Springer Series in Advanced Microelectronics

Volume 52

The Springer Series in Advanced Microelectronics provides systematic information on all the topics relevant for the design, processing, and manufacturing of microelectronic devices. The books, each prepared by leading researchers or engineers in their fields, cover the basic and advanced aspects of topics such as wafer processing, materials, device design, device technologies, circuit design, VLSI implementation, and subsystem technology. The series forms a bridge between physics and engineering and the volumes will appeal to practicing engineers as well as research scientists.

More information about this series at http://www.springer.com/series/4076

Souvik Mahapatra
Editor

Fundamentals of Bias Temperature Instability in MOS Transistors

Characterization Methods, Process and Materials Impact, DC and AC Modeling

Springer

Editor
Souvik Mahapatra
IIT Bombay
Mumbai
India

ISSN 1437-0387 ISSN 2197-6643 (electronic)
Springer Series in Advanced Microelectronics
ISBN 978-81-322-2507-2 ISBN 978-81-322-2508-9 (eBook)
DOI 10.1007/978-81-322-2508-9

Library of Congress Control Number: 2015943350

Springer New Delhi Heidelberg New York Dordrecht London

Printed on acid-free paper

Springer (India) Pvt. Ltd. is part of Springer Science+Business Media (www.springer.com)

To my parents
Krishna Chandra and Anjali Mahapatra

Preface

Bias Temperature Instability (BTI) is a serious reliability concern and continues to threaten the performance and lifetime of Complementary MOS (CMOS) devices and circuits. BTI affects both n- and p-channel Metal Oxide Semiconductor Field Effect Transistors (MOSFETs). Historically, Negative BTI (NBTI) became an important issue for p-MOSFETs as the semiconductor industry migrated from Silicon Oxide (SiO_2) to Silicon Oxynitride (SiON) gate insulators. NBTI continues to remain a serious issue for sub-22 nm technology nodes featuring High-K Metal Gate (HKMG) gate insulator based FinFET technologies. Positive BTI (PBTI) for n-MOSFETs remained negligible for SiON devices and emerged as a serious concern with the advent of HKMG gate insulator technology. However, latest reports from sub-22 nm Replacement Metal Gate (RMG) FinFET technologies have shown that PBTI is no longer a significant reliability concern.

Due to its technological significance, various research groups have extensively studied BTI in the last 15 years. In keeping with the technological trend, published reports initially focused on NBTI in SiON p-MOSFTs and later on both NBTI and PBTI in HKMG MOSFETs. These reports can be broadly classified into five major categories as explained hereinafter. Some reports have focused on process optimization for BTI mitigation and technology qualification, and studied the impact of various gate insulator and other processes and materials on BTI degradation. Other reports have focused on the development of novel methodologies for artifact-free characterization of BTI degradation and associated gate insulator defects. A significant majority of reports have focused on the understanding of BTI physical mechanism and development of numerical and compact models to explain measured data for DC and AC BTI degradation. Recently, some reports have focused on BTI variability, which is becoming important as device dimensions are scaled down. Finally, a significant majority of reports have also been published to study the impact of BTI on circuit and product degradation.

This book covers the first three aspects of BTI degradation as mentioned above, *viz.*, characterization methodologies, impact of gate insulator processes and materials, and physics-based models for DC and AC BTI. In the early years of research, no consensus existed among different groups on the experimental kinetics, process and materials dependence and physical mechanism of NBTI degradation. Therefore, development of suitable processes for NBTI mitigation and control became extremely difficult. Moreover, it became almost impossible to develop predictive models for estimating DC NBTI degradation at end-of-life, and to predict AC activity aware degradation for switching logic. Furthermore, although vast majority of published reports from different groups have dealt with understanding the physical mechanism and modeling of NBTI degradation, none can provide a comprehensive framework to explain measured results in SiON and HKMG p-MOSFETs and for DC and AC stress. Finally, some controversy also surrounded the physical mechanism and modeling of PBTI in HKMG n-MOSFETs.

The results presented in this book are extensively based on research carried out by our group in the past 13 years. Published results are also reported as and when necessary, to establish a comprehensive and universal picture of BTI degradation throughout the book. Our research has demonstrated that the lack of consensus in early NBTI reports arose due to the use of different nonstandard characterization methodologies and different gate insulator processes by different groups. Subsequently, proper artifact-free stress and measurement methodologies have been developed to characterize BTI degradation and also the underlying defects responsible for BTI degradation. The gate insulator process and material dependencies of NBTI degradation have been established for SiON and HKMG p-MOSFETs. The gate insulator process and material dependencies of PBTI degradation in HKMG n-MOSFETs are also established.

A consistent physical mechanism has been proposed to explain and model the experimentally observed process and material dependencies of NBTI in SiON and HKMG p-MOSFETs and PBTI in HKMG n-MOSFETs. The proposed framework is based on different mutually uncoupled underlying defect subcomponents, which are also assessed using independent measurement methods. Finally, a comprehensive DC and AC NBTI modeling framework is proposed and verified against experimental data in SiON and HKMG p-MOSFETs. The framework can successfully explain time evolution of measured NBTI degradation during and after DC stress and during multiple DC stress and recovery cycles, and during AC stress at different frequency and duty cycle. It can also predict extrapolated degradation at end of life for DC and AC NBTI stress.

Chapter 1 reviews NBTI and PBTI results from different published reports, and also shows measured DC and AC stress data using ultra-fast methods in SiON and HKMG devices under different experimental conditions. Different ultra-fast BTI characterization methods are discussed in Chap. 2, along with different characterization techniques for directly estimating gate insulator defects responsible for BTI. A consistent NBTI and PBTI mechanism is proposed and verified against experimental data in Chap. 3. Compact NBTI and PBTI models are developed in Chap. 4 based on the mechanism developed in Chap. 3, and the models are used to explain measured NBTI data in differently processed SiON and HKMG p-MOSFETs and PBTI data in different HKMG n-MOSFETs. The fundamentals of the

Reaction-Diffusion (RD) model are discussed in Chap. 5. Finally, the comprehensive DC and AC NBTI modeling framework is established in Chap. 6, and verified against ultra-fast measured data in SiON and HKMG p-MOSFETs.

Acknowledgments

This book has been made possible by active contribution and support from several people. I am grateful to my former and current students at IIT Bombay whose research contributions have been included in this book: Vrajesh Maheta, Gautam Kapila, P. Bharath Kumar, Dhanoop Varghese, E. Naresh Kumar, Shweta Deora, Kaustubh Joshi, Nirmal Nanaware, Vipul Chaudhary, Subhadeep Mukhopadhyay, Ankush Chaudhary and Nilesh Goel. I am grateful to Ahmad Ehteshamul Islam and Muhammad Ashraful Alam for collaboration in the area of NBTI modeling, Sandip De, Rajan Pandey and Kota Murali for DFT simulations, Bijesh Rajamohanan for providing flicker noise data in HKMG devices, Purushothaman Srinivasan, Jacopo Franco and Ben Kaczer for providing experimental data in SiGe devices. I wish to thank Steven Hung, Chris Olsen, Khaled Ahmed, Andreas Kerber, Anand Krishnan, Vincent Huard, Amr Haggag, Vijay Reddy, Sriram Kalpat, Srikanth Krishnan, Tanya Nigam, Eiichi Murakami, Hideki Aono, Tibor Grasser, Eduard Cartier, Giuseppe La-Rosa, Guido Groeseneken, Hans Reisinger, Sanjay Rangan, Stephen Ramey, Hiu Yung Wong, Jamil Kawa and Victor Moroz for many useful discussions. I am grateful to CEN IIT Bombay for measurement and computational facilities and Applied Materials for providing experimental devices and active support. I wish to thank my colleagues at IIT Bombay and friends for their understanding and support. Last but not least, I wish to profoundly thank my wife Sahana Mukherjee and daughter Adrija Mahapatra for their constant encouragement and support, which immensely helped me during the course of writing this book.

Mumbai, India Souvik Mahapatra

Contents

Editor and Contributors

About the Editor

Souvik Mahapatra received his Bachelors and Masters degrees in Physics from Jadavpur University, Calcutta, India in 1993 and 1995 respectively, and Ph.D. in Electrical Engineering from IIT Bombay, Mumbai, India in 1999. During 2000–2001, he was with Bell Laboratories, Lucent Technolgies, Murray Hill, NJ, USA. Since 2002 he is with the Department of Electrical Engineering at IIT Bombay and currently holds the position of full professor. His primary research interests are in the area of semiconductor device characterization, modeling and simulation, and in particular, MOS transistor and Flash memory device scaling and reliability. He has contributed in several technologically relevant research areas such as MOS gate insulator scaling, Channel Hot Carrier Degradation and Bias Temperature In-stability in CMOS devices, and CHISEL NOR Flash, SONOS NOR and NAND Flash and Metal Nanodot NAND Flash memory devices. He has published more than 150 papers in peer reviewed journals and conferences, delivered invited talks and tutorials in major international conferences including at the IEEE IEDM and IEEE IRPS, and served as a committee member and session chair in several IEEE conferences. He is a fellow of the Indian National Academy of Engineering, senior member of IEEE and a distinguished lecturer of IEEE EDS.

Contributors

Muhammad Ashraful Alam Department of Electrical and Computer Engineering, Purdue University, West Lafayette, IN, USA

Ankush Chaudhary Department of Electrical Engineering, Indian Institute of Technology Bombay, Mumbai, Maharashtra, India

Nilesh Goel Department of Electrical Engineering, Indian Institute of Technology Bombay, Mumbai, Maharashtra, India

Ahmad Ehteshamul Islam Materials and Manufacturing Directorate, Air Force Research Laboratory, Dayton, OH, USA; National Research Council, National Academy of Sciences, Washington DC, USA

Kaustubh Joshi Department of Electrical Engineering, Indian Institute of Technology Bombay, Mumbai, Maharashtra, India

Souvik Mahapatra Department of Electrical Engineering, Indian Institute of Technology Bombay, Mumbai, Maharashtra, India

Subhadeep Mukhopadhyay Department of Electrical Engineering, Indian Institute of Technology Bombay, Mumbai, Maharashtra, India

Chapter 1
Introduction: Bias Temperature Instability (BTI) in N and P Channel MOSFETs

Souvik Mahapatra, Nilesh Goel and Subhadeep Mukhopadhyay

Abstract In this chapter, the basic experimental signatures of NBTI and PBTI degradation respectively in p- and n-channel MOSFETs are discussed. Historical results from published reports are briefly reviewed for SiON and HKMG MOSFETs. Results obtained using ultra-fast characterization methods are shown for DC and AC BTI degradation in state-of-the-art SiON and HKMG MOSFETs. The impact of gate insulator processes on magnitude of NBTI degradation and its time, bias and temperature dependence is discussed for SiON p-MOSFETs. Time evolution of NBTI and PBTI degradation during and after DC stress and during AC stress is shown for HKMG MOSFETs, and the impact of stress bias and temperature, as well as that of AC pulse duty cycle and frequency are discussed. Impact of basic HKMG process variations, such as Nitrogen incorporation and interlayer thickness scaling are also discussed. Similarities and differences between NBTI and PBTI results are highlighted.

1.1 Introduction

Metal Oxide Semiconductor Field Effect Transistors (MOSFETs) are fundamental building blocks of Complementary MOS (CMOS) circuits. Figure 1.1 shows the schematic of p- and n-channel MOSFETs. It is now well known that for more than past 40 years, the gate length (L_G) of MOSFETs is reduced to increase its performance and keep up with Moore's scaling law [1]. According to MOSFET scaling principle, L_G scaling must be associated with gate insulator thickness (T_{OX}) scaling to keep different short channel effects under control [2]. Historically, conventional T_{OX} scaling based on reduction in Silicon Dioxide (SiO_2) thickness became unfeasible at 90 nm technology node, due to significant increase in gate

S. Mahapatra (✉) · N. Goel · S. Mukhopadhyay
Department of Electrical Engineering, Indian Institute of Technology Bombay, Mumbai 400076, India
e-mail: souvik@ee.iitb.ac.in; mahapatra.souvik@gmail.com

S. Mahapatra (ed.), *Fundamentals of Bias Temperature Instability in MOS Transistors*, Springer Series in Advanced Microelectronics 139,
DOI 10.1007/978-81-322-2508-9_1

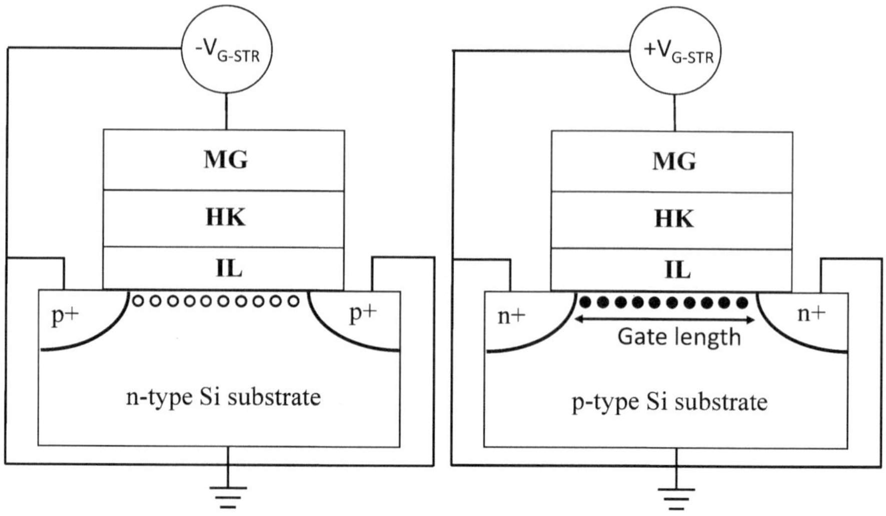

Fig. 1.1 Schematic of p- and n-channel MOSFETs having HKMG gate insulator stacks, configured respectively for NBTI and PBTI stress

leakage current (I_G) arising out of direct tunneling of inversion layer carriers across the gate insulator. Therefore, the industry has migrated first to Silicon Oxynitride (SiON) gate insulator at 90 nm technology node [3], and later to High-K Metal Gate (HKMG) gate insulator at 45 nm [4] or 28 nm [5] node, to continue T_{OX} scaling without significant increase in gate leakage. However, controlling the short channel effects has become a serious challenge at 20 nm node, and therefore, the industry has migrated from planar MOSFET to FinFET at 22 nm [6] or 16 nm (or 14 nm) [7] node.

Bias Temperature Instability (BTI) is a crucial reliability issue since it degrades the performance of p- and n-channel MOSFETs used in CMOS circuits. BTI results in a gradual shift in MOSFET characteristics, such as threshold voltage (V_T), transconductance (g_m), subthreshold slope (S), linear and saturation drain current (I_{DLIN} and I_{DSAT}), etc., over time [8], and hence in-turn degrades the performance of digital, memory and analog CMOS circuits [9]. As a result, the operating lifetime of CMOS devices and circuits gets reduced, and if unchecked, BTI can lead to premature failure of Integrated Circuits (IC) chips.

Figure 1.1 shows the BTI biasing conditions for p- and n-channel MOSFETs having HKMG gate insulators. BTI degradation of p-MOSFET devices, known as Negative Bias Temperature Instability (NBTI), occurs when the gate of the device is made negative w.r.t. other terminals. NBTI results in buildup of positive charges in the gate insulator and causes negative shift in V_T (ΔV_T) [8]. On the other hand, Positive Bias Temperature Instability (PBTI) occurs in n-MOSFETs when the gate of the device is made positive w.r.t. other terminals. PBTI causes negative charges in the gate insulator and results in positive ΔV_T [10].

NBTI was first observed half a century ago in SiO_2 MOSFETs [11], although it became a crucial reliability concern when SiON replaced SiO_2 as the gate insulator [12–14]. It continues to remain as a significant concern even today for planar and FinFET devices having HKMG based gate stacks [15–18]. Note that the physical mechanism of NBTI is presumed to be identical for SiON and HKMG p-MOSFETs [8, 19–23]. NBTI in SiO_2 or SiON p-MOSFETs is attributed to buildup of positive charges at the interface between the Silicon (Si) channel and SiO_2 (or SiON) dielectric, as well as in the SiO_2 (or SiON) dielectric bulk. Similarly, NBTI in HKMG p-MOSFETs is attributed to buildup of positive charges in the SiO_2 or SiON interlayer (IL), which separates the MOSFET channel and the Hafnium Dioxide (HfO_2) High-K layer, refer to Fig. 1.1. The interface between the MOSFET channel and IL as well as IL bulk becomes positively charged. The High-K layer presumably acts as a voltage divider.

PBTI remained negligible for SiO_2 and SiON n-MOSFETs and became important only with the introduction of HKMG technology [10, 16–18]. Generation of negative charges in HfO_2 High-K layer is believed to be responsible for PBTI, refer to Fig. 1.1. The SiO_2 (or SiON) IL presumably does not degrade for production quality HKMG MOSFETs and acts only as a voltage divider. Although both NBTI and PBTI impacts HKMG technology based circuits, recent results from Replacement Metal Gate (RMG) technology based HKMG FinFETs suggest PBTI to be much less of a concern than NBTI [17, 18, 24]. It is important to realize that the gate stack and other (such as backend) processes significantly influences BTI, and the magnitudes of NBTI and PBTI degradation depend on process optimization and are very much technology and industry specific. This aspect is discussed later in this chapter.

Although BTI is an important reliability issue, it is not the only mechanism that causes device degradation during operation of a CMOS circuit. Other mechanisms such as Channel Hot Carrier (CHC) degradation [25, 26], Stress Induced Leakage Current (SILC) [27, 28] and Time Dependent Dielectric Breakdown (TDDB) [29, 30] are also responsible for device and circuit performance degradation. As an example, Fig. 1.2 illustrates the interplay of different reliability mechanisms during the operation of a CMOS inverter circuit. The inverter is made of p- and n-channel MOSFETs connected in series between power and ground rails, and the output

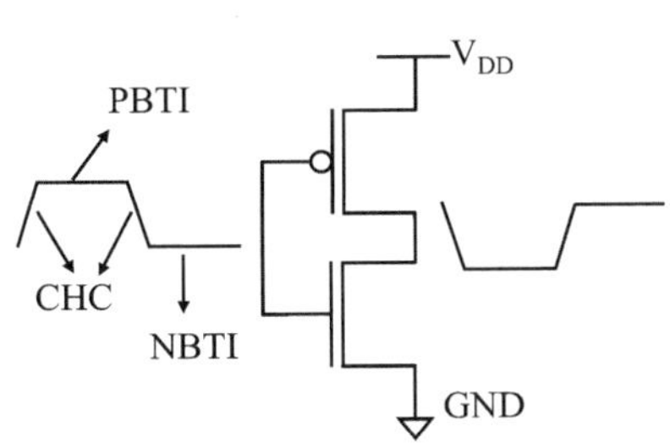

Fig. 1.2 Schematic of a CMOS inverter with input and output waveforms illustrating BTI and CHC reliability regimes. SILC and TDDB have identical regimes as BTI

voltage is complimentary to the input voltage. The p-MOSFET suffers from NBTI when input is low and output is high, i.e., the gate is negative w.r.t. all the other terminals of the device, while the n-MOSFET suffers from PBTI when input is high and output is low, i.e., the gate is positive w.r.t. all the other terminals of the device. SILC and TDDB regimes for p- and n-MOSFETs coincide with NBTI and PBTI regimes, respectively, while CHC degradation occurs during the transition periods when the input goes from low to high (output goes from high to low) and vice versa. Although a technology should be qualified for all these degradation regimes, BTI has become and remained as the dominant front-end reliability concern since 90 nm technology node and is the subject of this book.

1.2 Brief Overview of NBTI and PBTI

From a practical point of view, it is of utmost importance to estimate BTI degradation at the end of expected usage lifetime of devices (and hence circuits and products). The assessment of BTI induced MOSFET parametric degradation is usually done by stressing the device at an accelerated aging condition, using a gate bias (V_G) that is higher ($V_G = V_{G\text{-STR}}$) than that used during normal operation. MOSFET transfer *I–V* characteristics are measured before and after BTI stress, and the difference between pre- and post-stress values is used to access degradation in device parameters.

However, rather than a single value, it is of interest to estimate the time evolution of BTI degradation. In a typical BTI test scenario, transfer *I–V* measurements are performed by interrupting the stress at certain pre-defined times, the interruptions are usually spaced in logarithmic intervals of time, and the time evolution of BTI induced shift in MOSFET parameters is estimated. This scheme is defined as Measure-Stress-Measure (MSM) method. However, it was later realized that BTI degradation recovers as soon as stress is removed for measurement, and alternatively, the degradation in MOSFET drain current can be continuously monitored as the gate is being stressed, which is defined as On-The-Fly (OTF) method [31]. Different BTI characterization methods will be discussed in Chap. 2 of this book. The accelerated stress test is performed up to several 1000's of seconds or hours, or sometimes up to days, in wafer-level setup, although sometimes the test can go on for months in package-level setup. The measured time evolution of parametric degradation at accelerated stress condition ($V_G = V_{G\text{-STR}}$) is then extrapolated to expected end-of-life (e.g., 10 years) and to normal operating V_G by using suitable models. BTI modeling for lifetime estimation is a topic of great interest and is discussed later in this book.

It is necessary to follow certain cautionary measures for BTI stress and measurement. Note, BTI and TDDB have identical stress regimes as shown in Fig. 1.2. Since TDDB defects build up faster in time and has higher V_G acceleration rate as compared to BTI [32, 33], the use of very high stress V_G should be avoided to prevent TDDB related "corruption" of BTI data. In other words, although BTI dominates TDDB at relatively lower V_G, as stress V_G is increased, the degradation due to TDDB increases at a much faster rate than that for BTI. Therefore, the maximum limit of stress V_G should be carefully chosen such that the contribution from TDDB defects remains negligible as compared to BTI defects during accelerated stress test. This will be discussed later in Chap. 2. Moreover as mentioned before, it is now well known that BTI degradation starts to recover as soon as the magnitude of V_G is reduced for measurement [31, 34]. Although earlier NBTI measurements were done using slower methods, recent NBTI and PBTI measurements are performed in a way such that the impact of recovery is negligible. Several methods have been developed in recent past for "recovery artifact free" BTI measurements, and will be discussed in Chap. 2 of this book.

1.2.1 NBTI in SiO_2 and SiON p-MOSFETs

Early reports show that gate insulator quality plays an important role in determining NBTI degradation. Figure 1.3 plots time evolution of ΔV_T during NBTI stress in p-MOSFETs having SiO_2 gate insulators grown by wet and dry oxidation [12, 35]. Dry SiO_2 devices have superior gate insulator quality and hence show lower NBTI degradation.

As discussed in detail later in this book, NBTI degradation is due to uncorrelated contributions from interface trap generation (ΔN_{IT}) at Si/SiO_2 interface, hole trapping in process-related SiO_2 bulk traps (ΔN_{HT}) and generation of new SiO_2 bulk traps (ΔN_{OT}). In general, gate oxide process can impact any one or all of these components. However, ΔN_{HT} is a fast saturating phenomenon and is unlikely to be

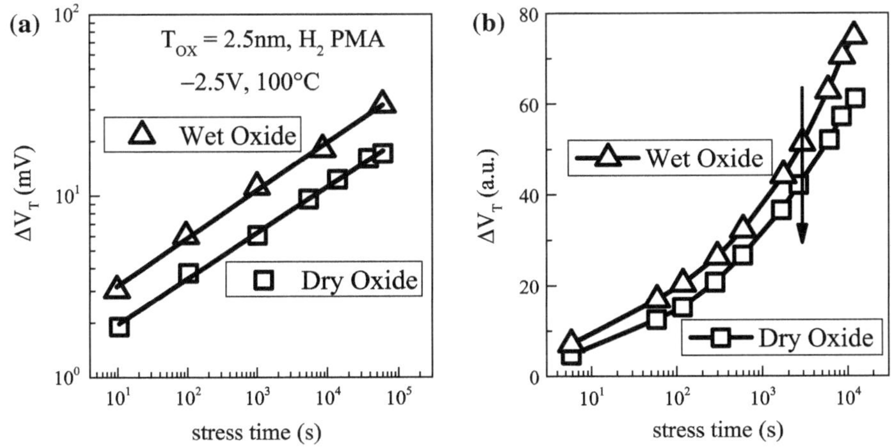

Fig. 1.3 Time evolution of measured ΔV_T for NBTI stress in dry and wet SiO_2 p-MOSFETs. Data from [12] (*left panel*) and [35] (*right panel*)

captured in slow MSM measurement method used in [12, 35]. Furthermore, ΔN_{OT} is associated with TDDB like trap generation process and therefore has strong voltage de-acceleration factor [32, 36]; ΔN_{OT} contribution is usually negligible unless high $V_{G\text{-}STR}$ is used. Therefore, the improvement is likely due to lower ΔN_{IT} in dry SiO_2 devices.

The role of generated interface traps is further ascertained by studying NBTI in SiO_2 p-MOSFETs having conventional Hydrogen (H_2) and Deuterium (D_2) based Post Metallization Anneal (PMA). Figure 1.4a plots time evolution of ΔV_T during NBTI stress in H_2 and D_2 PMA based SiO_2 p-MOSFETs [12]; lower degradation is observed for D_2 devices. Measured ΔV_T is likely dominated by ΔN_{IT} due to the reasons mentioned above. As discussed later in this book, interface trap generation is explained by Reaction-Diffusion (RD) model [37], which suggests breaking of Si–H bonds at Si/SiO_2 interface and subsequent diffusion of H_2 [8], refer to Chap. 5 for details. For D_2 devices, this would imply breaking of Si–D bonds and subsequent diffusion of D_2. Therefore, the difference in ΔN_{IT} and ΔV_T can be associated to difference in Si–D and Si–H bond strengths and/or different diffusivity of D_2 compared to H_2. As discussed in Chap. 2, ΔN_{IT} can be directly determined by using Charge Pumping (CP) measurements [38] before and after NBTI stress. Figure 1.4b shows measured ΔV_T and ΔN_{IT} for NBTI stress in SiO_2 devices having H_2 and D_2 PMA [35]. Lower ΔV_T for D_2 devices is due to lower ΔN_{IT} as shown.

As mentioned before, NBTI became a crucial reliability concern with the introduction of SiON gate insulator technology. Figure 1.5a plots the time evolution of ΔV_T for NBTI stress in pure SiO_2 and SiON p-MOSFETs having different Nitrogen (N) content in the gate stack [12]. Time evolution of ΔV_T shows power law dependence for both SiO_2 and SiON devices, i.e., $\Delta V_T = A * t^n$, and the magnitude of ΔV_T increases with increased N content. Time evolution of ΔV_T is measured in SiON p-MOSFETs having different N content [39] and fitted using the

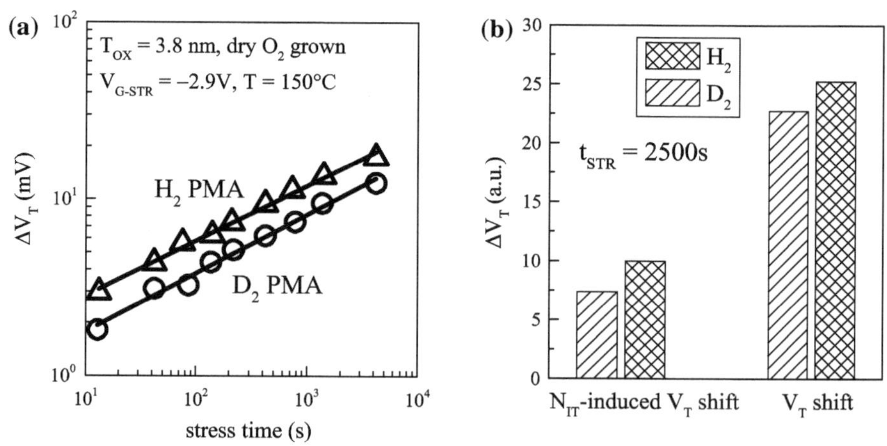

Fig. 1.4 **a** Time evolution of ΔV_T and **b** fixed time ΔV_T and ΔN_{IT}, measured for NBTI stress in dry SiO_2 p-MOSFETs having H_2 and D_2 PMA. Data from [12] (*left panel*) and [35] (*right panel*)

power-law expression shown above. Figure 1.5b plots pre-factor (A) as a function of atomic N content (N%) in the gate stack. Note that the pre-factor A increases with increased N%. A higher rate of increase is seen for $N > 8$ % for these particular devices. Figure 1.5 clearly indicates the emergence of NBTI as a crucial reliability issue with the advent of SiON technology. Once again, note that the impact of N% on ΔV_T is likely due to variations in ΔN_{IT}, since ΔN_{HT} is likely to be negligible due to the use of slow MSM measurement method, and ΔN_{OT} is also negligible as $V_{G\text{-}STR}$ is kept low during stress.

Note that N incorporation increases the permittivity (K) of gate insulator stack, and for identical SiO_2 physical thickness, it results in lower electrical thickness for SiON gate insulators. Different nitridation techniques have been studied to explore the possibility of obtaining SiON gate insulator without corresponding increase in NBTI. Figure 1.6 plots time evolution of ΔV_T for NBTI stress in differently processed SiON p-MOSFETs [40]. Since different process have different gate insulator thickness, $V_{G\text{-}STR}$ is suitably adjusted to perform NBTI stress at similar oxide field (E_{OX}) for fair comparison, as NBTI depends on E_{OX} and not $V_{G\text{-}STR}$ [32]. Note that Nitrogen Implanted Silicon Substrate (NISS) process shows highest ΔV_T, followed by Rapid Thermal Oxidation (RTO) in NO ambient and N_2O based thermal oxidation, while Remote Plasma Nitridation (RPN) process shows lowest NBTI. It is important to remark that the process impact on ΔV_T is likely coming from variations in ΔN_{IT} due to the reasons mentioned before.

Figure 1.7 plots the time evolution of ΔV_T and ΔN_{IT} obtained respectively from slow MSM I–V and CP measurements for NBTI stress in NO based thermal nitrided SiON and RPN-SiON p-MOSFETs [14]. RPN-SiON devices have lower ΔN_{IT} and hence have lower ΔV_T as shown. This once again verifies that ΔN_{HT} and ΔN_{OT}

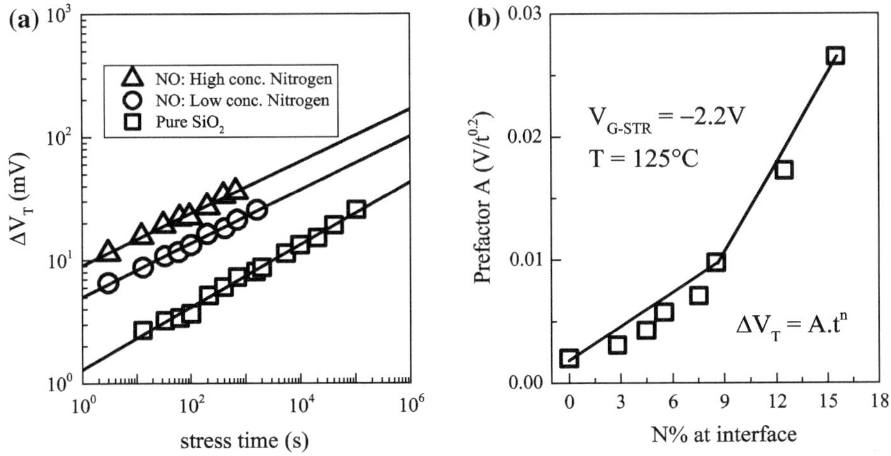

Fig. 1.5 **a** Time evolution of ΔV_T in SiO_2 and SiON p-MOSFETs having different N dose for NBTI stress; data fitted using power-law expression $A * t^n$. **b** Pre-factor A as a function of atomic N % for SiON devices having different N content. Data from [12] (*left panel*) and [39] (*right panel*)

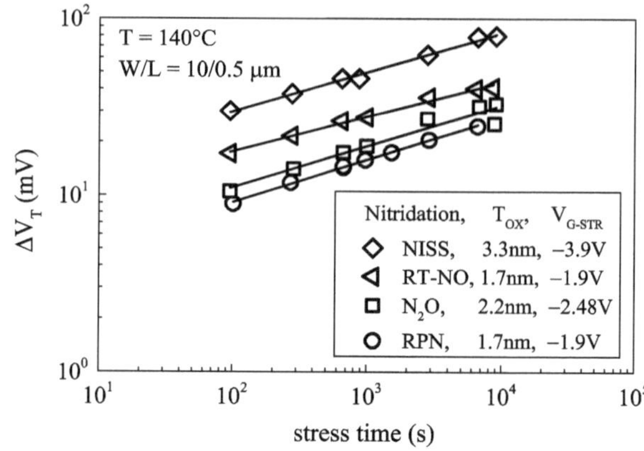

Fig. 1.6 Time evolution of ΔV_T for NBTI stress in SiON p-MOSFETs fabricated using different gate insulator nitridation processes. Data from [40]

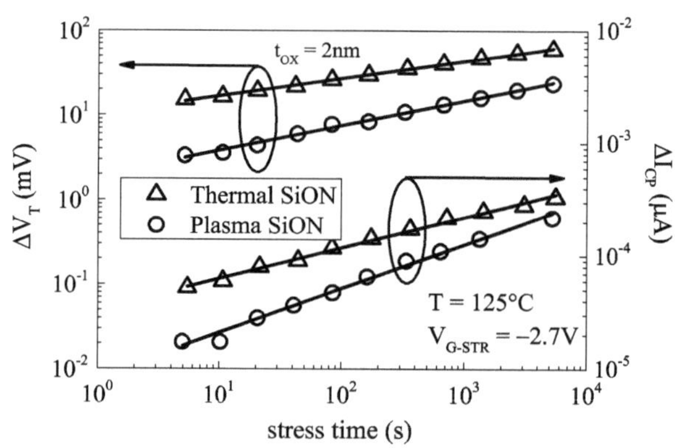

Fig. 1.7 Time evolution of ΔV_T and ΔN_{IT} for NBTI stress in plasma and thermal nitrided SiON p-MOSFETs. Data form [14]

contributions are negligible and ΔN_{IT} dominates ΔV_T for slow MSM measurement and low $V_{G\text{-}STR}$ stress as used in [14]. It is now well known that for a particular total N dose, RPN process results in lower N density at the Si channel and SiON interface compared to conventional NO-SiON process, since bulk of the N is placed closer to the SiON and poly-Si gate interface for RPN process [41]. Lower N density at Si/SiON interface results in lower NBTI for RPN compared to conventional thermal NO based SiON devices [42, 43]. Therefore, RPN-based nitridation technique has become widely adopted in the industry for gate insulator technology.

Figures 1.6 and 1.7 suggest that it is not the total N content in the gate insulator but rather the N density at or near the Si channel and SiON interface that controls NBTI. As a further verification, Fig. 1.8 plots N density distribution profile in the gate insulator and the corresponding NBTI degradation for different SiON gate insulator processes [44, 45]. N density profile can be obtained using Angle Resolved X-ray Photoelectron Spectroscopy (ARXPS) measurements [46]. In the top figure, NO first process has lower N density at Si/SiON interface in spite of having higher peak and overall integrated N content in the gate insulator as

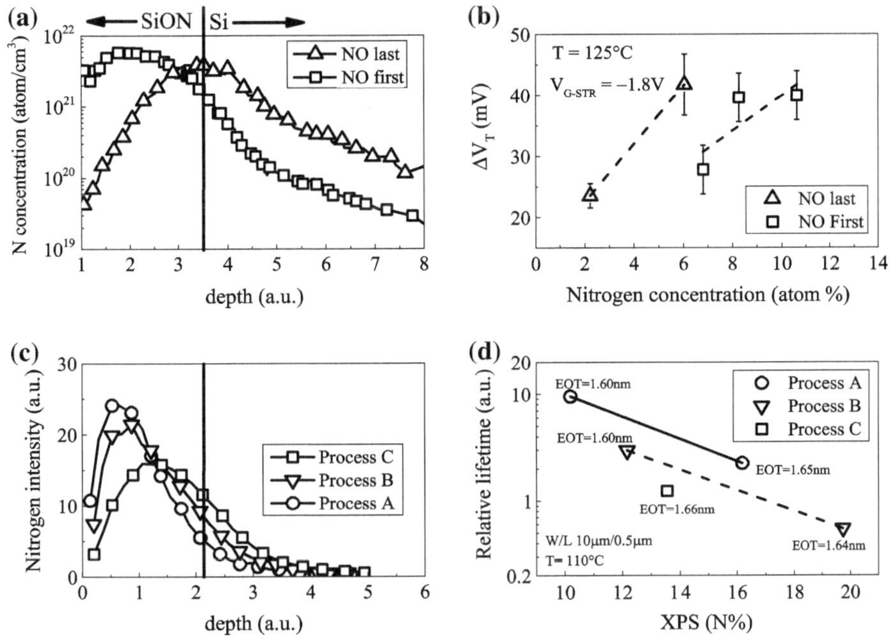

Fig. 1.8 Nitrogen distribution profile in SiON gate insulators (*left panels*) and corresponding NBTI degradation and lifetime (*right panels*) for NO first and NO last process (*top panels*) [44] and different plasma nitridation processes (*bottom panels*) [45]

compared to NO last process [44]. Therefore, the NO first process results in lower ΔV_T as shown. In the bottom plot, Si/SiON interfacial N density increases while the peak and total integrated N content in the gate insulator reduces from process A through C [45]. For a particular total N content in the gate stack, NBTI degradation is lowest, which results in highest extrapolated lifetime for process-A, and the degradation is highest and hence the extrapolated lifetime is lowest for process-C as shown.

Note that for RPN process having a particular total N dose, the Post Nitridation Anneal (PNA) condition [47] plays an important role in controlling NBTI degradation [48]. Figure 1.9 shows the time evolution of ΔV_T for NBTI stress in SiON p-MOSFETs having similar N content but different post-nitridation optimization [49]. Although the total N dose remains same, optimized devices show lower ΔV_T as shown. Therefore not only the RPN nitridation process but the PNA process is also equally important to control NBTI and is now widely adopted in the industry for gate insulator technology. More results on SiON devices having different RPN N content and PNA processes will be discussed later in this chapter.

Fluorine (F) incorporation in the gate insulator stack also impacts NBTI degradation. Figure 1.10 shows the time evolution of ΔV_T and ΔN_{IT} obtained, respectively, from slow MSM *I–V* and CP measurements for NBTI stress in devices without and with F in the gate stack [35, 50].

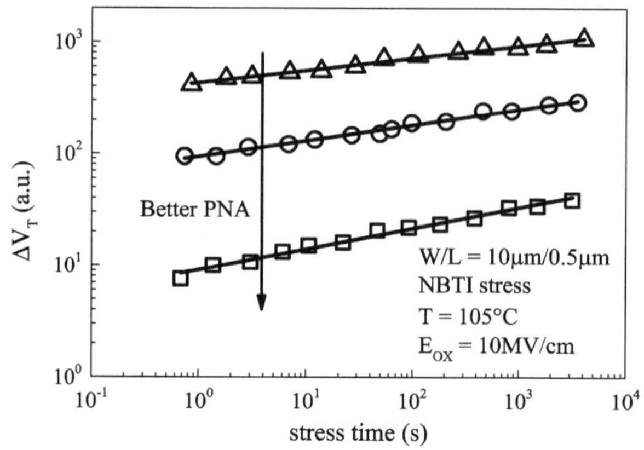

Fig. 1.9 Time evolution of ΔV_T for NBTI stress in SiON p-MOSFETs fabricated using plasma nitridation and different post nitridation anneals. Data from [49]

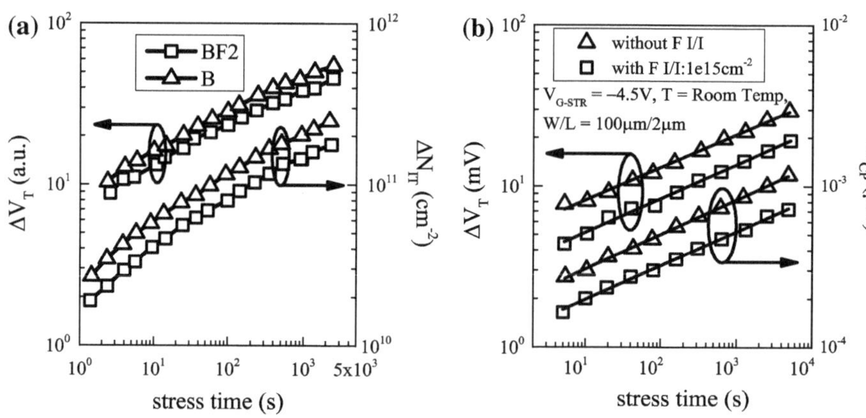

Fig. 1.10 Time evolution of ΔV_T and ΔN_{IT} ($\sim \Delta I_{CP}$) for NBTI stress in SiON p-MOSFETs having different fluorine content in the gate stack. Data from [35] (*left panel*) and [50] (*right panel*)

In Fig. 1.10a, F incorporation is achieved using BF_2 implanted Source and Drain junctions and subsequent activation anneal, while the reference device has Boron (B) implanted junctions [35]. F is implanted directly in the gate oxide in Fig. 1.10b [50]. Note that measurements were done using slow MSM method with low $V_{G\text{-}STR}$, and results in ΔN_{IT} domination of ΔV_T as discussed earlier. F incorporation reduces ΔN_{IT} and hence ΔV_T as shown. Once again, Fig. 1.10 proves the role of generated interface traps on NBTI degradation.

The results shown above are among the earliest reported that studied the impact of different SiO_2 and SiON gate insulator processes on NBTI degradation, refer to [51] for further details. Note that measurements were done using slower methods, as NBTI recovery and its implications [31, 34] were not fully understood at that time. Subsequently, methods having ~millisecond [31] and ~microsecond [52, 53] delay were developed and used to study the impact of gate insulator nitridation on NBTI [43, 48]. These results will be discussed later in this chapter.

1.2.2 NBTI and PBTI in HKMG MOSFETs

As mentioned before, similar to SiO_2 or SiON p-MOSFETs, HKMG devices are also affected by NBTI stress due to positive charge buildup in the SiO_2 (or SiON) IL that separates the channel and HfO_2 High-K layer. Similar to SiON devices, the optimization of HKMG process plays a significant role in reducing NBTI degradation. As an illustration, Fig. 1.11a plots measured ΔV_T time evolution for NBTI stress in p-MOSFETs having "initial" or premature and "final" or production quality HKMG gate insulator stacks [16]. Note that time evolution of ΔV_T shows power law dependence similar to SiON devices. The "initial" premature process shows much larger ΔV_T and very low time exponent n when compared to the production quality process and highlights the importance of process optimization. Figure 1.11b plots measured ΔV_T at fixed stress time (t_{STR}) as a function of gate oxide field E_{OX} for NBTI stress in production quality 65 nm SiON, as well as 45, 32 and 22 nm HKMG gate insulator processes [16, 17, 54]. Note that NBTI degradation reduces for migration from 65 nm SiON to 45 nm HKMG technology, presumably due to reduction in N in the gate stack; however, it gradually increases for migration from 45 to 32 nm planar and eventually to 22 nm FinFET technologies. Of importance is the power-law field acceleration factor (Γ_E) or the slope of

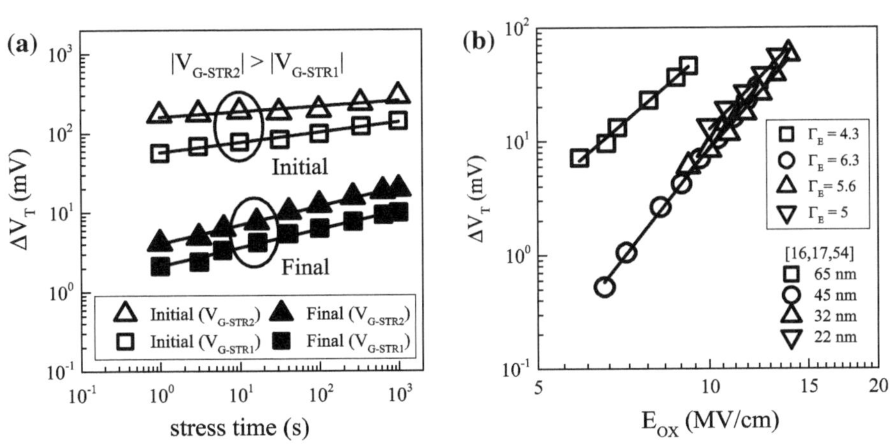

Fig. 1.11 **a** Time evolution of ΔV_T for NBTI stress in HKMG p-MOSFETs with different quality of gate insulator stacks. **b** Fixed time ΔV_T as a function of oxide field E_{OX} in a log–log plot, for different SiON and HKMG production quality devices. Data from [16, 17, 54]

ΔV_T versus E_{OX} plot, which is larger for 45 nm HKMG compared to 65 nm SiON process, although it reduces slightly for 32 and 22 nm HKMG processes.

On the other hand, PBTI was negligible for SiO_2 or SiON n-MOSFETs and become appreciable with the introduction of HKMG devices, as it is associated with the buildup of negative charges in HfO_2 High-K layer [10]. It will be shown later in this book that ΔV_T for PBTI stress is due to uncoupled contribution from trap generation at IL/High-K interface ($\Delta N_{IT\text{-}HK}$) and High-K bulk ($\Delta N_{OT\text{-}HK}$), as well as electron trapping in pre-existing process-related bulk High-K traps (ΔN_{ET}). PBTI measurements in early HKMG devices show very large degradation due to significant contribution from ΔN_{ET} in inferior quality and thick HfO_2 dielectrics [55, 56]. PBTI reduces with reduction in HfO_2 High-K thickness [10], while process optimization played a significant role in further reducing PBTI degradation [16], primarily by reducing the ΔN_{ET} contribution [57]. As an example, Fig. 1.12 plots ΔV_T determined at a particular charge flounce for PBTI stress at different $V_{G\text{-}STR}$ versus thickness of the High-K layer. ΔV_T reduces significantly with reduction in High-K thickness as shown, due to reduction in the volume for electron trapping.

Figure 1.13a plots the time evolution of measured ΔV_T for PBTI stress in HKMG n-MOSFETs having "initial" and "final" gate insulator processes [16]. Time evolution of ΔV_T shows power law dependence similar to NBTI stress. Moreover, similar to NBTI results, the "initial" or premature process shows much higher ΔV_T and lower time exponent n also for PBTI stress when compared to the "final" production quality process. Figure 1.13b plots measured ΔV_T at fixed t_{STR} as a function of E_{OX} for PBTI stress in production quality 45 and 22 nm HKMG gate insulator processes [16, 17, 54]. Unlike NBTI, ΔV_T for PBTI stress reduces with HKMG technology scaling as shown.

An important difference between NBTI and PBTI in HKMG MOSFETs can be observed by measuring the peak transconductance degradation (Δg_m). Figure 1.14 plots the correlation of peak Δg_m to ΔV_T for NBTI and PBTI stress in premature or "initial" and fully optimized or "final" production quality HKMG processes [16].

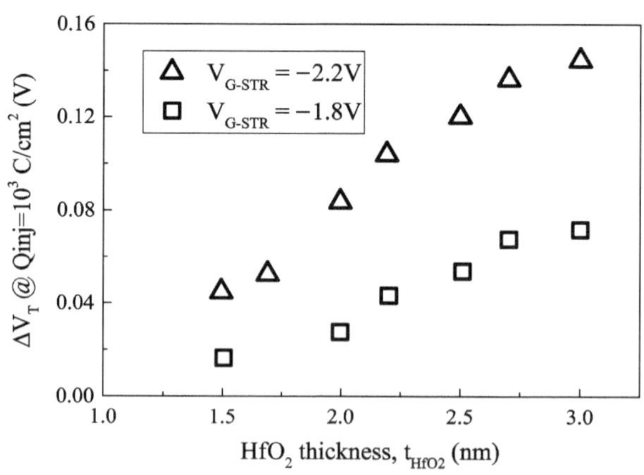

Fig. 1.12 Measured ΔV_T for PBTI stress in HKMG n-MOSFETs having different High-K layer thickness. Data from [10]

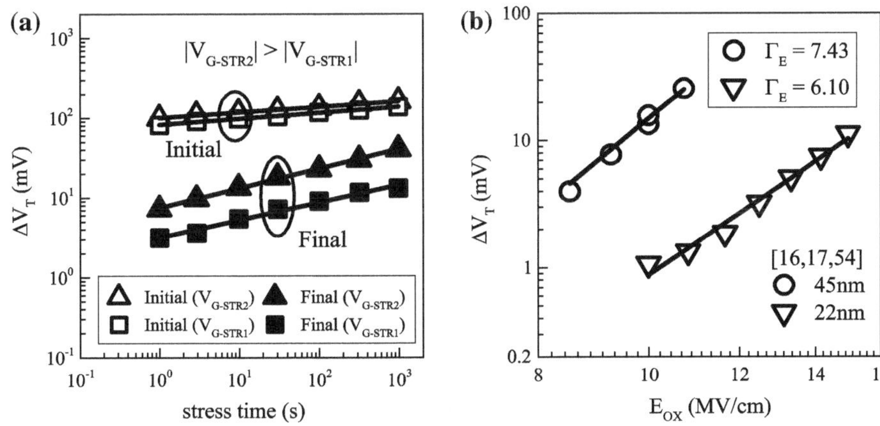

Fig. 1.13 **a** Time evolution of ΔV_{T} for PBTI stress in HKMG n-MOSFETs with different quality of gate insulator stacks. **b** Fixed time ΔV_{T} as a function of oxide field E_{OX} in a log–log plot, for different HKMG production quality devices. Data from [16, 17, 54]

The "initial" or premature process shows very large ΔV_{T} and Δg_m for both NBTI and PBTI. Process optimization helps in reducing ΔV_{T} for both NBTI and PBTI stress. Interestingly, the optimized process shows reduced but non-negligible Δg_m for NBTI but negligible Δg_m for PBTI stress. Note, peak g_m degradation can be related to channel mobility degradation and is due to enhanced Coulomb scattering by BTI induced charges in the gate insulator. The presence of g_m degradation for NBTI is consistent with the fact that NBTI causes charge buildup in the SiO_2 (or SiON) IL, which is nearer to the MOSFET channel. Process optimization helps in improving the IL quality and reduces buildup of charges during NBTI stress, and results in lower ΔV_{T} and Δg_m.

On the other hand, large Δg_m for PBTI stress in premature devices implies degradation of the IL in addition to the High-K layer. However, IL degradation is negligible for PBTI stress in well-optimized HKMG process, and charge buildup happens only in the High-K layer that is further away from the channel. Therefore, the "final" HKMG process exhibits negligible Δg_m as shown. It is now universally believed that PBTI stress does not cause significant IL degradation in production quality n-MOSFET devices.

The relative magnitude of NBTI and PBTI degradation is of interest, since both mechanisms add up and can result in enhanced degradation of CMOS circuits for HKMG technologies. As mentioned before, BTI is strongly influenced by the gate insulator and other processes and hence is industry specific. For example, Fig. 1.15 plots NBTI and PBTI degradation measured in several production-quality HKMG technologies [18]. NBTI increases from 32 nm planar through 14 nm FinFETs, although reduction in NBTI is observed for optimized 14 nm process. This is similar to Fig. 1.11b [16, 17, 54] that shows increase in NBTI from 45 nm planar through 22 nm FinFET processes, and latest results [24] that show a reduction in

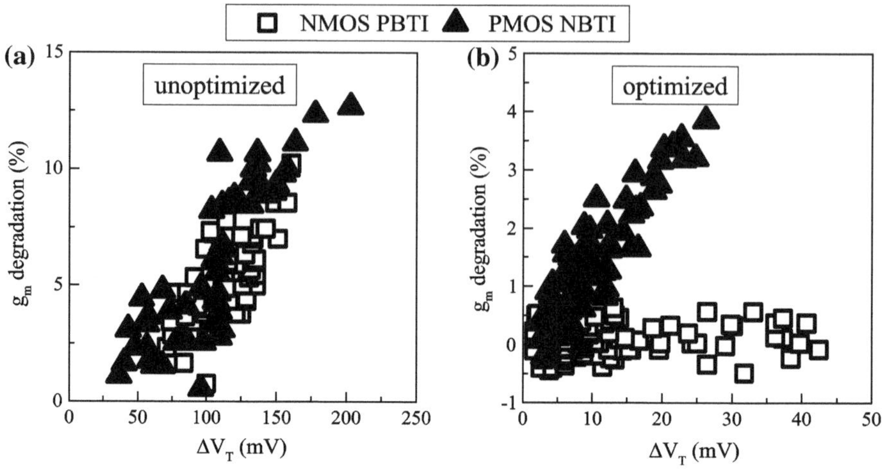

Fig. 1.14 Correlation of peak g_m degradation to V_T shift for NBTI and PBTI stress in HKMG MOSFETs, having **a** un-optimized and **b** optimized gate insulator stacks. Data from [16]

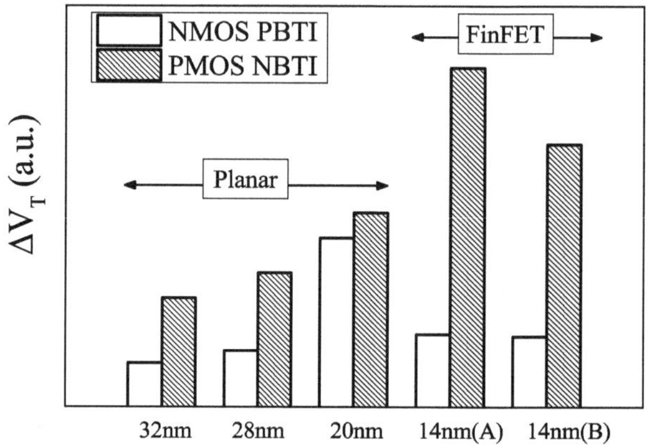

Fig. 1.15 Relative NBTI and PBTI degradation in different HKMG technology nodes. Data from [18]

NBTI for 14 nm FinFETs. Figure 1.15 also shows that 32 and 28 nm planar HKMG technologies have higher NBTI than PBTI. However, PBTI increases and comparable NBTI and PBTI degradation has been observed for the planar 20 nm node. This is contrary to Fig. 1.13b data obtained from [16, 17, 54], which show reduction in PBTI degradation with scaling from 45 nm through 22 nm nodes. A possible explanation for this discrepancy is while RMG HKMG process has been used for all nodes in [16, 17, 54], the processes in [18] used Gate First (GF) HKMG scheme for 32 and 28 nm and RMG scheme for 20 nm technology nodes [58]. However, PBTI has reduced for 14 nm RMG HKMG FinFETs [18], similar to results shown in [24]. The process impact and physical mechanism of BTI will be explained later in this book.

Note that the HKMG BTI results presented in this section have been obtained by using slower methods. Although fair conclusions can be made on relative magnitude of degradation for different processes, slower methods fail to capture accurate magnitude of BTI. Ultra-fast characterization of DC and AC BTI in HKMG MOSFETs is presented later in this chapter.

1.3 Ultra-Fast Characterization of NBTI in SiON p-MOSFETs

In this section, NBTI is characterized in differently processed SiON p-MOSFETs by using Ultra-Fast On-The-Fly (UF-OTF) I_{DLIN} method [52]. As described in detail in Chap. 2, in UF-OTF method, the drain current in linear regime (I_{DLIN}) is continuously monitored as the device is being stressed at $V_G = V_{G\text{-}STR}$, and time evolution of degradation is calculated from measured drain current degradation, ΔI_{DLIN} ($=I_{DLIN} - I_{DLIN0}$), using the relation $\Delta V = -\Delta I_{DLIN}/I_{DLIN0} * (V_{G\text{-}STR} - V_{T0})$, where V_{T0} is pre-stress V_T and I_{DLIN0} is the first data point recorded at $V_G = V_{G\text{-}STR}$, within a time t_0 of application of stress. The impact of t_0 delay will be discussed in Chap. 2; all measurements in this section have been done using $t_0 = 1$ μs. Mobility degradation ($\Delta\mu_{eff}$) needs to be considered for properly converting ΔV to threshold voltage shift, ΔV_T, which can be done using a post-processing correction algorithm described in [59], refer to Chap. 2 for further details.

Table 1.1 lists the gate insulator process details of different SiON p-MOSFETs studied in this section [43]. Plasma Nitrided Oxide (PNO) devices have been fabricated using Decoupled Plasma Nitridation (DPN) process [60] with suitable Post Nitridation Anneal (PNA) [47]. The "PNO with proper PNA" devices W1 through W5 have different starting SiO_2 base oxide thickness and were subjected to different N dose during subsequent plasma nitridation, which results in different Equivalent Oxide Thickness (EOT) and N content of the gate insulator stack. PNO with non-optimized improper PNA (W6), and N_2O based Rapid Thermal Nitrided

Table 1.1 Details of SiON devices for NBTI study using UF-OTF method

Device no.	Type	Base oxide thickness (Å)	N dose ($\times 10^{15}$ cm^{-2})	TXPS (Å)	EOT (Å)	% N
W1	PNO	15	0.0 + 2.8	18.48	14	23
W2	PNO	20	0.0 + 2.9	22.34	17.7	20
W3	PNO	20	0.0 + 5.3	23.16	15.6	35
W4	PNO	20	0.0 + 6.8	24.37	14.6	43
W5	PNO	25	0.0 + 3.1	28.3	23.5	17
W6	PNO (moderate PNA)	20	0.0 + 2.7	24.09	20.2	17
W7	RTNO	25	0.8 + 0.0	26.43	22.5	6

Oxide (RTNO) devices (W7) [40] are also fabricated. The atomic N content (N%) of SiON dielectrics is extracted using XPS measurements [46].

1.3.1 Process Impact on Threshold Voltage Degradation

Figure 1.16 shows time evolution of ΔV_T in (a) PNO and (b) RTNO SiON devices at different stress bias ($V_{G\text{-}STR}$) and temperature (T). Figure 1.16c, d re-plot the above data but for longer stress time, t_{STR}. The devices have similar EOT, and the oxide field during stress is estimated as $E_{OX} = (V_{G\text{-}STR} - V_{T0})/\text{EOT}$, EOT being the SiO_2 equivalent oxide thickness. In spite of lower N dose, RTNO device shows much larger NBTI degradation as compared to PNO device at identical E_{OX} and T. As mentioned before, although overall N content is high, N density at the Si/SiON interface is lower for PNO compared to RTNO process [14], and it is the Si/SiON N density that actually controls NBTI degradation. ΔV_T time evolution increases rapidly at shorter time (t_{STR} < 1 s), and shows power law dependence at

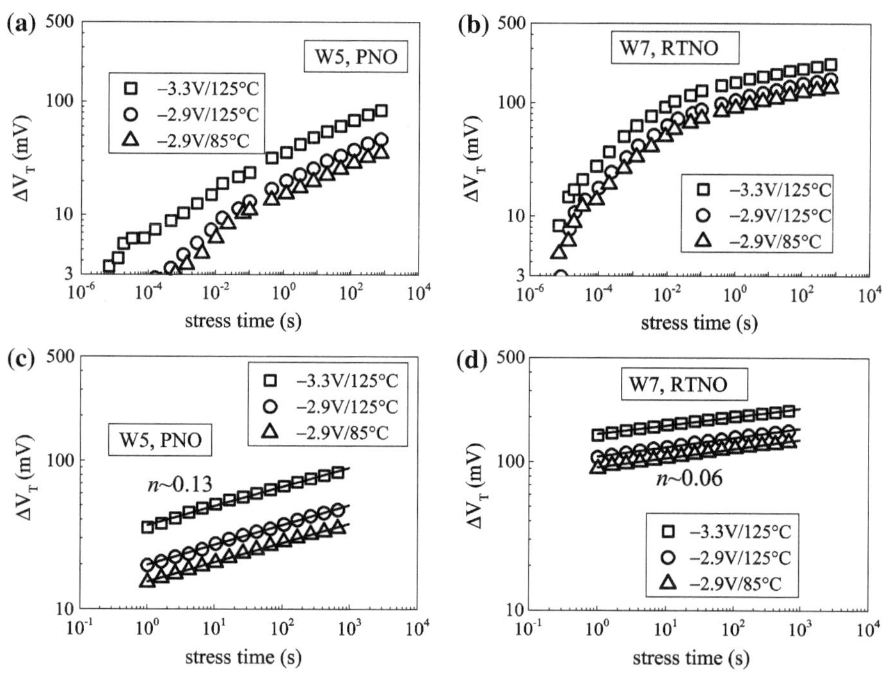

Fig. 1.16 Time evolution of ΔV_T from short to long stress time for **a** PNO and **b** RTNO SiON p-MOSFETs under NBTI stress at different $V_{G\text{-}STR}$ and T. Data obtained from UF-OTF measurements after mobility correction. Re-plot of long-time ΔV_T time evolution data for **c** PNO and **d** RTNO SiON p-MOSFETs

longer time (t_{STR} > 10 s), with an exponent (n) that is higher for PNO compared to RTNO devices. However for a particular PNO or RTNO device, similar n values are observed across different E_{OX} and T. It is interesting to note that appreciable degradation is observed in the sub-1 ms time scale for RTNO devices, which is not observed for PNO devices. Furthermore, the large, sub-1 ms degradation in RTNO devices has weak T activation but non-negligible E_{OX} dependence as shown. Overall, PNO devices show stronger E_{OX} dependence and T activation compared to RTNO devices.

Figure 1.17 shows time evolution of ΔV_T at different stress T, in PNO-SiON devices having (a) proper PNA process but different PNO N dose, as well as (b) proper and improper PNA processes and different N dose. In (b), the N content for the proper PNA process is higher than that for the improper PNA process; refer to Table 1.1. As expected, ΔV_T increases with increase in N dose for proper PNA devices as shown in Fig. 1.17a. The long-time power-law time exponent (n) reduces with increased N dose. Moreover, PNO devices with high N dose show very high, but weak T activated degradation in the sub-1 ms time scale. Therefore, PNO devices with proper PNA but very high N dose behave quite similar to RTNO devices, as shown in Fig. 1.16b. On the other hand, Fig. 1.17b shows that in spite of having relatively lower N dose, improper PNA devices show very large overall ΔV_T, as well as very high, weak T activated degradation in the sub-1 ms time scale. The power-law time exponent (n) at longer t_{STR} is much lower for improper PNA devices. Therefore, in spite of lower N dose, improper PNA-PNO devices also behave similar to RTNO devices, and Fig. 1.17b shows the importance of utilizing proper PNA [47] after N incorporation using the DPN process [60]. Figures 1.16

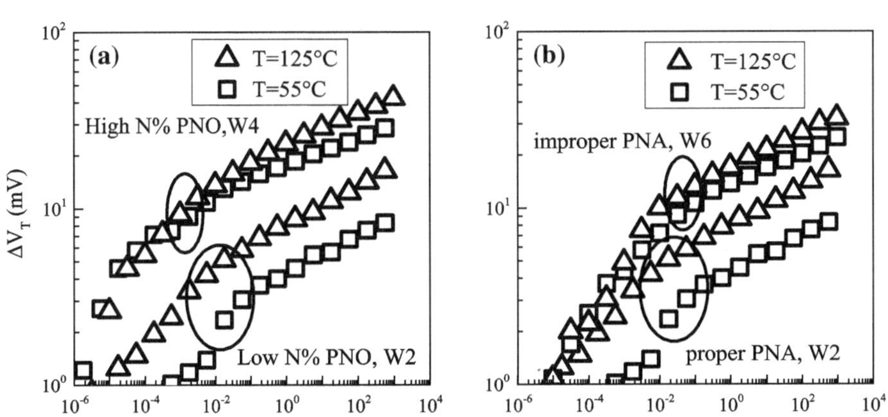

Fig. 1.17 Time evolution of ΔV_T from short to long stress time for PNO SiON p-MOSFETs having **a** different N dose and **b** different PNA, under NBTI stress at different T. Data obtained from UF-OTF measurements after mobility correction

and 1.17 clearly demonstrate that SiON gate insulator process has significant impact on NBTI degradation. Studying gate insulator process dependence helps understand the underlying NBTI physical mechanism, which is explained and modeled later in this book.

1.3.2 Process Impact on NBTI Parameters

The power-law time exponent (n) corresponding to longer time BTI degradation is of significant interest, since it determines the extrapolation of measured data to end-of-life. Figure 1.18 plots the exponent (n) as a function of (a) stress E_{OX} and (b) stress T for RTNO-SiON and different PNO-SiON devices; n is extracted from linear regression of ΔV_T time evolution data in t_{STR} range of 10 s to 1 Ks. For all devices, similar values of n (within experimental error) have been obtained for different stress E_{OX} (or $V_{G\text{-}STR}$) and T. PNO devices having different EOT and low to moderately high N dose have similar n, which is much larger compared to RTNO devices. Note that the magnitudes of ΔV_T and n are always reciprocal to each other. In spite of having lower N dose, RTNO device shows higher ΔV_T and lower n compared to PNO devices, as evident from Figs. 1.16 and 1.18.

Figure 1.19 plots stress E_{OX} dependence of time exponent (n) for PNO-SiON devices having (a) proper PNA process but different N dose, as well as (b) proper and improper PNA processes respectively with relatively higher and lower N dose. Increase in N dose for proper PNA process and the improper PNA process result in lower n values as shown. Once again, lower n values are exactly reciprocal to higher ΔV_T values for these processes as shown in Fig. 1.17.

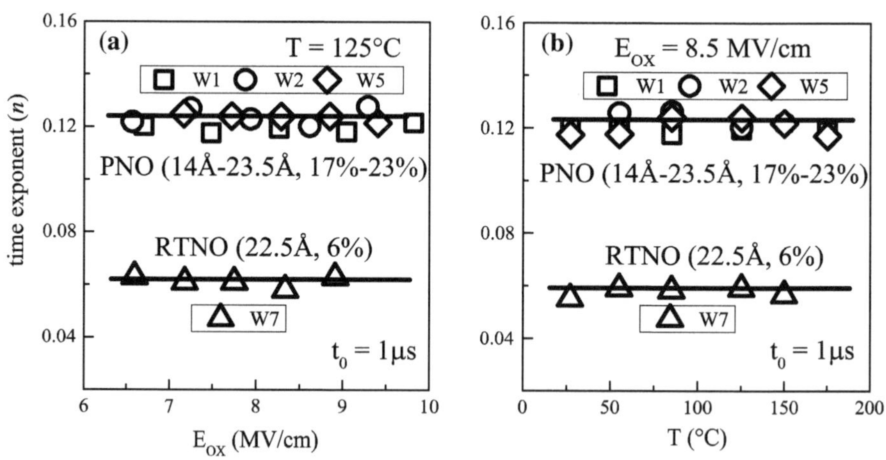

Fig. 1.18 Extracted long-time power law time exponent n versus **a** stress E_{OX} and **b** stress T, for NBTI stress in p-MOSFETs having PNO and RTNO SiON gate insulators. PNO devices have different N content and EOT

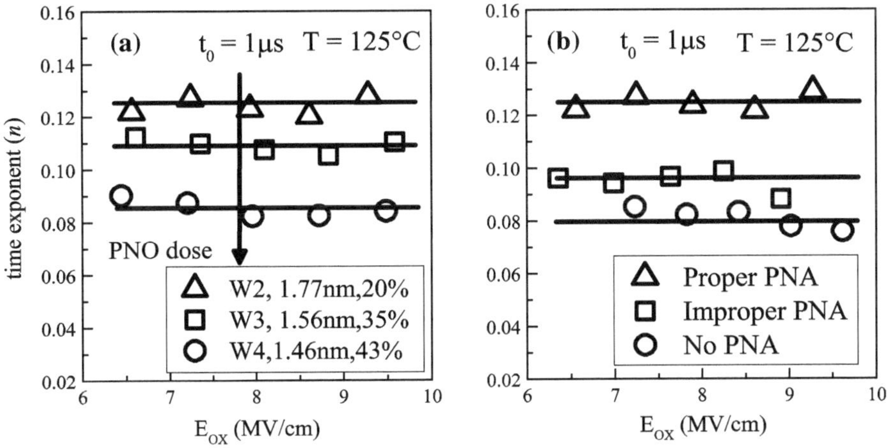

Fig. 1.19 Extracted long-time power law time exponent n versus stress E_{OX} for NBTI stress in PNO SiON p-MOSFETs having **a** different N dose and **b** different PNA

Before proceeding further, it is important to note that the observed values of n is dependent on the t_0 delay for OTF measurements, and will be discussed further in Chap. 2. It is also important to highlight that several stress and measurement artifacts can impact the stress E_{OX} (or $V_{G\text{-}STR}$) and T independence of measured n as observed in Figs. 1.18 and 1.19, and they are listed as follows:

(a) Additional contribution from TDDB like bulk trap generation ΔN_{OT} can result in increased n values at longer t_{STR} for higher stress V_G and/or T, since ΔN_{OT} demonstrates higher time exponent, $V_{G\text{-}STR}$ acceleration and T activation than generated interface traps ΔN_{IT} as mentioned before [32, 33, 61]; refer to Chap. 4 for additional details. Corruption of NBTI data by TDDB process will be discussed further in Chap. 2.
(b) Recovery related artifacts associated with relatively slower BTI measurements can result in increased n values at higher T [62], which is explained in detail elsewhere [63], and will be discussed in Chap. 2.
(c) In some situations, ΔV_T magnitude can become high enough to significantly reduce the effective stress E_{OX} [64]. This negative feedback would cause saturation in ΔV_T at higher stress time, resulting in lower n values. In such situations, n would reduce at higher stress V_G and T.
(d) Finally, as discussed later in this book, BTI induced V_T shift has two mutually uncorrelated components, generation of new defects and charging of process induced pre-existing defects; while the former evolves as power-law with stress time and has larger T activation, the later saturates quickly and has lower T activation. In certain situations, especially if BTI experiments are carried out over a wide range of T, charging of pre-existing defects would dominate at lower T and the generation of new defects would dominate at higher T, hence the time exponent (n) would increase with increase in T.

Therefore, BTI stress and measurements must be carefully done to obtain similar values of time exponent (n) across different stress E_{OX} (or $V_{G\text{-}STR}$) and T as shown in Figs. 1.18 and 1.19.

The observation of similar power-law time exponent (n) for different E_{OX} and T implies that longer-time ΔV_T versus t_{STR} data are parallel to each other when plotted in a log–log scale, as shown in Fig. 1.16c, d. Therefore, ΔV_T at fixed t_{STR} can be extracted for different E_{OX} at constant stress T, and different stress T at constant E_{OX}, and the extracted E_{OX} and T dependence of ΔV_T would therefore be independent of t_{STR}. The E_{OX} dependence of BTI can be used to extrapolate measured accelerated stress data to normal operating condition. The T dependence can be used to understand the underlying BTI physical mechanism, as discussed later in this book.

Figure 1.20 shows ΔV_T at fixed t_{STR} as a function of (a) stress E_{OX} and (b) stress T for PNO with proper PNA devices having different N dose and RTNO devices. In spite of having lower N dose, RTNO devices show higher ΔV_T when compared to PNO devices for a particular E_{OX} and T. Moreover, power-law field acceleration (Γ_E) and Arrhenius T activation energy (E_A) values are lower for RTNO compared to PNO devices. Note that since ΔV_T shows power-law time dependence for longer stress time and the time exponent (n) is similar across different E_{OX} and T, extracted Γ_E and E_A would be independent of the t_{STR} value used for ΔV_T measurement. It is important to remark that unless n remains independent of E_{OX} and T, extracted Γ_E and E_A will have not much meaning as their magnitude would be dependent on the exact value of t_{STR} used for extraction. Therefore, it is important to ensure that the artifacts mentioned above are avoided while determining field acceleration and temperature activation of BTI degradation.

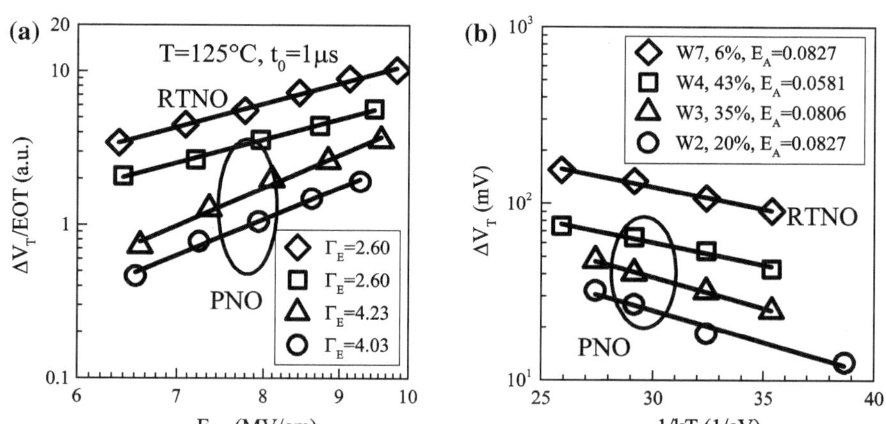

Fig. 1.20 Fixed time ΔV_T versus **a** stress E_{OX} and **b** stress T, for NBTI stress in SiON p-MOSFETs having PNO and RTNO gate insulators. E_{OX} dependence is shown in a log–log plot, T dependence in a semi-log plot. PNO devices have different N content and EOT

It is interesting to note from results presented in Figs. 1.16, 1.17, 1.18, 1.19, 1.20 that a particular SiON process that results in higher ΔV_T magnitude also results in lower NBTI parameters n, E_A and Γ_E. As a further demonstration of process dependence of NBTI parameters, Fig. 1.21 plots (a) power-law time exponent (n) obtained in t_{STR} range of 10 s to 1 Ks, as well as (b) Arrhenius T activation energy (E_A) and (c) power-law field acceleration factor (Γ_E) as a function of atomic N% measured using XPS for different SiON processes [36, 43]. The control SiO_2 data are also shown, although it should be noted that the device suffers from Boron penetration from the poly-Si gate due to very low N content in the gate stack [65]. The PNO with proper PNA devices have different starting base oxide thickness and were subjected to different N dose, which resulted in different N% in the gate stack [43] as shown in Table 1.1. For these devices, n, E_A and Γ_E reduce with increase in N%, and rapid changes are observed for N > 30 %, presumably due to large increase in N density at or near the Si/SiON interface. Furthermore, for a particular N%, RTNO and PNO without proper PNA devices show lower n, E_A and Γ_E compared to the PNO with proper PNA devices. This is again attributed to relatively larger N concentration at the Si/SiON interface for RTNO and improper

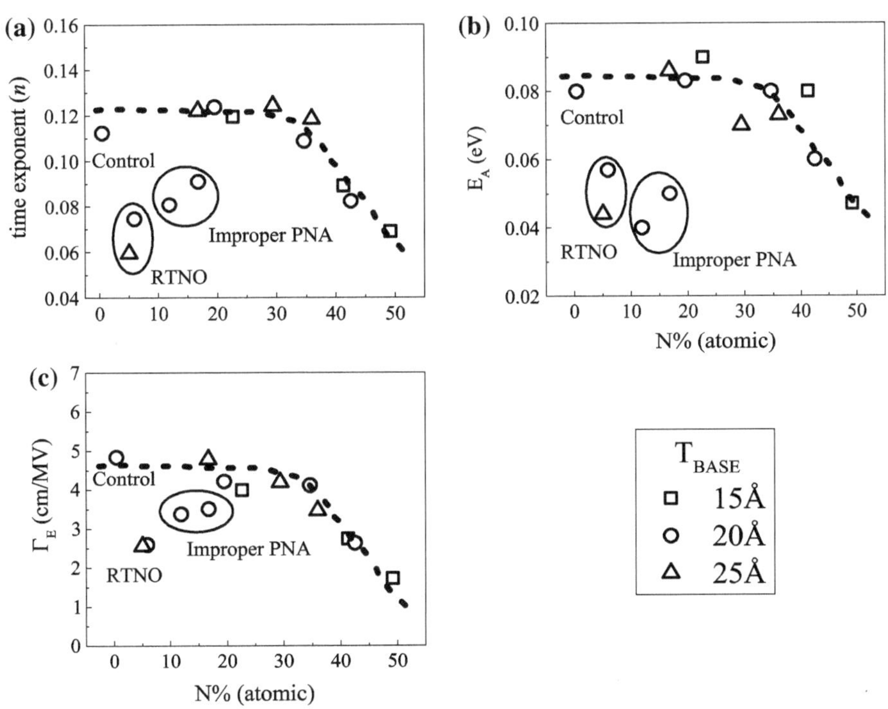

Fig. 1.21 **a** Power-law time exponent n, **b** T activation energy E_A and **c** E_{OX} acceleration factor Γ_E versus XPS measured atomic N% for NBTI stress in different SiON processes. PNO devices have different starting base oxide thickness and different N dose, resulting in different N%. Control SiO_2 data are also shown

PNA-PNO devices. The impact of Nitrogen on NBTI parameters will be explained later in this book.

1.4 Ultra-Fast DC BTI Characterization in HKMG MOSFETs

Results presented in this and later sections of this chapter are obtained by Ultra-Fast Measure-Stress-Measure (UF-MSM) method, where transfer I_{DLIN} versus V_G characteristics are measured in approximately 10 μs before and during logarithmically spaced intervals of BTI stress [66]. As illustrated in Fig. 1.22 and discussed in detail in Chap. 2, transfer *I–V* curves are measured for V_G sweeps ranging from ±1.2 to ±0.6 V, the "+" and "−" signs are for PBTI and NBTI measurements, respectively. V_T is extracted from measured *I–V* sweep data using the conventional "max-g_m" method [2].

Table 1.2 lists HKMG MOSFETs studied in this and following sections. The devices were fabricated using Gate First integration scheme having different Rapid Thermal Process (RTP) based ultra-scaled thermal IL layers. All devices have identical HfO_2 High-K thickness. EOT scaling has been achieved by fabricating RTP thermal IL having thickness of 5 Å (D1) and 3 Å (D2), post-High-K nitridation

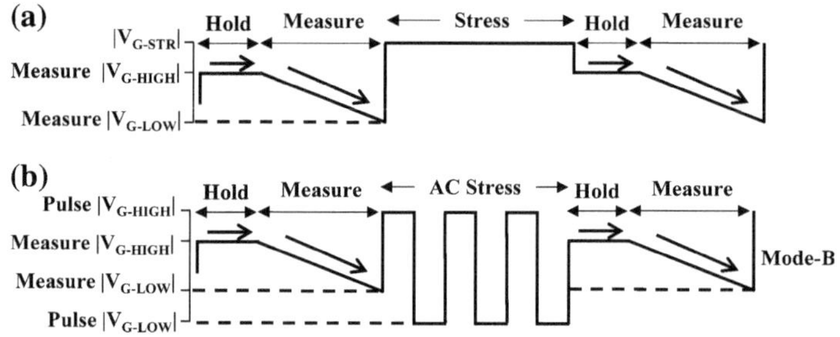

Fig. 1.22 Applied gate waveforms for (*top*) DC and (*bottom*) mode-B AC stress and UF-MSM measurements. Default hold time is 50 ns and sweep time is 10 μs. Refer to Chap. 2 for additional details

Table 1.2 Details of HKMG devices for NBTI and PBTI study using UF-MSM method

Device no.	IL thickness (Å)	High-K thickness (Å)	Remarks
D1	5	23	Low-T RTP
D2	3	23	Low-T RTP
D3	2.5	23	Low-T RTP, PHKN after high-K deposition
D4	1.5	23	Low-T RTP in N ambient

of the 3 Å IL device (D3), and sub 2 Å IL using RTP on Nitrogen (N) surface passivated Si surface (D4), refer to [67] and Chap. 3, Fig. 3.2 for further details. All results shown in Sects. 1.4 and 1.5 are obtained from device D2 unless specifically mentioned otherwise.

1.4.1 Drain Current Degradation

Figure 1.23 shows transfer I_{DLIN} versus V_G characteristics measured before and after NBTI and PBTI stress respectively in (a) p-channel and (b) n-channel HKMG MOSFETs. Reduction in I_{DLIN} is observed for both NBTI and PBTI stress. NBTI results in negative shift and skew of the *I–V* characteristics after stress, while PBTI only results in positive shift and negligible skew of the *I–V* characteristics. Shift in *I–V* characteristics after stress is a measure of ΔV_T due to gate insulator charges; positive charges for NBTI causing negative ΔV_T, while negative charges for PBTI causing positive ΔV_T. The g_m versus V_G characteristics are also plotted in Fig. 1.23 before and after (c) NBTI and (d) PBTI stress. The presence of skew in *I–V* characteristics implies transconductance degradation (Δg_m) for NBTI stress, while negligible Δg_m is observed for PBTI stress, consistent with results shown

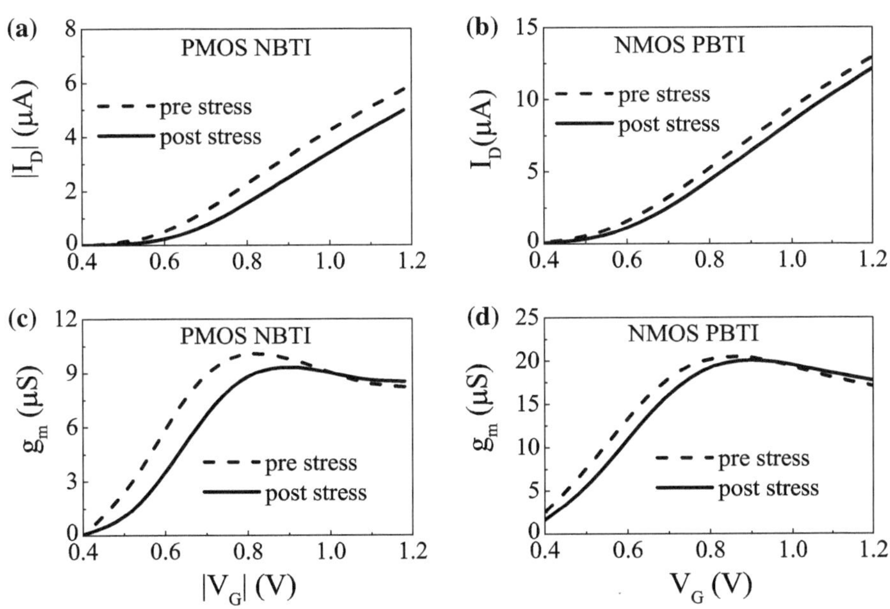

Fig. 1.23 **a**, **b** I_{DLIN} versus V_G and **c**, **d** g_m versus V_G characteristics measured using the UF-MSM method before and after **a**, **c** NBTI and **b**, **d** PBTI stress respectively in p- and n-channel HKMG MOSFETs

earlier in Fig. 1.14. Note that the term "transconductance degradation" after NBTI stress is applicable for peak g_m and for g_m corresponding to low V_G values, while interestingly, an apparent "improvement" is observed in g_m for high V_G values. The implication of this aspect will be discussed later in Chap. 2.

Figure 1.24 shows the correlation of peak Δg_m to ΔV_T during NBTI and PBTI stress respectively in p- and n-channel HKMG MOSFETs. Two different devices D1 and D2 having different IL thickness of the HKMG gate insulator have been used. Note that these IL thicknesses are much smaller than the IL thickness of the devices shown in Fig. 1.14 [16]. The values of Δg_m and ΔV_T are obtained from *I–V* characteristics for different intervals of BTI stress, the stress being performed at different $V_{G\text{-STR}}$. For a given ΔV_T, NBTI shows much larger Δg_m than PBTI stress. Identical Δg_m to ΔV_T correlation is observed for different IL thickness for NBTI stress. On the other hand, PBTI stress on devices having thicker IL has negligible Δg_m, while higher Δg_m for a given ΔV_T is observed for PBTI stress on the gate stack with thinner IL. This is consistent with the fact that NBTI results in IL degradation while PBTI causes degradation of the HfO_2 High-K layer. The increase in Δg_m for a given ΔV_T for PBTI stress in thinner IL stack is due to increase in Coulomb scattering, since High-K charges come closer to the channel as IL is scaled. Once again, Fig. 1.24 shows absence of IL degradation during PBTI stress in well-optimized devices.

1.4.2 Threshold Voltage Degradation

Figure 1.25 shows time evolution of ΔV_T from very short to long stress time (t_{STR}) during (a) NBTI and (b) PBTI stress in p- and n-channel HKMG MOSFETs respectively, the stress being done at different $V_{G\text{-STR}}$ and T. Figure 1.25c, d re-plot the same data for longer t_{STR}. As expected, enhanced BTI degradation is observed when stress is done at higher $V_{G\text{-STR}}$ and T, which is true for both n- and p-channel

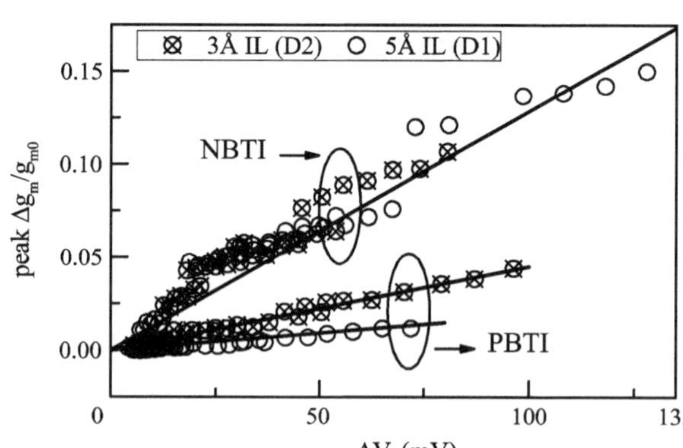

Fig. 1.24 Correlation of Δg_m versus ΔV_T for NBTI and PBTI stress respectively in p- and n-channel MOSFETs having different IL thickness of HKMG gate insulator stack

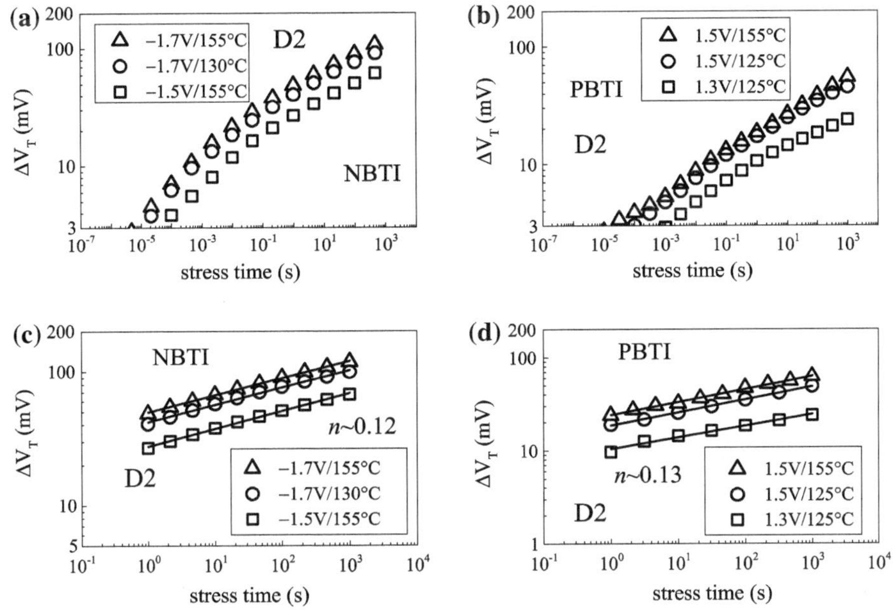

Fig. 1.25 Time evolution of ΔV_T from short to long stress time for **a** NBTI and **b** PBTI stress respectively in p- and n-channel HKMG MOSFETs at different $V_{G\text{-}STR}$ and T. Data obtained from UF-MSM measurements. Re-plot of long-time ΔV_T time evolution data for **c** NBTI and **d** PBTI stress

devices. Note that similar to NBTI in SiON p-MOSFETs, ΔV_T for both NBTI and PBTI in HKMG MOSFETs also show rapid increase in time during the initial period of stress ($t_{STR} < 1$ s), while power law time dependence ($\Delta V_T \sim t^n$) is observed for relatively longer stress time ($t_{STR} > 1$ s).

Figure 1.26 plots long-time power law time exponent (n) as a function of (a, b) stress E_{OX} and (c, d) stress T, for both (a, c) NBTI and (b, d) PBTI stress respectively in p- and n-channel HKMG MOSFETs. The exponent is obtained by linear regression of measured ΔV_T time-evolution data in t_{STR} interval of 10 s to 1 Ks. For this particular HKMG process, NBTI and PBTI show similar values of n; moreover, the values of n are also similar across different stress E_{OX} and T. Note that the stress E_{OX} and T independence of n is similar to that observed for NBTI in SiON p-MOSFETs, and is obtained only when the artifacts mentioned in Sect. 1.3 do not influence BTI stress and measurements.

To evaluate the impact of stress E_{OX} and T, BTI stress tests were performed at three sets of T; for each T, three sets of E_{OX} have been used, and the time evolution of ΔV_T has been obtained. Figure 1.27 plots the (a, b) stress E_{OX} and (c, d) stress T dependence of ΔV_T at a fixed t_{STR}, obtained from measured ΔV_T time evolution data for (a, c) NBTI and (b, d) PBTI stress in p- and n-channel HKMG MOSFETs, respectively. The E_{OX} dependence of ΔV_T is used to extract the power-law field acceleration factor (Γ_E), which is found to be independent of T, as shown in

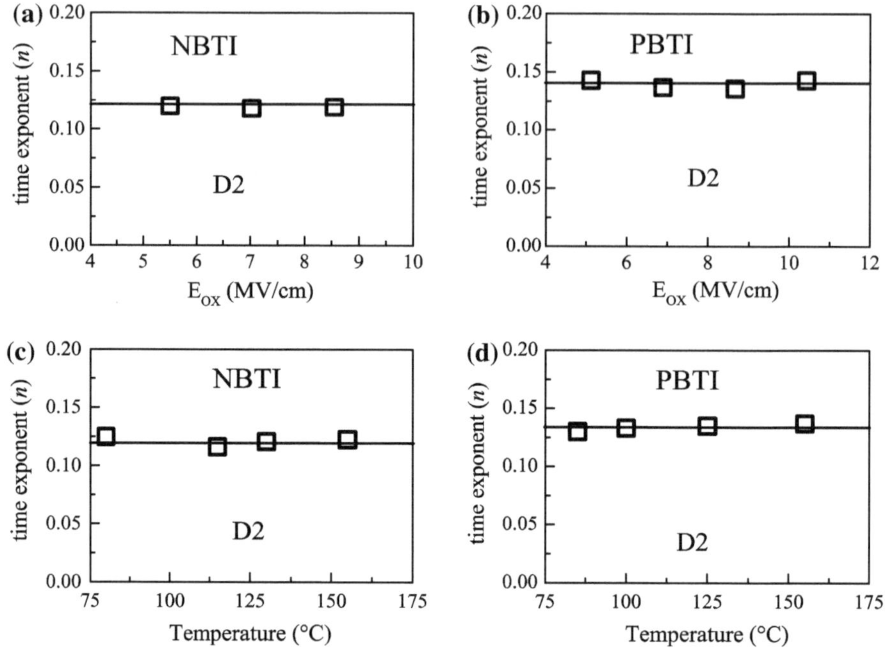

Fig. 1.26 Extracted long-time power law time exponent n versus **a**, **b** stress E_{OX} and **c**, **d** stress T, for **a**, **c** NBTI and **b**, **d** PBTI stress respectively in p- and n-channel HKMG MOSFETs

Fig. 1.27a, b respectively for NBTI and PBTI stress. On the other hand, T dependence of ΔV_T for three sets of E_{OX} is used to extract Arrhenius T activation energy (E_A), which is found to be independent of E_{OX}, as shown respectively in Fig. 1.27c, d for NBTI and PBTI stress. Note that it is important to obtain parallel ΔV_T versus t_{STR} curves when plotted in a log–log scale, or similar time exponent n across different stress E_{OX} and T, to obtain similar Γ_E and E_A at different t_{STR} as mentioned before. Therefore, the experiments must be free from different stress and measurement artifacts as mentioned in Sect. 1.3. Moreover, note that T independence of Γ_E and E_{OX} independence of E_A is also obtained only when BTI stress and measurements are free from the artifacts mentioned in Sect. 1.3.

As a counter example, lower than expected ΔV_T would be measured at higher E_{OX} and/or higher T when BTI is impacted by stress saturation. In such situations, lower exponent n would be obtained at higher stress E_{OX} and T, the E_{OX} acceleration factor Γ_E would be lower at higher T, while the T activation energy E_A would be lower at higher E_{OX}, and the independence observed in Fig. 1.27 would be lost. In a similar way, recovery impact or TDDB traps would also corrupt extraction of Γ_E and E_A, since these effects would result in non-parallel ΔV_T versus t_{STR} curves in a log–log plot or different n for different stress E_{OX} and T. Therefore, BTI stress and measurements must be carefully done to obtain mutually independent Γ_E and E_A values.

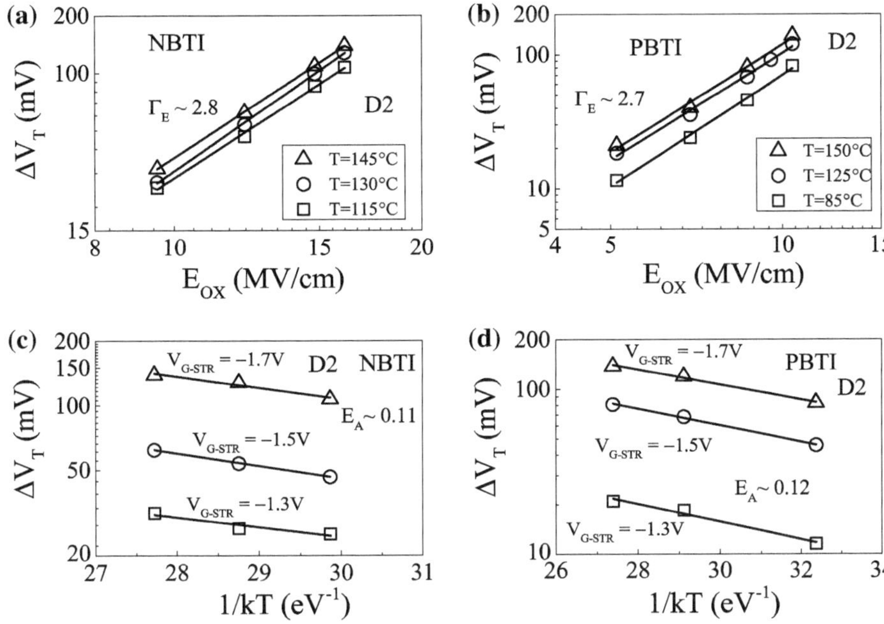

Fig. 1.27 Fixed time UF-MSM measured ΔV_T as a function of **a**, **b** stress E_{OX} for different stress T (log–log plot) and **c**, **d** stress T for different stress E_{OX} (semi-log plot) obtained under **a**, **c** NBTI and **b**, **d** PBTI stress respectively in p- and n-channel HKMG MOSFETs

1.5 Ultra-Fast Characterization of BTI Recovery and AC Degradation in HKMG MOSFETs

As mentioned before, it is now well known that BTI degradation recovers after the reduction of stress V_G [31, 34]. On one hand, recovery poses a serious challenge to BTI measurements, as experimental data must be obtained to capture the true effect of stress with negligible impact of recovery. This is discussed in Chap. 2. On the other hand, recovery results in lower BTI degradation for AC stress, as buildup of ΔV_T during pulse on phase partially recovers during pulse off phase. This "relief" in degradation obtained for AC BTI is of interest to all switching CMOS circuits, and is modeled in Chap. 6.

1.5.1 Recovery Measurements

Figure 1.28 shows the time evolution of ΔV_T measured after DC (a) NBTI and (b) PBTI stress respectively in p- and n-channel HKMG MOSFETs. The devices were first stressed at a fixed $V_{G\text{-}STR}$ and for a fixed duration of t_{STR}, and subsequently the V_G is reduced to a low value ($V_G = V_{G\text{-}REC}$), and the recovery of ΔV_T is

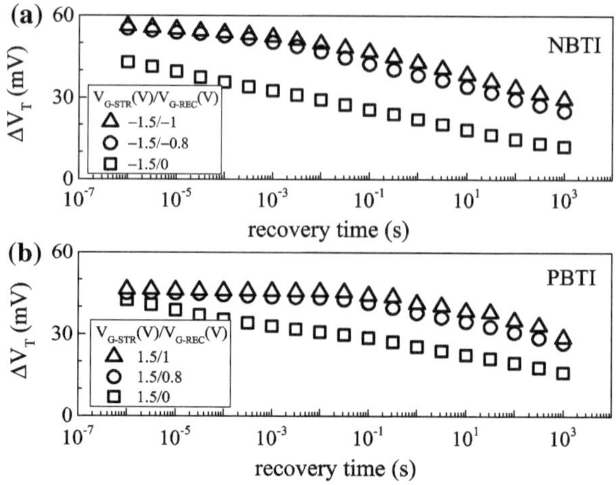

Fig. 1.28 Time evolution of ΔV_T recovery measured at different $V_{G\text{-}REC}$ following **a** NBTI and **b** PBTI stress respectively in p- and n-channel HKMG MOSFETs. Data obtained from UF-MSM measurements

measured using 10 μs UF-MSM method in logarithmic intervals of time. Three different values of $V_{G\text{-}REC}$ have been used as shown, while MSM *I–V* sweeps were always performed in the V_G range of ±1.1 to ±0.6 V, the "+" and "−" signs respectively represent PBTI and NBTI measurements. Note that ΔV_T starts to recover immediately after stress when the devices are recovered at $V_{G\text{-}REC} = 0$ V. However, a delayed start of recovery is observed and overall magnitude of recovery is lower when recovery is done at higher $|V_{G\text{-}REC}|$. Interestingly, note that both NBTI and PBTI demonstrate very similar impact of recovery V_G as shown.

As a further illustration, Fig. 1.29 shows magnitude of ΔV_T recovery as a function of recovery V_G for different recovery time (t_{REC}). The recovery values are obtained from measured time evolution of ΔV_T recovery after DC (a) NBTI and (b) PBTI stress in p- and n-channel HKMG MOSFETs respectively, while stress V_G and stress time were kept constant across all experiments. As shown, the magnitude of ΔV_T recovery is governed by both $V_{G\text{-}REC}$ and t_{REC}. In these HKMG devices, for

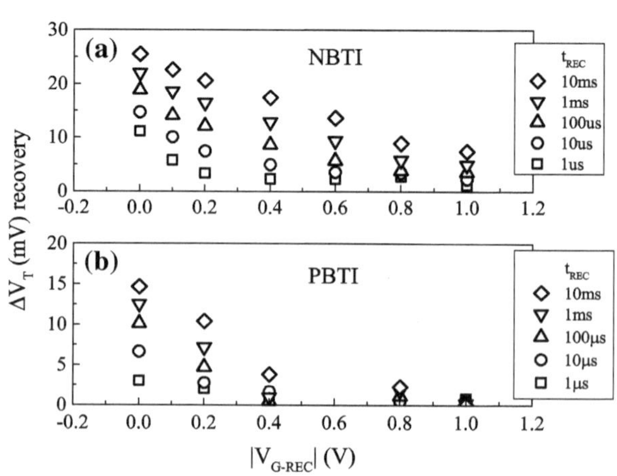

Fig. 1.29 Magnitude of ΔV_T recovery as a function of $V_{G\text{-}REC}$, for different recovery time following **a** NBTI and **b** PBTI stress respectively in HKMG p- and n-channel MOSFETs. Data obtained from UF-MSM measurements

a particular t_{REC}, lower ΔV_T recovery is observed for $|V_{G\text{-}REC}| > 0.6$ V, while the magnitude of ΔV_T recovery significantly increases for $|V_{G\text{-}REC}| \leq 0.4$ V, and similar behavior is observed for both NBTI and PBTI stress.

The information provided in Fig. 1.29 can be used as a guideline to determine the correct procedure for "recovery artifact free" BTI measurements. As discussed in detail in Chap. 2, BTI measurements are often performed by reducing V_G from stress, either by using full I_{DLIN} versus V_G sweep [66] or using I_{DLIN} measurement at a fixed V_G [62]. The measurement time and measurement V_G (fixed or sweep range) would determine the magnitude of BTI recovery. It is obvious that recovery would be higher for higher measurement or stress-off time, while recovery would also be higher if relatively lower $|V_G|$ is used for end of sweep or one spot measurements. The sweep V_G range and sweep time for UF-MSM measurements used in these sections are chosen to keep BTI recovery at a minimum. Finally, it is important to note that many industry level BTI measurements are done using 1 ms delay where I_{DLIN} is recorded at a fixed, low V_G (close to pre-stress V_T) value [62]. Such measurements would likely suffer from recovery related artifacts and need careful attention. This aspect is discussed in Chap. 2.

1.5.2 AC BTI Measurements

AC stress waveform having a particular pulse duty cycle (PDC) and frequency (*f*) is applied to the gate, as shown in Fig. 1.22 and discussed in detail in Chap. 2, and full *I–V* measurements are performed in 10 μs before and during stress, by interrupting the stress at logarithmic intervals of time. The stress time, t_{STR} is recorded till the end of last full cycle as shown, and the net pulse on time would therefore depend on the value of PDC. Similar to DC stress, *I–V* sweeps are done in the V_G range of ±1.1 to ±0.6 V, the "+" and "−" signs represent PBTI and NBTI measurements, respectively. Note that it is necessary to use identical measurement delay and sweep bias range for DC and AC experiments, to determine proper AC/DC BTI ratio. This aspect will be discussed in Chap. 2.

Figure 1.30 shows time evolution of ΔV_T for AC (a) NBTI and (b) PBTI stress respectively in p- and n-channel HKMG MOSFETs. AC gate pulse with different PDC and *f* but identical $V_{G\text{-}STR}$ and *T* have been used; note that $V_{G\text{-}STR}$ for AC stress denotes the pulse high value ($V_{G\text{-}HIGH}$). Pulse low value ($V_{G\text{-}LOW}$) is kept at 0 V unless mentioned otherwise. As mentioned before, t_{STR} for AC stress is the time during which AC pulse is applied to gate, while the actual stress happens only during the pulse on time, given by the value of PDC * t_{STR}. Only long t_{STR} values of ΔV_T are plotted, which shows power-law time evolution with similar time exponent *n* across different PDC and *f*. As shown, ΔV_T increases at higher PDC but remains independent of *f* for both NBTI and PBTI stress. The DC stress data are plotted for reference, which show higher ΔV_T when compared to AC stress. As mentioned before, AC BTI is lower than its DC counterpart due to recovery; ΔV_T increases with increase in PDC due to reduction in pulse off or recovery time.

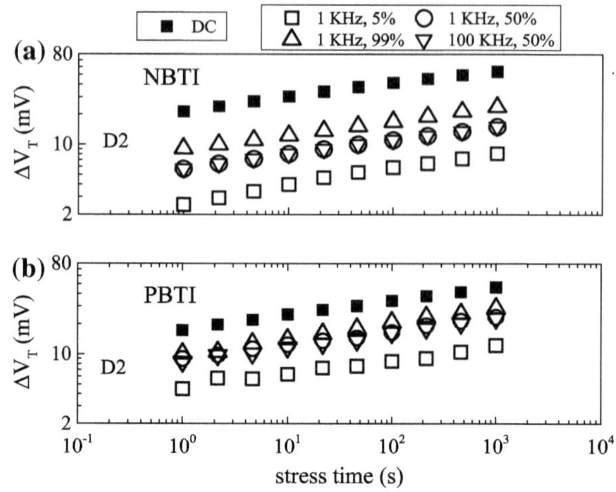

Fig. 1.30 Time evolution of ΔV_T for AC **a** NBTI and **b** PBTI stress respectively in p- and n-channel HKMG MOSFETs at different pulse duty cycle and frequency. Time evolution of ΔV_T for DC NBTI and PBTI stress is also shown. Data obtained from UF-MSM measurements

On the other hand, for a particular PDC, the net pulse on and off times remain identical for AC pulses having different *f* and AC BTI is independent of *f*. The PDC and *f* dependence of BTI will be explained and modeled later in this book.

Figure 1.31 plots the measured time exponent *n* for AC (a, c) NBTI and (b, d) PBTI stress for different (a, b) PDC and (c, d) *f* of the gate pulse. All experiments were performed at identical $V_{G\text{-}STR}$ and *T*, and as before, the exponent *n* is extracted by linear regression of measured data in t_{STR} range of 10 s to 1 Ks. The DC value is also shown as reference. Note that *n* for AC stress remains independent of PDC and *f*, and has higher value when compared to DC stress. Moreover, similar values of *n* are obtained for NBTI and PBTI for both DC and AC stress. The physical mechanism responsible for different *n* for AC and DC BTI will be discussed later in this book.

Figure 1.32 plots ΔV_T at fixed t_{STR} for AC (a, c) NBTI and (b, d) PBTI stress in p- and n-channel HKMG MOSFETs, respectively. Experiments were performed at different PDC and constant *f*, see Fig. 1.32a, b, and at different *f* and constant PDC, see Fig. 1.32c, d, for identical $V_{G\text{-}STR}$ and *T*. In addition to the default pulse low ($V_{G\text{-}LOW}$) value of 0 V, the PDC dependent experiments were also done at higher $|V_{G\text{-}LOW}|$ of 0.8 V, while $V_{G\text{-}STR}$ or pulse high value is kept constant. On the other hand, the *f* dependence is studied at different PDC. As mentioned before, t_{STR} for AC stress is the time for which the AC pulse is applied, while the actual stress time will be governed by PDC and hence the pulse on time. The corresponding DC data at identical *T*, $V_{G\text{-}STR}$ and t_{STR} are also shown.

The PDC dependence of ΔV_T demonstrates a typical "S" curve for both NBTI and PBTI stress as shown respectively in Fig. 1.32a, b. Note that ΔV_T increases with increase in PDC and a "kink" or "jump" in ΔV_T magnitude is observed for both NBTI and PBTI, between high PDC AC and DC stress. For a particular PDC,

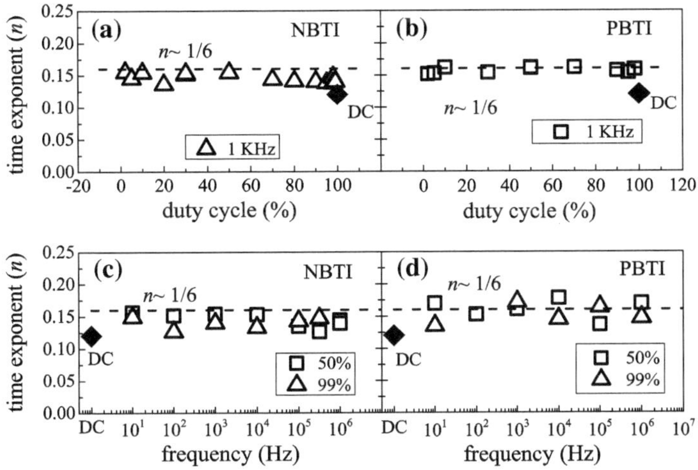

Fig. 1.31 Extracted long-time power law time exponent n versus **a**, **b** pulse duty cycle and **c**, **d** frequency, for AC **a**, **c** NBTI and **b**, **d** PBTI stress respectively in p- and n-channel HKMG MOSFETs. DC data are shown as reference

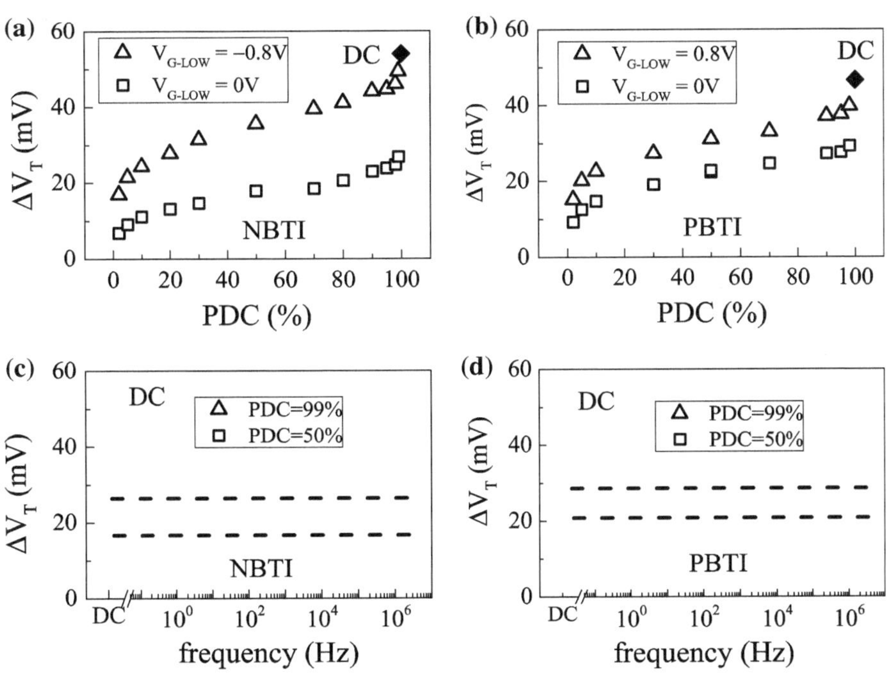

Fig. 1.32 Fixed time UF-MSM measured ΔV_T versus **a**, **b** pulse duty cycle and **c**, **d** frequency, obtained for AC **a**, **c** NBTI and **b**, **d** PBTI stress respectively in p- and n-channel HKMG MOSFETs. DC data are shown as reference. PDC dependence is shown for different $V_{G\text{-LOW}}$, f dependence shown for different PDC

the AC/DC ratio depends on the pulse low value; higher ΔV_T for AC stress is observed when AC stress pulse has relatively higher $|V_{G\text{-}LOW}|$. This is observed for both NBTI and PBTI and is fully consistent with lower ΔV_T recovery for higher $|V_{G\text{-}REC}|$ as shown in Fig. 1.29. On the other hand, f independence of ΔV_T is observed for different PDC and for both NBTI and PBTI stress, as shown, respectively, in Fig. 1.32c, d. The dependence of BTI degradation on PDC, f and $V_{G\text{-}LOW}$ of stress AC gate pulse will be explained and modeled later in this book.

1.6 Ultra-Fast Characterization of the Impact of HKMG EOT Scaling on BTI

As shown in Fig. 1.1, HKMG gate insulator stack consists of two layers; SiO_2 (or SiON) based IL and High-K dielectrics. The present state-of-the-art HKMG gate insulator stacks used in sub 20 nm technology nodes have Chemical Oxide (Chem-Ox) IL and HfO_2 High-K layers, having thicknesses of ~6–7 and ~15–18 Å, respectively [15–18]. It is difficult to reduce the HfO_2 High-K thickness from its present value of ~15–18 Å due gate leakage, and further EOT scaling can be achieved by either nitridation of the gate stack after High-K deposition or by reducing the IL thickness, as discussed in detail elsewhere [67]. Note that reducing the High-K thickness by 1 Å reduces EOT of the gate stack only by 0.2 Å, but increases the gate leakage current (I_G) by 10× [68], and is not beneficial. On the other hand, reduction of IL thickness by 1 Å reduces the gate stack EOT by the same amount, while the gate leakage increases by 10× for every 2 Å reduction in IL thickness. Therefore, IL thickness scaling is more beneficial than High-K thickness scaling. Oxygen scavenging method has been used for Chem-Ox IL scaling [69]; the scaled stacks have good leakage and mobility but suffer from severe increase in BTI degradation [58]. On the other hand, Rapid Thermal Process (RTP) based thermal IL has been successfully scaled down to a thickness of less than 2 Å [67], as shown in Table 1.2, and the scaled gate stacks demonstrate good leakage, mobility as well as BTI reliability. In this section, the impact of EOT scaling of thermal IL stacks, achieved by gate insulator nitridation following High-K deposition as well as IL thickness scaling, on DC BTI is studied. As mentioned before, BTI degradation is characterized using 10 μs UF-MSM method.

1.6.1 Effect of Nitrogen (N) Incorporation

Figure 1.33 shows ΔV_T time evolution during (a, b) NBTI and (c, d) PBTI stress in p- and n-channel HKMG MOSFETs, respectively. The gate insulator stacks of these devices were fabricated (a, c) with nitridation after High-K deposition (D3) and (b, d) RTP IL growth on N passivated Si surface (D4), as shown in Table 1.2 and described in detail in Chap. 3, refer to Fig. 3.2. Since nitridation changes the

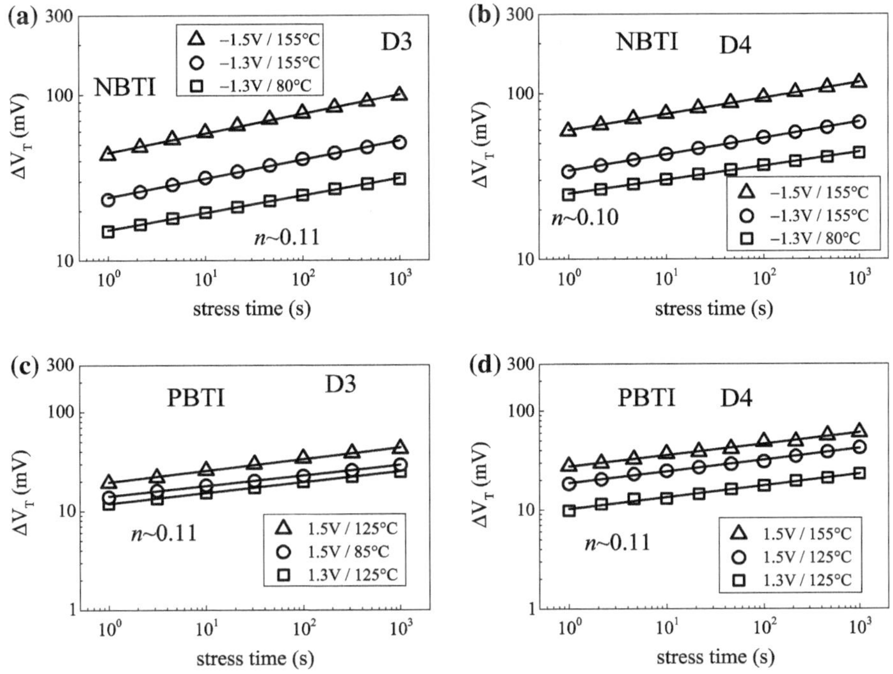

Fig. 1.33 Time evolution of ΔV_T for **a**, **b** NBTI and **c**, **d** PBTI stress respectively in p- and n-channel HKMG MOSFETs with gate stacks having **a**, **c** post High-K nitridation (D3) and **b**, **d** nitrided IL (D4); refer to Table 1.2 for device details. Data obtained from UF-MSM measurements

EOT of the gate insulator, $V_{G\text{-}STR}$ has been adjusted to obtain identical E_{OX} during stress. Once again, E_{OX} is obtained using a simple formula of $(V_{G\text{-}STR} - V_{T0})$/EOT, where V_{T0} is pre-stress V_T.

Only the long t_{STR} data measured at different stress E_{OX} and T are plotted. Similar to non-nitrided devices shown in Fig. 1.25, measured ΔV_T data in different type of nitrided devices also demonstrate power law time dependence; the time exponent n, extracted in t_{STR} range of 10 s to 1 Ks, is always lower for nitrided devices as compared to their non-nitrided counterparts for both NBTI and PBTI stress. Similar to the non-nitrided device, measured n for nitrided device is also independent of stress E_{OX} and T for both NBTI and PBTI stress, as long as the artifacts mentioned in Sect. 1.3 are avoided. For identical stress E_{OX} and T, NBTI shows higher ΔV_T for the nitrided device D3 compared to its not nitrided counterpart D2, refer to Figs. 1.25 and 1.33. Note that the impact of N on NBTI magnitude and time exponent n is exactly identical for SiON and HKMG p-MOSFETs. However device D3 shows lower ΔV_T compared to device D2 for PBTI stress, and therefore, N has just the opposite impact on PBTI magnitude when compared to NBTI stress. However, similar to NBTI, N also results in lower n for PBTI stress as shown. Process impact of NBTI and PBTI is modeled in Chap. 4.

Figure 1.34 plots (a, b) stress E_{OX} and (c, d) stress T dependence of measured ΔV_T, obtained at fixed t_{STR} for (a, c) NBTI and (b, d) PBTI stress respectively in p- and n-channel HKMG MOSFETs. Results are shown for thermal IL based gate insulator stacks without (D2) and with (D3) post High-K nitridation, and also with N passivated IL (D4), refer to Table 1.2. The E_{OX} dependence experiments were performed at constant T, while T dependence is studied at constant E_{OX}. Note that for any E_{OX} and T, devices having N in the gate stack show higher ΔV_T for NBTI but lower ΔV_T for PBTI when compared to non-nitrided device. Moreover, the nitrided devices demonstrate lower power-law field acceleration Γ_E and T activation E_A for both NBTI and PBTI stress, when compared to the non-nitrided device. Note that similar to non-nitrided devices, obtained Γ_E and E_A values are independent of t_{STR} since ΔV_T shows power law time dependence with similar n across stress E_{OX} and T. Moreover, Γ_E is independent of stress T and E_A is independent of stress E_{OX}, not explicitly shown here for brevity, as the artifacts mentioned in Sect. 1.3 do not play a significant role.

Therefore, expect for the magnitude of ΔV_T that increases for NBTI but reduces for PBTI, other BTI parameters such that time exponent (n), field acceleration (Γ_E)

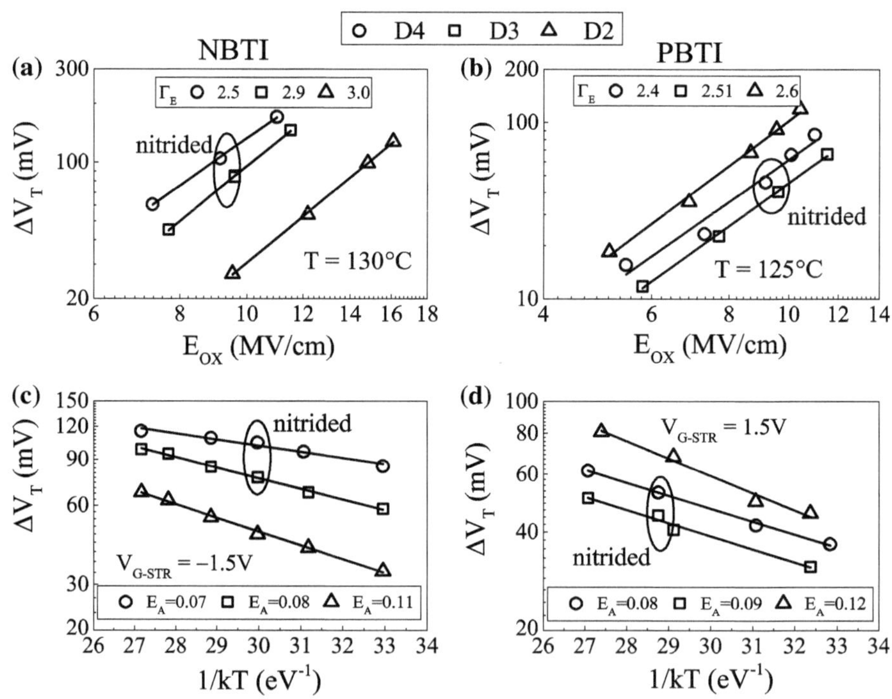

Fig. 1.34 Fixed time UF-MSM measured ΔV_T as a function of **a**, **b** stress E_{OX} and **c**, **d** stress T, obtained under **a**, **c** NBTI and **b**, **d** PBTI stress respectively in p- and n-channel MOSFETs having different HKMG gate insulator processes. E_{OX} dependence is shown in a log–log plot, T dependence in a semi-log plot

and Arrhenius T activation energy (E_A) reduce when N is introduced in HKMG stacks. The mechanism responsible for the impact of N on NBTI and PBTI will be discussed later in this book.

1.6.2 *Effect of Interlayer (IL) Scaling*

Time evolution of ΔV_T is measured during NBTI and PBTI stress respectively in p- and n-channel HKMG MOSFETs having RTP thermal IL based HKMG gate stacks

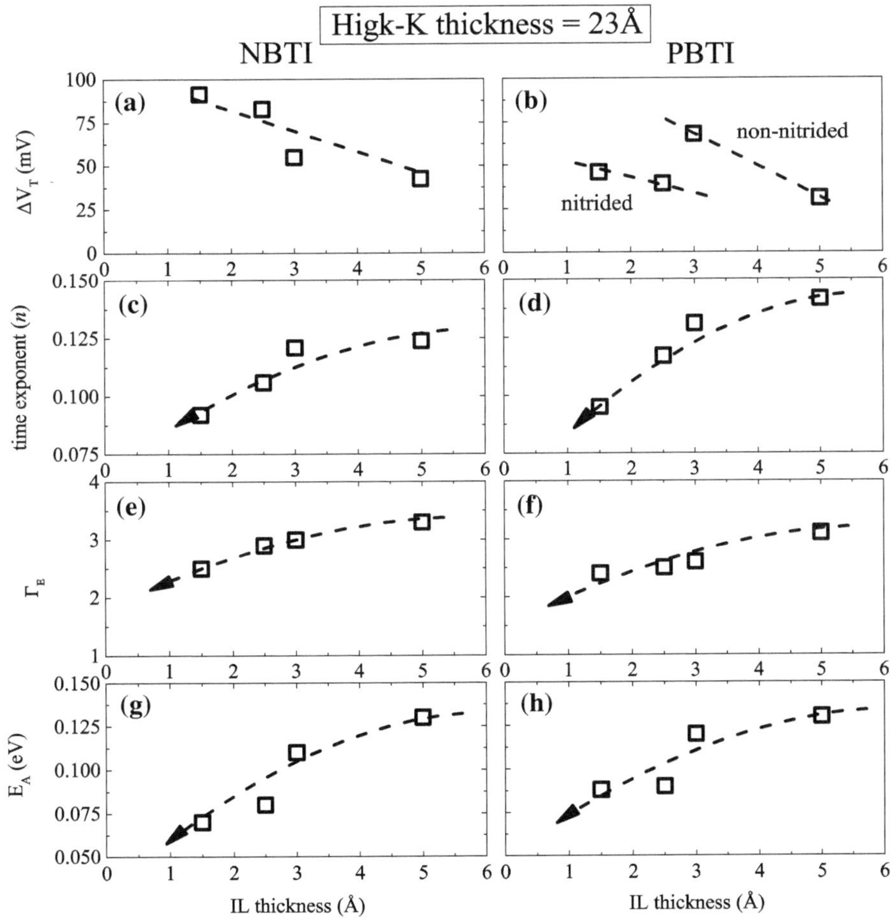

Fig. 1.35 Interlayer (IL) thickness dependence of UF-MSM measured **a**, **b** ΔV_T at fixed overdrive, **c**, **d** power-law time exponent n, **e**, **f** power-law E_{OX} acceleration factor Γ_E, and **g**, **h** T activation energy E_A for NBTI (*left panels*) and PBTI (*right panels*) stress respectively in p- and n-channel HKMG MOSFETs

of varying EOT as listed in Table 1.2, refer to Chap. 3 for additional details [67]. BTI stress has been performed and time evolution of ΔV_T is measured at different $V_{G\text{-}STR}$ and constant T, and different T at constant $V_{G\text{-}STR}$. The magnitude of ΔV_T is extracted at identical gate overdrive ($V_{G\text{-}STR} - V_{T0}$), T and t_{STR}, while the time exponent n, which is found to be independent of $V_{G\text{-}STR}$ and T for all devices, is extracted by linear regression in the t_{STR} range of 10 s to 1 Ks. The power-law E_{OX} acceleration Γ_E and Arrhenius T activation E_A of ΔV_T is extracted at fixed t_{STR} of 1 Ks. However as mentioned before, similar Γ_E and E_A values would be obtained at different t_{STR} as the devices show similar power-law time exponent n for different stress E_{OX} and T. Moreover, Γ_E has been found to be independent of stress T and E_A is independent of stress E_{OX} for all devices.

The measured data are plotted in Fig. 1.35 to show the IL scaling impact on (a, b) ΔV_T, (c, d) time exponent n, (e, f) power-law field acceleration Γ_E and (g, h) Arrhenius T activation E_A for both NBTI and PBTI stress. Note that extreme EOT scaling is achieved by either post-High-K nitridation or by using N surface passivation before IL growth, and would imply higher N concentration for the two lowest EOT gate stacks.

It can be readily observed that ΔV_T increases as EOT is scaled for NBTI stress. Although ΔV_T increases with EOT scaling also for PBTI stress, the nitrided devices have lower ΔV_T than non-nitrided devices as shown. Since experiments are performed at constant overdrive, EOT scaling results in higher E_{OX}. NBTI degradation increases with nitridation and increase in E_{OX} as EOT is scaled. On the other hand, PBTI degradation reduces due to nitridation but increases with increase in E_{OX} as EOT is scaled. The parameters n, Γ_E and E_A reduce with EOT scaling for both NBTI and PBTI stress. The observed reduction in BTI parameters n, Γ_E and E_A with IL scaling is consistent with the impact of N shown earlier in Figs. 1.33 and 1.34. The impact of EOT scaling on BTI magnitude and its parameters will be explained and modeled later in this book.

1.7 Summary

To summarize, NBTI is shown to degrade the performance of SiON and HKMG p-MOSFETs, while only HKMG n-MOSFETs suffer from PBTI issues. The gate insulator process is shown to significantly impact BTI degradation for both SiON and HKMG MOSFETs. The role of gate oxide quality, post metallization anneal, Nitrogen and Fluorine incorporation on NBTI degradation in SiO_2 and SiON p-MOSFETs and the impact of gate insulator processes on NBTI and PBTI degradation in HKMG MOSFETs are reviewed from published reports. The relative magnitude of NBTI and PBTI degradation in recent HKMG technologies is shown to be process and industry specific, although the latest RMG HKMG based FinFETs demonstrate higher NBTI and negligible PBTI degradation universally across the semiconductor industry.

An ultra-fast OTF method has been used to study gate insulator process impact of NBTI for SiON p-MOSFETs. It is observed that PNO devices having relatively moderate N dose and proper PNA show lower overall V_T degradation and negligible ΔV_T in sub-1 ms stress time, higher power-law time exponent n at longer t_{STR}, and stronger E_{OX} acceleration Γ_E and T activation E_A. However, PNO with proper PNA devices having large N dose, PNO with improper PNA devices, and RTNO devices show very large overall degradation and significant ΔV_T in sub-1 ms time scale, lower power-law time exponent n at longer t_{STR}, and weaker E_{OX} and T dependence. Rather than total N dose, the N density at the Si/SiON interface controls NBTI.

An ultra-fast I–V characterization method has been used to measure NBTI and PBTI degradation respectively in p- and n-channel MOSFETs having state-of-the-art HKMG gate insulator stacks. The impact of post-High-K Nitrogen incorporation and IL scaling has been studied. Time evolution of BTI during and after DC stress and during AC stress is characterized, and the impact of stress bias, temperature, duty cycle and frequency have been analyzed. The following observations are made:

(a) The magnitude of ΔV_T for NBTI stress increases while that for PBTI stress reduces with N incorporation in the gate insulator stack, when compared at identical stress E_{OX} and T. The impact of N incorporation on NBTI has been found to be similar for SiON and HKMG p-MOSFETs.
(b) The magnitude of ΔV_T for NBTI increases with IL scaling for HKMG devices when compared at constant overdrive bias. The magnitude of ΔV_T for PBTI also increases with IL scaling, but nitrided devices show lower ΔV_T compared to non-nitrided devices.
(c) Time evolution of ΔV_T shows power-law dependence having exponent n at longer stress time. Similar n has been observed for NBTI and PBTI stress, which does not vary with stress E_{OX} and T, but reduces with higher N concentration in the gate stack and IL scaling. The impact of N incorporation on n has been found to be similar for SiON and HKMG p-MOSFETs. The exponent n is higher for AC compared to DC stress, and does not vary with PDC and f for AC stress.
(d) Power-law field dependence Γ_E and Arrhenius T activation energy E_A values have been found to be independent of stress T and E_{OX} respectively for both NBTI and PBTI stress. The values of Γ_E and E_A reduce with increase in N concentration and IL scaling. Once again, the impact of N incorporation on Γ_E and E_A for NBTI has been found to be similar for SiON and HKMG p-MOSFETs.
(e) Reduction of ΔV_T during recovery following NBTI and PBTI stress shows strong dependence on gate bias during recovery experiments.
(f) AC NBTI and PBTI degradation show similar "S" shaped dependence on pulse duty cycle but show frequency independence. A kink or jump in ΔV_T magnitude is observed between large PDC AC and DC stress.

These features will be explained and modeled in this book.

Acknowledgments The authors would like to acknowledge Vrajesh Maheta for NBTI measurements in SiON devices, Applied Materials for providing SiON and HKMG devices and Ankush Chaudhary for editorial support.

References

1. G.E. Moore, Cramming more components onto integrated circuits. Proc. IEEE **86**, 82 (1998)
2. Y. Taur, T.H. Ning, *Fundamentals of Modern VLSI Devices* (Cambridge University Press, Cambridge, 2009)
3. T. Ghani, M. Armstrong, C. Auth, M. Bost, P. Charvat, G. Glass, T. Hoffmann, K. Johnson, C. Kenyon, J. Klaus, B. McIntyre, K. Mistry, A. Murthy, J. Sandford, M. Silberstein, S. Sivakumar, P. Smith, K. Zawadzki, S. Thompson, M. Bohr, A 90 nm high volume manufacturing logic technology featuring novel 45 nm gate length strained silicon CMOS transistors, in *IEEE International Electron Devices Meeting Technical Digest* (2003), p. 11.6
4. K. Mistry, C. Allen, C. Auth, B. Beattie, D. Bergstrom, M. Bost, M. Brazier, M. Buehler, A. Cappellani, R. Chau, C.-H. Choi, G. Ding, K. Fischer, T. Ghani, R. Grover, W. Han, D. Hanken, M. Hattendorf, J. He, J. Hicks, R. Huessner, D. Ingerly, P. Jain, R. James, L. Jong, S. Joshi, C. Kenyon, K. Kuhn, K. Lee, H. Liu, J. Maiz, B. McIntyre, P. Moon, J. Neirynck, S. Pae, C. Parker, D. Parsons, C. Prasad, L. Pipes, M. Prince, P. Ranade, T. Reynolds, J. Sandford, L. Shifren, J. Sebastian, J. Seiple, D. Simon, S. Sivakumar, P. Smith, C. Thomas, T. Troeger, P. Vandervoorn, S. Williams, K. Zawadzki, A 45 nm logic technology with High-k+ metal gate transistors, strained silicon, 9 Cu interconnect layers, 193 nm dry patterning, and 100 % Pb-free packaging, in *IEEE International Electron Devices Meeting Technical Digest* (2007), p. 247
5. X. Chen, S. Samavedam, V. Narayanan, K. Stein, C. Hobbs, C. Baiocco, W. Li, D. Jaeger, M. Zaleski, H.S. Yang, N. Kim, Y. Lee, D. Zhang, L. Kang, J. Chen, H. Zhuang, A. Sheikh, J. Wallner, M. Aquilino, J. Han, Z. Jin, J. Li, G. Massey, S. Kalpat, R. Jha, N. Moumen, R. Mo, S. Kirshnan, X. Wang, M. Chudzik, M. Chowdhury, D. Nair, C. Reddy, Y.W. Teh, C. Kothandaraman, D. Coolbaugh, S. Pandey, D. Tekleab, A. Thean, M. Sherony, C. Lage, J. Sudijono, R. Lindsay, J.H. Ku, M. Khare, A. Steegen, A cost effective 32 nm high-K/metal gate CMOS technology for low power applications with single-metal/gate-first process, in *Symposium on VLSI Technology: Digest of Technical Papers* (2008), p. 88
6. C. Auth, C. Allen, A. Blattner, D. Bergstrom, M. Brazier, M. Bost, M. Buehler, V. Chikarmane, T. Ghani, T. Glassman, R. Grover, W. Han, D. Hanken, M. Hattendorf, P. Hentges, R. Heussner, J. Hicks, D. Ingerly, P. Jain, S. Jaloviar, R. James, D. Jones, J. Jopling, S. Joshi, C. Kenyon, H. Liu, R. McFadden, B. McIntyre, J. Neirynck, C. Parker, L. Pipes, I. Post, S. Pradhan, M. Prince, S. Ramey, T. Reynolds, J. Roesler, J. Sandford, J. Seiple, P. Smith, C. Thomas, D. Towner, T. Troeger, C. Weber, P. Yashar, K. Zawadzki, K. Mistry, A 22 nm high performance and low-power CMOS technology featuring fully-depleted tri-gate transistors, self-aligned contacts and high density MIM capacitors, in *Symposium on VLSI Technology: Digest of Technical Papers* (2012), p. 131
7. S.-Y. Wu, C.Y. Lin, M.C. Chiang, J.J. Liaw, J.Y. Cheng, S.H. Yang, S.Z. Chang, M. Liang, T. Miyashita, C.H. Tsai, C.H. Chang, V.S. Chang, Y.K. Wu, J.H. Chen, H.F. Chen, S.Y. Chang, K. H. Pan, R.F. Tsui, C.H. Yao, K.C. Ting, T. Yamamoto, H.T. Huang, T.L. Lee, C.H. Lee, W. Chang, H.M. Lee, C.C. Chen, T. Chang, R. Chen, Y.H. Chiu, M.H. Tsai, S.M. Jang, K.S. Chen, Y. Ku, An enhanced 16 nm CMOS technology featuring 2nd generation FinFET transistors and advanced Cu/low-k interconnect for low power and high performance applications, in *IEEE International Electron Devices Meeting Technical Digest* (2014), p. 3.1.1
8. S. Mahapatra, N. Goel, S. Desai, S. Gupta, B. Jose, S. Mukhopadhyay, K. Joshi, A. Jain, A.E. Islam, M.A. Alam, A comparative study of different physics-based NBTI models. IEEE Trans. Electron Devices **60**, 901 (2013)

9. S. Mahapatra, V. Huard, A. Kerber, V. Reddy, S. Kalpat, A. Haggag, Universality of NBTI—from devices to circuits and products, in *IEEE International Reliability Physics Symposium Proceedings* (2014), p. 3B.1.1
10. A. Kerber, E.A. Cartier, Reliability challenges for CMOS technology qualifications with hafnium oxide/titanium nitride gate stacks. IEEE Trans. Device Mater. Reliab. **9**, 147 (2009)
11. B.E. Deal, M. Sklar, A.S. Grove, E.H. Snow, Characteristics of the surface-state charge (Qss) of thermally oxidized silicon. J. Electrochem. Soc. **114**, 266 (1967)
12. N. Kimizuka, K. Yamaguchi, K. Imai, T. Iizuka, C.T. Liu, R.C. Keller, T. Horiuchi, NBTI enhancement by nitrogen incorporation into ultrathin gate oxide for 0.10-μm gate CMOS generation, in *Symposium on VLSI Technology: Digest of Technical Papers* (2000), p. 92
13. Y. Mitani, M. Nagamine, H. Satake, A. Toriumi, NBTI mechanism in ultra-thin gate dielectric —nitrogen-originated mechanism in SiON, in *IEEE International Electron Devices Meeting Technical Digest* (2002), p. 509
14. Y. Mitani, H. Satake, A. Toriumi, Influence of nitrogen on negative bias temperature instability in ultrathin SiON. IEEE Trans. Device Mater. Reliab. **8**, 6 (2008)
15. S. Krishnan, V. Narayanan, E. Cartier, D. Ioannou, K. Zhao, T. Ando, U. Kwon, B. Linder, J. Stathis, M. Chudzik, A. Kerber, K. Choi, Bias temperature instability in high-κ/metal gate transistors—gate stack scaling trends, in *IEEE International Reliability Physics Symposium Proceedings* (2012), p. 5A.1.1
16. S. Pae, M. Agostinelli, M. Brazier, R. Chau, G. Dewey, T. Ghani, M. Hattendorf, J. Hicks, J. Kavalieros, K. Kuhn, M. Kuhn, J. Maiz, M. Metz, K. Mistry, C. Prasad, S. Ramey, A. Roskowski, J. Sandford, C. Thomas, J. Thomas, C. Wiegand, J. Wiedemer, BTI reliability of 45 nm high-K+ metal-gate process technology, in *IEEE International Reliability Physics Symposium Proceedings* (2008), p. 352
17. S. Ramey, A. Ashutosh, C. Auth, J. Clifford, M. Hattendorf, J. Hicks, R. James, A. Rahman, V. Sharma, A. St. Amour, C. Wiegand, Intrinsic transistor reliability improvements from 22 nm tri-gate technology, in *IEEE International Reliability Physics Symposium Proceedings* (2013), p. 4C.5.1
18. K.T. Lee, K. Wonchang, C. Eun-Ae, G. Kim, H. Shin, H. Lee, H. Kim, M. Choe, N.-I. Lee, A. Patel, J. Park, P. Jongwoo, Technology scaling on high-K and metal-gate FinFET BTI reliability, in *IEEE International Reliability Physics Symposium Proceedings* (2013), p. 2D.1.1
19. S. Deora, V.D. Maheta, G. Bersuker, C. Olsen, K.Z. Ahmed, R. Jammy, S. Mahapatra, A comparative NBTI study of $HfO_2/HfSiO_X$, and SiON p-MOSFETs using UF-OTF I_{DLIN} technique. IEEE Electron Device Lett. **30**, 152 (2009)
20. K. Joshi, S. Mukhopadhyay, N. Goel, S. Mahapatra, A consistent physical framework for N and P BTI in HKMG MOSFETs, in *IEEE International Reliability Physics Symposium Proceedings* (2012), p. 5A.3.1
21. S. Desai, S. Mukhopadhyay, N. Goel, N. Nanaware, B. Jose, K. Joshi, S. Mahapatra, A comprehensive AC / DC NBTI model: Stress, recovery, frequency, duty cycle and process dependence, in *IEEE International Reliability Physics Symposium Proceedings* (2013), p. XT.2.1
22. N. Goel, K. Joshi, S. Mukhopadhyay, N. Nanaware, S. Mahapatra, A comprehensive modeling framework for gate stack process dependence of DC and AC NBTI in SiON and HKMG p-MOSFETs. Microelectron. Reliab. **54**, 491 (2014)
23. N. Goel, S. Mukhopadhyay, N. Nanaware, S. De, R.K. Pandey, K.V.R.M. Murali, S. Mahapatra, A comprehensive DC/AC model for ultra-fast NBTI in deep EOT scaled HKMG p-MOSFETs, in *IEEE International Reliability Physics Symposium Proceedings* (2014), p. 6A.4.1
24. S. Novak, S. Parker, C. Becher, D. Agostinelli, M. Chahal, M. Lui, M. Nakapani, P. Packan, P. Natarajan, Transistor aging and reliability in 14 nm tri-gate technology, in *IEEE International Reliability Physics Symposium Proceedings* (2015), p. 2F.2
25. S. Mahapatra, D. Saha, D. Varghese, P.B. Kumar, On the generation and recovery of interface traps in MOSFETs subjected to NBTI, FN, and HCI stress. IEEE Trans. Electron Devices **53**, 1583 (2006)

26. Y.M. Randriamihaja, X. Federspiel, V. Huard, A. Bravaix, P. Palestri, New hot carrier degradation modeling reconsidering the role of EES in ultra short N-channel MOSFETs, in *IEEE International Reliability Physics Symposium Proceedings* (2013), p. XT.1.1
27. S. Takagi, M. Takayanagi, A. Toriumi, Experimental examination of physical model for direct tunneling current in unstressed/stressed ultrathin gate oxides, in *IEEE International Electron Devices Meeting Technical Digest* (1999), p. 461
28. E. Cartier, A. Kerber, Stress-induced leakage current and defect in nFETs with HfO_2/TiN gate stacks during positive-bias temperature stress, in *IEEE International Reliability Physics Symposium Proceedings* (2009), p. 486
29. M.A. Alam, J. Bude, A. Ghetti, Field acceleration for oxide breakdown-can an accurate anode hole injection model resolve the E vs. 1/E controversy?, in *IEEE International Reliability Physics Symposium Proceedings* (2000), p. 21
30. T. Nigam, P. Peumans, TDDB in the presence of interface states: implications for the PMOS reliability margin, in *IEEE International Electron Devices Meeting Technical Digest* (2008). doi:10.1109/IEDM.2008.4796814
31. S. Rangan, N. Mielke, E.C.C. Yeh, Universal recovery behavior of negative bias temperature instability [PMOSFETs], in *IEEE International Electron Devices Meeting Technical Digest* (2003), p. 14.3.1
32. S. Mahapatra, P. Bharath Kumar, M.A. Alam, Investigation and modeling of interface and bulk trap generation during negative bias temperature instability of p-MOSFETs. IEEE Trans. Electron Devices **51**, 1371 (2004)
33. M.A. Alam, S. Mahapatra, A comprehensive model of PMOS NBTI degradation. Microelectron. Reliab. **45**, 71 (2005)
34. H. Reisinger, O. Blank, W. Heinrigs, A. Muhlhoff, W. Gustin, C. Schlunder, Analysis of NBTI degradation- and recovery-behavior based on ultra fast VT-measurements, in *IEEE International Reliability Physics Symposium Proceedings* (2006), p. 448
35. V. Huard, M. Denais, F. Perrier, N. Revil, C. Parthasarathy, A. Bravaix, E. Vincent, A thorough investigation of MOSFETs NBTI degradation. Microelectron. Reliab. **45**, 83 (2005)
36. S. Mahapatra, A.E. Islam, S. Deora, V.D. Maheta, K. Joshi, A. Jain, M.A. Alam, A critical re-evaluation of the usefulness of R-D framework in predicting NBTI stress and recovery, in *IEEE International Reliability Physics Symposium Proceedings* (2011), p. 6A.3.1
37. K.O. Jeppson, C.M. Svensson, Negative bias stress of MOS devices at high electric fields and degradation of MNOS devices. J. Appl. Phys. **48**, 2004 (1977)
38. G. Groeseneken, H.E. Maes, N. Beltran, R.F. De Keersmaecker, A reliable approach to charge-pumping measurements in MOS transistors. IEEE Trans. Electron Devices **31**, 42 (1984)
39. S.S. Tan, T.P. Chen, C.H. Ang, L. Chan, Atomic modeling of nitrogen neighboring effect on negative bias temperature instability of pMOSFETs. IEEE Electron Device Lett. **25**, 504 (2004)
40. C.H. Liu, M.T. Lee, J. Chen, K. Schruefer, J. Brighten, N. Rovedo, T.B. Hook, M.V. Khare, C. Wann, T.H. Ning, Mechanism and process dependence of negative bias temperature instability (NBTI) for pMOSFETs with ultrathin gate dielectrics, in *IEEE International Electron Devices Meeting Technical Digest* (2001), p. 39.2.1
41. T.B. Hook, R. Bolam, W. Clark, J. Burnham, N. Rovedo, L. Schutz, Negative bias temperature instability on three oxide thicknesses (1.4/2.2/5.2 nm) with nitridation variations and deuteration. Microelectron. Reliab. **45**, 47 (2005)
42. S. Mahapatra, K. Ahmed, D. Varghese, A.E. Islam, G. Gupta, L. Madhav, D. Saha, M.A. Alam, On the physical mechanism of NBTI in silicon oxynitride p-MOSFETs: can differences in insulator processing conditions resolve the interface trap generation versus hole trapping controversy?, in *IEEE International Reliability Physics Symposium Proceedings* (2007), p. 1
43. V.D. Maheta, E.N. Kumar, S. Purawat, C. Olsen, K. Ahmed, S. Mahapatra, Development of an ultrafast on-the-fly I_{DLIN} technique to study NBTI in plasma and thermal oxynitride p-MOSFETs. IEEE Trans. Electron Devices **55**, 2614 (2008)

44. T. Sasaki, K. Kuwazawa, K. Tanaka, J. Kato, Engineering of nitrogen profile in an ultrathin gate insulator to improve transistor performance and NBTI. IEEE Electron Device Lett. **24**, 150 (2003)
45. M. Terai, K. Watanabe, S. Fujieda, Effect of nitrogen profile and fluorine incorporation on negative-bias temperature instability of ultrathin plasma-nitrided SiON MOSFETs. IEEE Trans. Electron Devices **54**, 1658 (2007)
46. Angle Resolved X-RAY Photoelectron Spectroscopy. [Online]. Available: http://goliath.emt.inrs.ca/surfsci/arxps/introcss.html
47. C. Olsen, Two-step post nitridation annealing for lower EOT plasma nitrided gate dielectrics, WO2004081984 A2, (2004)
48. V.D. Maheta, C. Olsen, K. Ahmed, S. Mahapatra, The impact of gate dielectric nitridation methodology on NBTI of SiON p-MOSFETs as studied by UF-OTF technique, in *IEEE International Symposium on the Physical and Failure Analysis of Integrated Circuits* (2008). doi:10.1109/IPFA.2008.4588198
49. K. Sakuma, D. Matsushita, K. Muraoka, Y. Mitani, Investigation of nitrogen-originated NBTI mechanism in SiON with high-nitrogen concentration, in *IEEE International Reliability Physics Symposium Proceedings* (2006), p. 454
50. Y. Mitani, T. Yamaguchi, H. Satake, A. Toriumi, Reconsideration of hydrogen-related degradation mechanism in gate oxide, in *IEEE International Reliability Physics Symposium Proceedings* (2007), p. 226
51. S. Mahapatra, FEOL and BEOL process dependence of NBTI, in *Bias Temperature Instability for Devices and Circuits*, ed. by T. Grasser (Springer, New York, 2014)
52. E.N. Kumar, V.D. Maheta, S. Purawat, A.E. Islam, C. Olsen, K. Ahmed, M.A. Alam, S. Mahapatra, Material dependence of NBTI physical mechanism in silicon oxynitride (SiON) p-MOSFETs: a comprehensive study by ultra-fast on-the-fly (UF-OTF) I_{DLIN} technique, in *IEEE International Electron Devices Meeting Technical Digest* (2007), p. 809
53. C. Shen, M.-F. Li, C.E. Foo, T. Yang, D.M. Huang, A. Yap, G.S. Samudra, Y.-C. Yeo, Characterization and physical origin of fast Vth transient in NBTI of pMOSFETs with SiON dielectric, in *IEEE International Electron Devices Meeting Technical Digest* (2006). doi:10.1109/IEDM.2006.346776
54. S. Pae, A. Ashok, T. Ghani, K. Lemay, M. Liu, R. Lu, P. Packan, C. Parker, R. Purser, A. St. Amour, B. Woolery, Reliability characterization of 32 nm high-K and metal-gate logic transistor technology, in *IEEE International Reliability Physics Symposium Proceedings* (2010), p. 287
55. J. Mitard, X. Garros, L.P. Nguyen, C. Leroux, G. Ghibaudo, F. Martin, G. Reimbold, Large-scale time characterization and analysis of PBTI in HFO_2/metal gate stacks, in *IEEE International Reliability Physics Symposium Proceedings* (2006), p. 174
56. D. Heh, C.D. Young, G. Bersuker, Experimental evidence of the fast and slow charge trapping/detrapping processes in high-k dielectrics subjected to PBTI stress. IEEE Electron Device Lett. **29**, 180 (2008)
57. J. Yang, M. Masuduzzaman, K. Joshi, S. Mukhopadhyay, J. Kang, S. Mahapatra, M.A. Alam, Intrinsic correlation between PBTI and TDDB degradations in nMOS HK/MG dielectrics, in *IEEE International Reliability Physics Symposium Proceedings* (2012), p. 5D.4.1
58. E. Cartier, A. Kerber, T. Ando, M.M. Frank, K. Choi, S. Krishnan, B. Linder, K. Zhao, F. Monsieur, J. Stathis, V. Narayanan, Fundamental aspects of HfO_2-based high-k metal gate stack reliability and implications on t inv-scaling, in *IEEE International Electron Devices Meeting Technical Digest* (2011), p. 18.4.1
59. A.E. Islam, V.D. Maheta, H. Das, S. Mahapatra, M.A. Alam, Mobility degradation due to interface traps in plasma oxynitride PMOS devices, in *IEEE International Reliability Physics Symposium Proceedings* (2008), p. 87
60. P.A. Kraus, K.Z. Ahmed, C.S. Olsen, F. Nouri, Physical models for predicting plasma nitrided Si-O-N gate dielectric properties from physical metrology. IEEE Electron Device Lett. **24**, 559 (2003)

61. Y.M. Lin, C.J. Wang, K. Wu, A new finding on NBTI lifetime model and an investigation on NBTI degradation characteristic for 1.2 nm ultra thin oxide, in *IEEE International Reliability Physics Symposium Proceedings* (2005), p. 704
62. B. Kaczer, V. Arkbipov, R. Degraeve, N. Collaert, G. Groeseneken, M. Goodwin, Disorder-controlled-kinetics model for negative bias temperature instability and its experimental verification, in *IEEE International Reliability Physics Symposium Proceedings* (2005), p. 381
63. D. Varghese, D. Saha, S. Mahapatra, K. Ahmed, F. Nouri, M. Alam, On the dispersive versus arrhenius temperature activation of NBTI time evolution in plasma nitrided gate oxides: measurements, theory, and implications, in *IEEE International Electron Devices Meeting Technical Digest* (2005), p. 684
64. A.E. Islam, H. Kufluoglu, D. Varghese, S. Mahapatra, M.A. Alam, Recent issues in negative-bias temperature instability: initial degradation, field dependence of interface trap generation, hole trapping effects, and relaxation. IEEE Trans. Electron Devices **54**, 2143 (2007)
65. J.R. Pfiester, F.K. Baker, T.C. Mele, H.-H. Tseng, P.J. Tobin, J.D. Hayden, J.W. Miller, C.D. Gunderson, L.C. Parrillo, The effects of boron penetration on p+ polysilicon gated PMOS devices. IEEE Trans. Electron Devices **37**, 1842 (1990)
66. N. Goel, N. Nanaware, S. Mahapatra, Ultrafast AC–DC NBTI characterization of deep IL scaled HKMG p-MOSFETs. IEEE Electron Device Lett. **34**, 1476 (2013)
67. K. Joshi, S. Hung, S. Mukhopadhyay, V. Chaudhary, N. Nanaware, B. Rajamohnan, T. Sato, M. Bevan, A. Wei, A. Noori, B. McDougal, C. Ni, G. Saheli, C. Lazik, P. Liu, D. Chu, L. Date, S. Datta, A. Brand, J. Swenberg, S. Mahapatra, HKMG process impact on N, P BTI: role of thermal IL scaling, IL/HK integration and post HK nitridation, in *IEEE International Reliability Physics Symposium Proceedings* (2013), p. 4C.2.1
68. W. Tsai, L.-A. Ragnarsson, L. Pantisano, P.J. Chen, B. Onsia, T. Schram, E. Cartier, A. Kerber, E. Young, M. Caymax, S. De Gendt, M. Heyns, Performance comparison of sub 1 nm sputtered TiN/HfO/sub 2/nMOS and pMOSFETs, in *IEEE International Electron Devices Meeting Technical Digest* (2003), p. 13.2.1
69. K. Choi, H. Jagannathan, C. Choi, L. Edge, T. Ando, M. Frank, P. Jamison, M. Wang, E. Cartier, S. Zafar, J. Bruley, A. Kerber, B. Linder, A. Callegari, Q. Yang, S. Brown, J. Stathis, J. Iacoponi, V. Paruchuri, V. Narayanan, Extremely scaled gate-first high-k/metal gate stack with EOT of 0.55 nm using novel interfacial layer scavenging techniques for 22 nm technology node and beyond, in *Symposium on VLSI Technology: Digest of Technical Papers* (2009), p. 138

Chapter 2
Characterization Methods for BTI Degradation and Associated Gate Insulator Defects

Souvik Mahapatra, Nilesh Goel, Ankush Chaudhary, Kaustubh Joshi and Subhadeep Mukhopadhyay

Abstract In this chapter, different characterization methods are discussed to determine BTI degradation of MOSFET parameters and to directly estimate the pre-existing and generated gate insulator defects responsible for BTI. V_T shift is obtained from full I_D–V_G sweeps and also from spot I_D measurements at fixed V_G; one spot measurements are performed either on-the-fly at stress V_G or by dropping down to a lower V_G from stress. Impact of measurement delay and mobility degradation on V_T extracted from different methods is discussed. Flicker noise method is used to access the density of pre-existing defects for different gate insulator processes. Gated diode or DCIV, charge pumping (CP) and low voltage SILC methods are used to determine trap generation at or near the interface between Si channel and gate insulator. Conventional SILC is used to estimate generation of bulk gate insulator defects. Different artifacts related to improper choice of stress bias and measurement delay are discussed.

2.1 Introduction

As discussed in Chap. 1, Bias Temperature Instability (BTI) in Metal Oxide Semiconductor Field Effect Transistor (MOSFET) is normally assessed by stressing the device at higher than nominal gate bias ($V_G = V_{G\text{-}STR}$), and recording the shift in device parameters in logarithmic intervals of stress time (t_{STR}). Degradation in device parameters, such as threshold voltage shift (ΔV_T), is extracted from drain current (I_D) measured before and after stress. I_D is usually measured in the linear operating regime (I_{DLIN}), with a low drain bias (V_D) to keep oxide field (E_{OX}) almost uniform across the channel during stress.

S. Mahapatra (✉) · N. Goel · A. Chaudhary · K. Joshi · S. Mukhopadhyay
Department of Electrical Engineering, Indian Institute of Technology Bombay, Mumbai 400076, India
e-mail: souvik@ee.iitb.ac.in; mahapatra.souvik@gmail.com

S. Mahapatra (ed.), *Fundamentals of Bias Temperature Instability in MOS Transistors*, Springer Series in Advanced Microelectronics 139,
DOI 10.1007/978-81-322-2508-9_2

In earlier days, complete I_D–V_G sweeps were taken to extract V_T before and during BTI stress, the later was achieved by interrupting the stress for few seconds for measurements [1]. It was soon realized that BTI degradation recovers substantially after stress is paused for measurement [2, 3], and for certain situations, as much as 50 % of degradation recovers in just 1 s after stoppage of stress [4]. This makes the conventional, slow I_D–V_G sweep based Measure-Stress-Measure (MSM) technique unsuitable for BTI characterization, since it cannot capture the correct degradation magnitude. To circumvent this recovery issue, three different ultra-fast measurement techniques have been proposed in the literature, i.e., On-The-Fly (OTF) I_{DLIN}, MSM and One Spot Drop Down (OSDD), and are described in this chapter.

All three classes of measurements described in this chapter can be implemented using the setup illustrated in Fig. 2.1 [5–7]. A pulse generator is connected to the gate terminal of a MOSFET and is used to apply the desired stress and measurement biases. Source and substrate terminals are grounded, and the drain terminal is connected to a current–voltage converter (IVC) and subsequently to a digital storage oscilloscope (DSO). The combination of IVC and DSO helps in ultra-fast I_{DLIN} measurement in ~microseconds measurement time (t_M). All methods are described in this chapter are for Negative BTI (NBTI) stress in p-MOSFETs, although these can be readily adopted for Positive BTI (PBTI) stress in n-MOSFETs.

As described in detail later in this book, BTI is due to uncorrelated contribution from charging of pre-existing defects as well as generation of new defects during stress in the gate insulator of a MOSFET, and measured ΔV_T can be decomposed into these underlying sub-components. Therefore, it is important to independently and directly access the pre-existing and generated gate insulator traps for different gate insulator processes, to verify the accuracy of the decomposition method. Such assessment can be done using different trap characterization methods, i.e., Flicker (1/*f*) Noise, Charge Pumping (CP), Gated Diode (DCIV), Stress Induced Leakage Current (SILC) and Low Voltage (LV) SILC, and are described in this chapter.

Finally, it is important to remark that several stress and measurement artifacts can corrupt BTI experiments and are also discussed in this chapter. Since BTI and TDDB stress regimes are essentially the same, shown in Chap. 1, Fig. 1.2, and since

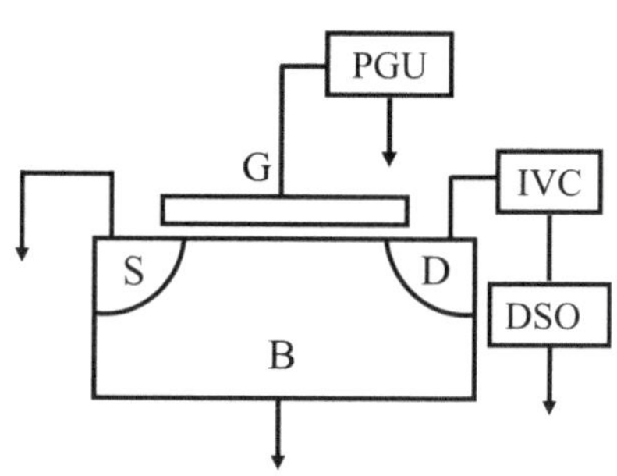

Fig. 2.1 Schematic of a representative setup for ultra-fast BTI measurements

TDDB defects have significantly different t_{STR} and V_G dependence, caution must be used to avoid TDDB corruption of BTI experiments. Several artifacts related to BTI recovery are also discussed.

2.2 Ultra-Fast On-the-Fly (UF-OTF) Method

Figure 2.2 shows the schematic of OTF measurement method. BTI induced I_{DLIN} degradation is measured on-the-fly at stress ($V_G = V_{G\text{-}STR}$) without reduction in V_G, and hence the technique does not suffer from recovery issues [3, 5]. However, the first data point (I_{DLIN0}) measured immediately after the application of $V_{G\text{-}STR}$ is assumed to be unstressed, and extracted ΔI_{DLIN} ($=I_{DLIN} - I_{DLIN0}$) therefore depends on the time-zero delay (t_0 delay) between application of $V_{G\text{-}STR}$ and measurement of I_{DLIN0} [5, 8]. I_{DLIN0} has been measured by using $t_0 = 1$ ms in conventional OTF [3, 9] and $t_0 = 1$ µs in Ultra-Fast OTF (UF-OTF) [5, 8] setups. Furthermore, the OTF technique needs to be modified to account for mobility degradation ($\Delta\mu_{eff}$), as-measured ΔI_{DLIN} is influenced by both ΔV_T and $\Delta\mu_{eff}$, as shown in Chap. 1, especially for NBTI stress in p-channel MOSFETs [10]. In [11], the OTF method is modified presumably to account for mobility degradation, where a small gate pulse is superimposed on $V_{G\text{-}STR}$ to measure transconductance (g_m), and ΔV_T is calculated using $\Delta V_T = \Delta I_{DLIN}/g_m$. However, it will be shown later that this method is not effective in providing correct ΔV_T, and also suffers from complications associated with application of a complex gate waveform. An alternative approach is to compute $\Delta V = -\Delta I_{DLIN}/I_{DLIN0} * (V_{G\text{-}STR} - V_{T0})$ by applying only a DC stress bias and measuring I_{DLIN} at $V_G = V_{G\text{-}STR}$, where V_{T0} is pre-stress V_T of the device. A post-processing correction procedure involving additional I_D–V_G sweep measurements is used to correct for the impact of $\Delta\mu_{eff}$ and convert ΔV to ΔV_T [12], which is discussed later in this chapter. OTF and UF-OTF methods have been

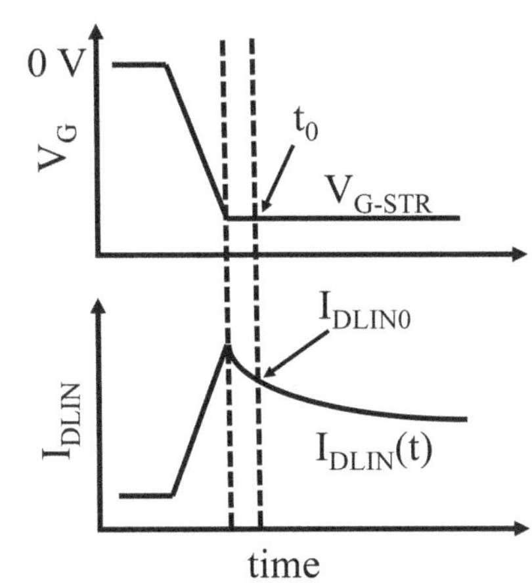

Fig. 2.2 Schematic of OTF measurement of BTI degradation during stress. Default time-zero delay is 1 µs unless mentioned otherwise

extensively used to characterize the gate insulator process dependence of NBTI degradation especially in Silicon Oxynitride (SiON) p-MOSFETs [9, 13], which will be discussed and analyzed later in this book.

2.2.1 Impact of Measurement Delay for DC Stress

Accuracy of the OTF technique depends on time-zero or t_0 delay as mentioned before, as the first data point measured immediately after the application of $V_{G\text{-}STR}$, i.e., I_{DLIN0}, is assumed as unstressed. As shown in Fig. 2.2, t_0 delay is the time lag between application of $V_{G\text{-}STR}$ and measurement of I_{DLIN0}. For a particular t_0 delay, a device that degrades more rapidly soon after the initiation of stress would show larger t_0 delay impact on measured ΔI_{DLIN} and vice versa. OTF measurements with t_0 delay down to 1 μs have been made using the setup shown in Fig. 2.1 [5, 8], and relevant results are discussed in this section. Although the implementation and results are shown for the case of NBTI in SiON p-MOSFETs, OTF method can be used to study NBTI and PBTI in HKMG devices as well.

Figure 2.3a plots I_{DLIN} measured before, during and immediately after application of $V_{G\text{-}STR}$ for OTF measurement in SiON p-MOSFET having Plasma Nitrided Oxide (PNO) gate insulator; refer to Chap. 1, Table 1.1 for SiON device details. In the beginning, the pulse generator shown in Fig. 2.1 supplies a low DC V_G to the gate terminal of the MOSFET, and the DSO is triggered to start sampling of I_{DLIN} in 1 μs intervals. The pulse generator is triggered next and V_G ramps up to $V_{G\text{-}STR}$, and DSO sampling captures the associated rise in I_{DLIN} as the MOSFET is driven to strong inversion. The first data point, i.e., I_{DLIN0}, is captured within 1 μs of the time of V_G becoming equal to $V_{G\text{-}STR}$. Measured I_{DLIN} then starts to decrease with increase in stress time due to BTI degradation.

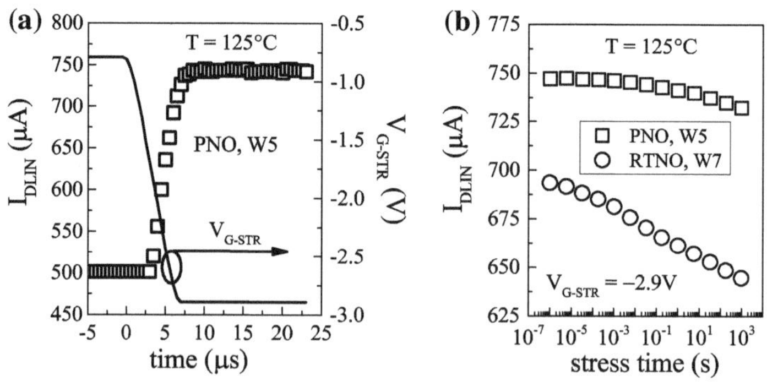

Fig. 2.3 OTF stress measurement: **a** application of gate stress voltage and corresponding measured I_{DLIN}, and **b** time evolution of I_{DLIN} degradation for PNO and RTNO SiON p-MOSFETs under NBTI stress

Figure 2.3b shows measured I_{DLIN} from short to longer stress time for PNO and Rapid Thermal Nitrided Oxide (RTNO) SiON p-MOSFETs. Measured I_{DLIN} at $V_{G\text{-}STR}$ starts to degrade due to BTI degradation. Although the devices have similar Equivalent Oxide Thickness (EOT), RTNO device shows lower starting I_{DLIN} and larger rate of I_{DLIN} degradation with increase in stress time, t_{STR}. As mentioned before, ΔI_{DLIN} is calculated from measured I_{DLIN} degradation using $I_{DLIN} - I_{DLIN0}$, and although the first data point captured within 1 µs can be used as I_{DLIN0}, it is possible to choose measured I_{DLIN} at different t_{STR} as I_{DLIN0} to study the impact of different t_0 delay on ΔI_{DLIN}. BTI degradation is estimated using $\Delta V = -\Delta I_{DLIN}/I_{DLIN0} * (V_{G\text{-}STR} - V_{T0})$ as mentioned above, where V_{T0} is pre-stress V_T of the device, and mobility correction has been used to convert ΔV to ΔV_T using the method described in [12] and discussed later in this chapter.

Figure 2.4 plots ΔV time evolution measured in (a) PNO and (b) RTNO devices using different t_0 delay. Higher degradation is obtained using lower t_0 delay as it captures a less degraded I_{DLIN0} and vice versa. Significantly, larger impact of t_0 delay is observed for RTNO compared to PNO device, as RTNO device has larger rate of I_{DLIN} degradation just after the initiation of stress as shown in Fig. 2.3b.

In spite of having similar EOT and much lower Nitrogen (N) content as mentioned in Chap. 1, Table 1.1, the RTNO device has much larger BTI magnitude compared to the PNO device when stressed at identical electric field, E_{OX}, and temperature (T), and measured using identical t_0 delay; E_{OX} is calculated using $(V_{G\text{-}STR} - V_{T0})$/EOT. Interestingly, the RTNO device shows large degradation in the sub-1 ms time scale when measured using t_0 delay of 1 µs, which is not observed in the PNO device, or with OTF measurements with higher t_0 delay. This observation has significant implication in understanding the underlying physical mechanism of BTI, and will be discussed in detail in Chap. 6. Therefore, Fig. 2.4 clearly demonstrates the importance of ultra-fast OTF measurements to accurately capture device degradation during BTI stress. Time evolution of NBTI degradation

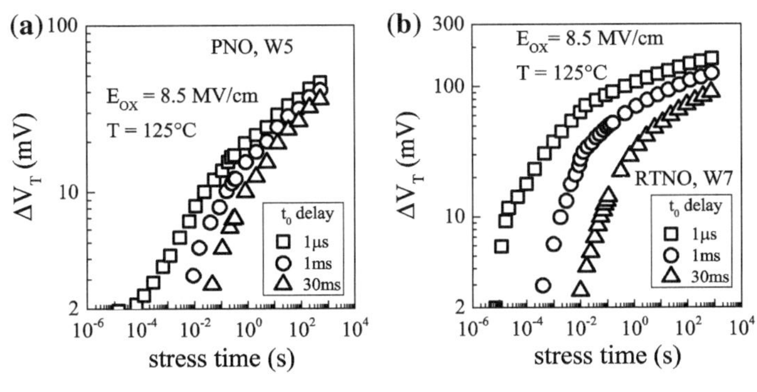

Fig. 2.4 Time evolution of OTF measured and mobility corrected ΔV_T for different time-zero delay, for **a** PNO and **b** RTNO SiON p-MOSFETs under NBTI stress. See Sect. 2.3 for mobility correction

at different stress E_{OX} and T measured using the ultra-fast OTF method in differently processed SiON p-MOSFETs has been shown in Chap. 1 and further analyzed later in this book.

As mentioned in Chap. 1, Fig. 1.16, BTI degradation shows power-law time dependence at longer stress time. Measured data are plotted in a log-log plot and the power-law time exponent (n) is obtained using linear regression in a specified time range. To verify the robustness of power-law time dependence, Fig. 2.5a plots n calculated from ΔV_T time evolution data as a function of $V_{G\text{-}STR}$ for NBTI stress in PNO-SiON p-MOSFET. Extraction is done for different ranges of t_{STR} over which regression has been performed. Note that within measurement error, similar values of n are observed for different t_{STR} range and across different stress V_G, as long as ΔV_T is measured using a small t_0 delay such that recovery artifacts remain negligible. It is important to note that the V_G independence of n shown in Fig. 2.5a does not hold all the time. As mentioned in Chap. 1, measured n can reduce at higher $V_{G\text{-}STR}$ due to saturation effects, which is caused by reduction in stress E_{OX} at longer t_{STR} when ΔV_T becomes large. On the other hand, n can also increase at higher $V_{G\text{-}STR}$ especially for thicker gate insulator devices, due to additional contribution from TDDB like bulk insulator trap generation [14], which is discussed later in this chapter and also in Chap. 4.

Figure 2.5b illustrates the impact of t_0 delay on extracted n for p-MOSFETs having different EOT and N content of PNO and RTNO based SiON gate stacks. Note that n is extracted over a fixed range of t_{STR} and reduces with reduction in t_0 delay, although PNO devices having different EOT and low to moderate N content show higher magnitude of n when compared to the RTNO device. It is important to note that a smaller t_0 captures I_{DLIN0} that is relatively less degraded, and results in larger BTI degradation and lower time exponent n. Similarly for a particular t_0 delay, I_{DLIN0} is less degraded for PNO compared to RTNO device, refer to Fig. 2.3b, and results in smaller degradation magnitude and higher n for the former device. Therefore, it is important to use ultra-fast OTF measurements to properly

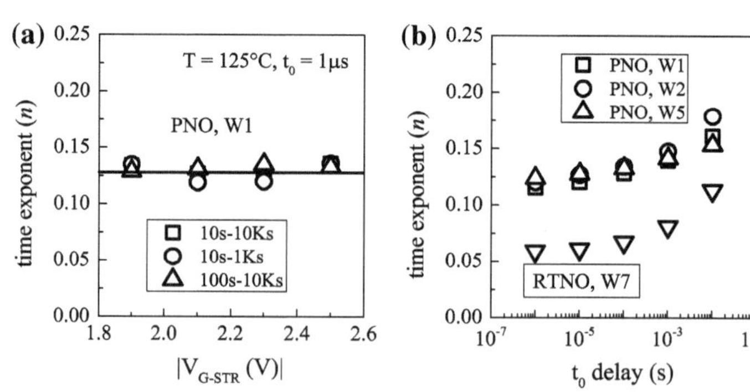

Fig. 2.5 Power-law time exponent n of OTF measured ΔV_T, **a** plotted versus $V_{G\text{-}STR}$ and extracted for different stress time ranges for linear regression, and **b** versus time-zero delay, for PNO and RTNO SiON p-MOSFETs under NBTI stress

measure BTI degradation in wide variety of gate insulator devices. More results of OTF measured time exponent n for different devices and stress conditions are presented and analyzed in other chapters of this book.

2.2.2 Recovery After DC Stress

As mentioned before, it is now well known that BTI degradation recovers substantially after the reduction of stress V_G. Figure 2.6a illustrates the gate bias sequence for OTF measurement of BTI recovery at $V_G = V_{G\text{-}REC}$ following BTI stress at $V_G = V_{G\text{-}STR}$ for a time t_{STR}. The pulse generator applies $V_G = V_{G\text{-}REC}$ to the gate of the MOSFET and the DSO is triggered to measure pre-stress I_{DLIN}, which is recorded as I_{DLIN0}. The pulse generator is then triggered to output $V_G = V_{G\text{-}STR}$ for a duration t_{STR}, during which the device degrades due to BTI, after which V_G drops back to $V_{G\text{-}REC}$. The DSO is synchronously triggered again with the falling edge of the gate pulse, and recovery of degraded I_{DLIN} is measured. Once again, the voltage shift for BTI recovery can be calculated using $\Delta V = -\Delta I_{DLIN}/I_{DLIN0} * (V_{G\text{-}REC} - V_{T0})$, where ΔI_{DLIN} is $I_{DLIN} - I_{DLIN0}$ and I_{DLIN0} is recorded at $V_G = V_{G\text{-}REC}$ before stress. Mobility correction is used to convert ΔV to ΔV_T using the method of [12] as discussed in the following section.

Figure 2.6b plots time evolution of ΔV_T recovery measured using the OTF technique after NBTI stress in PNO-SiON p-MOSFETs for different t_{STR} values. It is important to note that ΔV_T starts to recover as soon as V_G is reduced from $V_{G\text{-}STR}$

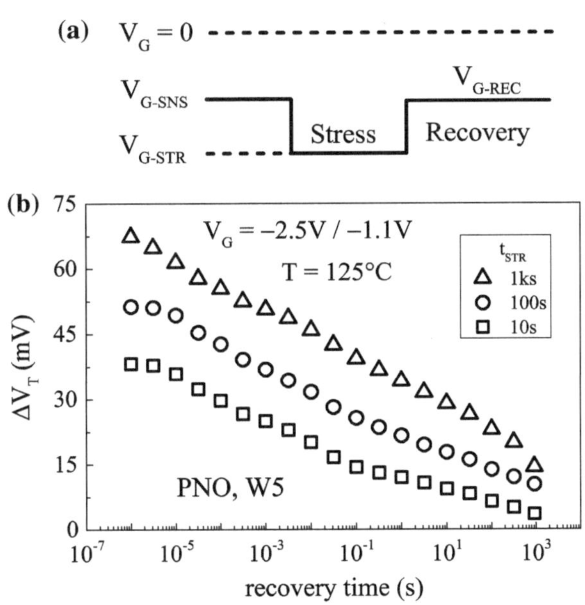

Fig. 2.6 OTF recovery measurement: **a** applied gate voltage sequence, **b** time evolution of measured and mobility corrected ΔV_T recovery after NBTI stress for different stress time in PNO SiON p-MOSFET

to $V_{G\text{-REC}}$. This suggests the necessity of using ultra-fast setup for BTI characterization using MSM or OSDD methods, and is discussed later in this chapter. More BTI recovery results measured using OTF method for different devices and stress conditions are presented and analyzed in Chap. 6 of this book.

2.3 Mobility Degradation

As discussed in Chap. 1, I_{DLIN} degradation especially during NBTI stress is caused by degradation in both V_T and μ_{eff}, the later results in transconductance (g_m) degradation, refer to Chap. 1, Fig. 1.16. Therefore, the contribution of $\Delta\mu_{eff}$ on ΔI_{DLIN} needs to be determined for accurate determination of time evolution of ΔV_T from OTF technique.

The drain current of a MOSFET in above threshold and linear regime of operation, I_{DLIN}, is given by the following equation [15]:

$$I_{DLIN} = \mu_{eff} C_{OX} \frac{W}{L}\left(V_G - V_T - m\frac{V_D}{2}\right) V_D \tag{2.1}$$

In (2.1), μ_{eff} is effective mobility, C_{OX} is gate capacitance, W and L are device width and channel length respectively, V_G and V_D are gate and drain biases respectively, and m is calculated using the subthreshold I_D versus V_G characteristics. For higher gate overdrive ($V_G - V_T$), the effective mobility can be expressed using a simple first order relation as follows [10, 12]:

$$\mu_{eff} = \frac{\mu_0}{1 + \theta(V_G - V_T)} \tag{2.2}$$

In (2.2), μ_0 is low field mobility and θ is high field reduction factor. The variation in I_{DLIN} due to BTI can be obtained by differentiating (2.1) as follows [12]:

$$-\left|\frac{\Delta I_{DLIN}}{I_{DLIN0}}\right| = \frac{\Delta\mu_{eff}}{\mu_{eff0}} \frac{V_G - V_T - m\frac{V_D}{2}}{V_G - V_{T0} - m_0\frac{V_D}{2}} + \frac{\mu_{eff}}{\mu_{eff,0}} \frac{-|\Delta V_T| - \frac{V_D}{2}|\Delta m|}{V_G - V_{T0} - m_0\frac{V_D}{2}} \tag{2.3}$$

Note that besides ΔV_T, ΔI_{DLIN} is also impacted by $\Delta\mu_{eff}$ ($=\mu_{eff} - \mu_{eff,0}$) and therefore by $\Delta\mu_0$ ($=\mu_0 - \mu_{0,0}$) and $\Delta\theta$ ($=\theta - \theta_0$), as well as by Δm ($=m - m_0$), where $\mu_{eff,0}$, θ_0 and m_0 are corresponding pre-stress values. The simplified expression used in Sect. 2.2, $\Delta V = -\Delta I_{DLIN}/I_{DLIN0} * (V_{G\text{-STR}} - V_{T0})$, is derived from (2.3) by ignoring $\Delta\mu_{eff}$ and Δm, and therefore, ΔV needs to be corrected, especially for NBTI stress, to obtain proper, "mobility corrected" ΔV_T [12].

The mobility and m factor degradation can be assessed by measuring I_{DLIN} versus V_G sweeps before and during logarithmically spaced intervals of NBTI stress. Pre- and post-stress I_{DLIN} versus V_G data in above threshold at high gate overdrive can be used to obtain μ_{eff} versus ($V_G - V_T$) data, where the mobility expression of (2.2) remains valid, and can be fitted to obtain μ_0 and θ, and hence, time evolution of $\Delta\mu_0$ and $\Delta\theta$ can be determined. Time evolution of Δm can be obtained from subthreshold slope degradation and that of ΔV_T from V_T extracted using conventional "peak g_m" method.

Figure 2.7 shows normalized $\Delta\mu_0$, $\Delta\theta$ and Δm as a function of ΔV_T for NBTI stress in SiON p-MOSFETs. The magnitude of μ_0 and θ reduce and that of m increases after NBTI stress. Moreover, $\Delta\mu_0$, $\Delta\theta$ and Δm are linearly correlated to ΔV_T, and this correlation can be used to construct a "mobility correction" algorithm as discussed later. Note that for a particular ΔV_T, much larger variation is usually observed for $\Delta\mu_0$ and $\Delta\theta$ than that for Δm, and therefore, correction due to Δm has negligible impact and can be ignored as discussed in [12].

It is important to remark that the correlation of $\Delta\mu_{eff}$ to ΔV_T depends on θ_0, the pre-stress slope of μ_{eff} versus $(V_G - V_T)$ data, refer to (2.2). Note that when compared at fixed $(V_G - V_T)$, obtained μ_{eff} always reduces after NBTI stress, and therefore, $\Delta\mu_{eff} < 0$. However, actual NBTI measurements are done at fixed V_G, and for devices having high θ_0, reduction in μ_0 can be overcompensated by reduction in θ_0 and $(V_G - V_T)$ after NBTI stress, which can result in $\Delta\mu_{eff} > 0$, refer to (2.2) [12].

As an example, Fig. 2.8 shows normalized $\Delta\mu_{eff}$ as a function of ΔV_T for two different SiON p-MOSFETs having (a) low and (b) high pre-stress θ_0. Note that μ_{eff} is extracted using (2.1) from pre- and post-stress I_{DLIN} versus V_G characteristics at fixed sense bias of $V_G = V_{G\text{-}SNS}$, while V_T is extracted by using the peak g_m method. The magnitude of $|\Delta\mu_{eff}|$ increases with increase in ΔV_T due to NBTI stress for both devices. However, the device with low θ_0 shows $\Delta\mu_{eff} < 0$ as expected, while the device with high θ_0 shows $\Delta\mu_{eff} > 0$ due to overcompensation effect discussed above. It is also important to note that for a particular ΔV_T, $|\Delta\mu_{eff}|$ reduces with

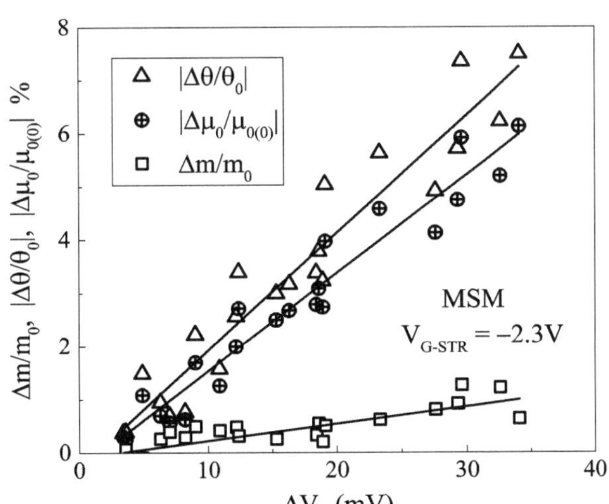

Fig. 2.7 Correlation of degradation in parameters m, μ_0 and θ shown in (2.1) and (2.2) to ΔV_T for NBTI stress in PNO SiON p-MOSFET. Data obtained using I–V sweeps in slow MSM method

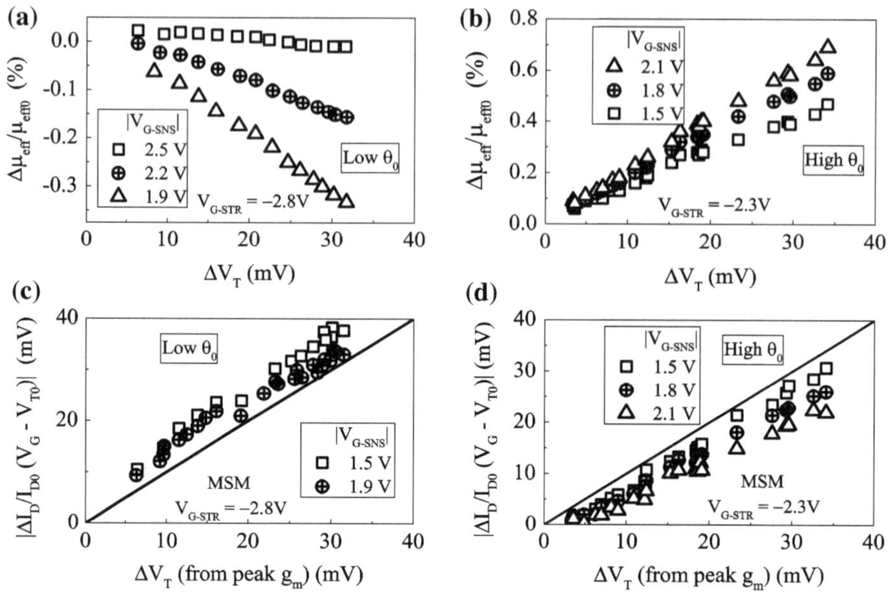

Fig. 2.8 Correlation of normalized mobility degradation measured using different sense V_G to ΔV_T for NBTI stress in PNO SiON p-MOSFET having **a** low θ_0 and **b** high θ_0. Correlation of corresponding mobility uncorrected voltage shift using simple I_{DLIN} expression to ΔV_T for **c** low θ_0 and **d** high θ_0 devices. ΔV_T is obtained using peak g_m method

increase in $|V_{G\text{-}SNS}|$ for low θ_0 device, while for high θ_0 device, $|\Delta\mu_{eff}|$ increases with increase in $|V_{G\text{-}SNS}|$ as shown in Fig. 2.8a, b.

Note that the magnitude and sign of $\Delta\mu_{eff}$ versus ΔV_T correlation would impact ΔI_{DLIN} versus ΔV_T correlation as per (2.3), since ΔI_{DLIN} is also obtained from I_{DLIN} measurement before and after stress at a fixed V_G. As an illustration, Fig. 2.8 plots the correlation of $\Delta V = -\Delta I_{DLIN}/I_{DLIN0} * (V_{G\text{-}SNS} - V_{T0})$ versus ΔV_T at different sense bias $V_{G\text{-}SNS}$ for SiON p-MOSFETs with (c) low and (d) high pre-stress θ_0; the $\Delta\mu_{eff}$ versus ΔV_T correlation of these devices are shown, respectively in Fig. 2.8a, b. Pre- and post-stress V_T and hence ΔV_T are obtained using the peak g_m method, and 1:1 correlation is shown by a solid line. As expected, ΔV is not equal to ΔV_T due to mobility degradation effects. For a particular ΔV_T, the device with lower θ_0 shows higher ΔV than 1:1 correlation line, see Fig. 2.8c, and the error in ΔV increases with reduction in $|V_{G\text{-}SNS}|$. On the other hand, the device with higher θ_0 has lower ΔV than 1:1 correlation line, see Fig. 2.8d, while the error in ΔV increases with increase in $|V_{G\text{-}SNS}|$. Therefore, if mobility correction is not taken into consideration, OTF measurements can result in higher or lower than actual NBTI degradation, and the error will be dependent on $V_{G\text{-}SNS}$, which is $V_{G\text{-}STR}$ for OTF method. Therefore, not only the NBTI magnitude but also the $V_{G\text{-}STR}$ or E_{OX} dependence of measured degradation will get corrupted, which will result in incorrect extrapolated degradation at use condition.

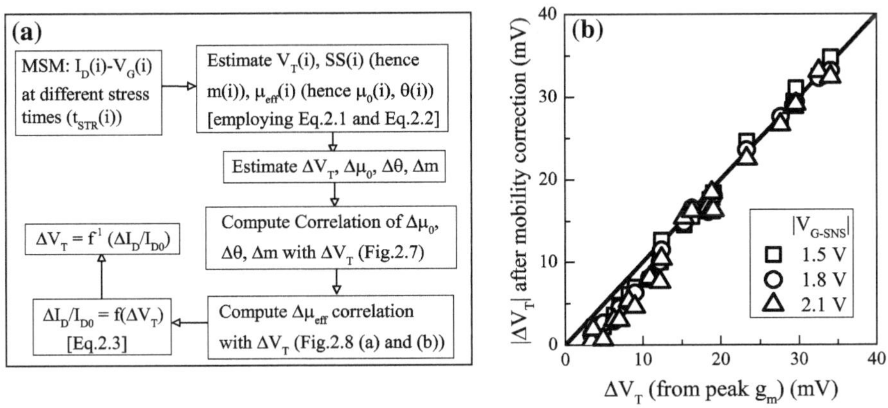

Fig. 2.9 **a** Algorithm for mobility correction of OTF data. **b** Correlation of mobility corrected calculated ΔV_T from measured I_{DLIN} at different sense V_G to measured ΔV_T using peak g_m method for NBTI stress in PNO SiON p-MOSFET

I_{DLIN} versus V_G measurements before and during NBTI stress interruptions can be used to calculate pre-stress $\mu_{0,0}$, θ_0 and m_0 values, as well as stress induced $\Delta\mu_0$, $\Delta\theta$ and Δm versus ΔV_T correlation as shown in Fig. 2.7. Moreover, (2.2) can be used to calculate $\Delta\mu_{eff}$ versus ΔV_T correlation as shown in Fig. 2.8a, b, and finally, (2.3) can be used to calculate "mobility corrected" ΔV_T [12]. Figure 2.9a illustrates this correction algorithm, and Fig. 2.9b plots "mobility corrected" ΔV_T from one spot ΔI_{DLIN} measurements at different $V_{G\text{-SNS}}$ as a function of ΔV_T obtained using the conventional peak g_m method. Note that once mobility degradation is taken into account using (2.3), 1:1 correlation is observed between "peak g_m" ΔV_T and ΔV_T obtained from ΔI_{DLIN} measured at different $V_{G\text{-SNS}}$. All UF-OTF NBTI data presented elsewhere in this book are corrected for mobility degradation using the method proposed in this section.

2.4 Ultra-Fast Measure-Stress-Measure (UF-MSM) Method

Ultra-Fast MSM (UF-MSM) method, illustrated in Fig. 2.10, relies on performing complete I_{DLIN} versus V_G sweeps in few microseconds before and during logarithmically spaced interruptions of BTI stress [6, 7]. V_T is extracted before and after stress from measured I_D–V_G data by using the conventional "peak g_m" method [15], and hence time evolution of ΔV_T can be obtained. Note that similar to conventional MSM method, the UF-MSM method also suffers from potential recovery issues since V_G is lowered from stress for measurement, although the recovery would be substantially lower, if not almost negligible, as stress interruption time is drastically minimized. Note that a tradeoff exists between measurement speed and accuracy,

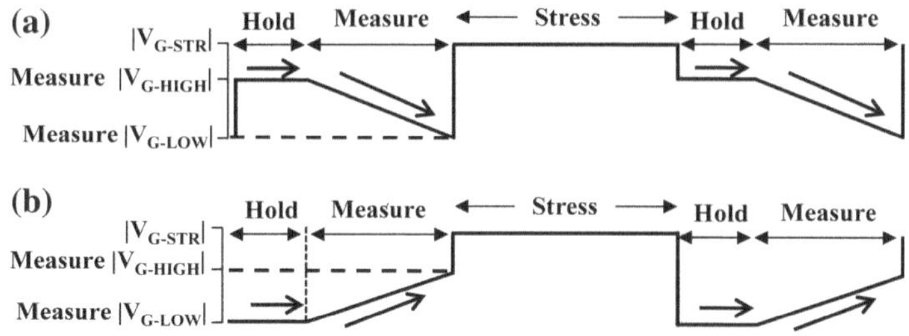

Fig. 2.10 Schematic of UF-MSM method for DC NBTI measurements. Arrows represent directions of V_G sweep for I–V measurement. Measurement delay is sum of hold time and sweep time. Default hold time is 50 ns and sweep time is 10 μs unless mentioned otherwise

and commercial setups are now available to reliably perform both DC and AC BTI experiments in MSM mode with ~10 μs delay [16]. The UF-MSM method has been extensively used to characterize DC and AC degradation during NBTI and PBTI stress in p- and n-channel High-K Metal Gate (HKMG) MOSFETs, respectively [16–19]; results are shown in Chap. 1 and further analyzed in later chapters of this book. This section describes the implementation details of the UF-MSM method. Although early implementation of UF-MSM method was made using custom setups [6], all results presented in this book are measured using commercially available instrument [16]. Although UF-MSM method has been used to characterize NBTI stress in both SiON and HKMG devices and PBTI stress in HKMG devices, results presented in this section are obtained from NBTI stress in p-channel HKMG MOSFETs.

2.4.1 Measurements During DC Stress

As shown in Fig. 2.10, V_G sweeps for I_{DLIN} measurements can be made either from high to low (H to L) or from low to high (L to H) values,[1] however, note that it is important to keep identical sweep direction for pre- and post-stress measurements. Figure 2.11 plots pre- and post-stress (a) I_{DLIN} and (b) g_m versus V_G data obtained using H to L and L to H sweeps. Note that due to recovery issues, the H to L and L to H sweep measurements are done on identical but separate devices, stressed at identical V_G and T. Identical pre-stress measured values for different sweep directions (small discrepancy is due to device to device variation) verify that the measurement is free from hysteresis issues, which can be either due to inaccurate setup or due to bad gate insulator quality of the device under test. The degradation, i.e., the difference between pre- and post-stress data is different for different sweep directions and is discussed below.

[1]For NBTI stress, low and high values correspond to the magnitude of gate pulse.

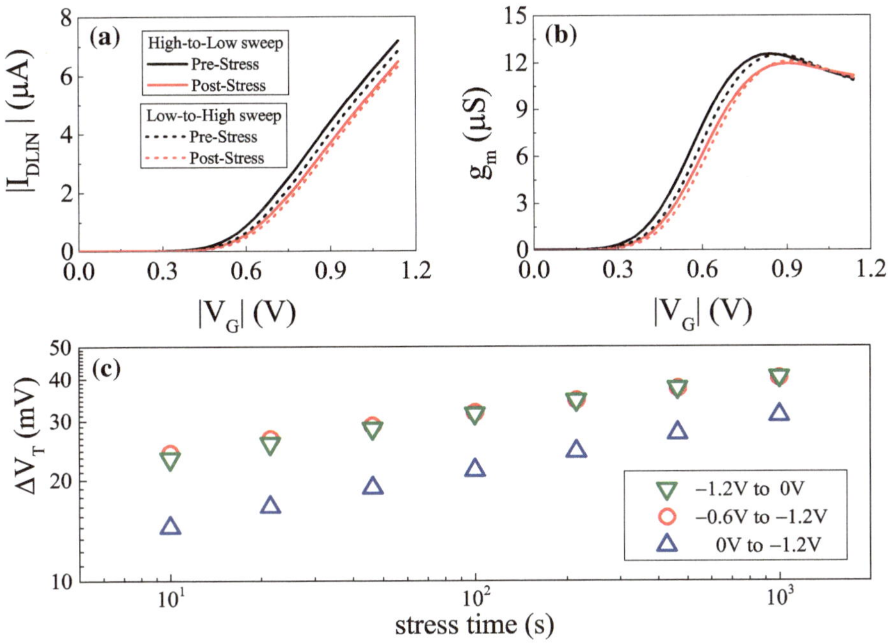

Fig. 2.11 UF-MSM measured **a** I_{DLIN} versus V_{G} and **b** g_m versus V_{G} characteristics before and after NBTI stress in HKMG p-MOSFET. Data shown for different sweep directions for *I*–*V* measurements. **c** Time evolution of UF-MSM measured ΔV_{T} for different sweep directions and ranges of sweep bias

As mentioned before, V_{T} is extracted using the conventional "peak g_m" method, where a tangent in drawn on the I_{DLIN} versus V_{G} curve at the V_{G} value corresponding to the peak g_m point. *I*–*V* characteristics are measured before and during logarithmically spaced interruptions in BTI stress, and ΔV_{T} time evolution is obtained. Figure 2.11c plots time evolution of ΔV_{T} obtained at identical NBTI stress conditions but using different sweep directions. Note that ΔV_{T} shows power-law time dependence with exponent *n* at longer t_{STR}. However, the magnitude of ΔV_{T} and its exponent *n* show different values for H to L and L to H sweeps for identical hold and sweep time. Higher ΔV_{T} and lower *n* are observed for H to L when compared to L to H sweep as shown; *n* being obtained using linear regression of measured data in t_{STR} range to 10 s to 1 Ks.

Although H to L and L to H sweeps are taken using identical voltage range (0 to −1.2 V) and sweep time, the hold before sweep is at higher $|V_{\mathrm{G}}|$ for H to L and at 0 V for L to H sweep, see Fig. 2.10. Since recovery accelerates at lower $|V_{\mathrm{G}}|$, refer to Chap. 1, Fig. 1.22, hold at 0 V for L to H sweep has larger recovery as compared to hold at higher $|V_{\mathrm{G}}|$ for H to L sweep, and therefore, the former results in lower ΔV_{T} magnitude and higher exponent *n* as shown. Note that for L to H sweep, it is sufficient to record I_{DLIN} from V_{G} values just below the peak g_m point for extraction of V_T using the tangent method, and it is not necessary to record I_{DLIN} all the way from $V_G = 0$ V. Figure 2.11c also plots the time evolution of ΔV_T for L to H sweep from −0.6 to −1.2 V, which, for identical $V_{G\text{-}STR}$ and T, shows larger ΔV_T and lower n compared to L to H sweep from 0 to −1.2 V, due to relatively higher hold V_G and correspondingly lower recovery. Note that the H to L sweep can also be done from −1.2 V up to −0.6 V instead of going all the way down to 0 V to reduce measurement time, t_M. It should be remarked that the non-zero lower limit of V_G for L to H and/or H to L sweep should be suitably chosen such that it crosses the V_G value corresponding to the peak g_m point, through which a tangent needs to be drawn to estimate V_T.

Figure 2.12 shows the impact of hold time on time evolution of ΔV_T for (a) H to L sweep from −1.2 to −0.6 V, as well as L to H sweep from (b) −0.6 to −1.2 V and (c) 0 to −1.2 V. The power-law time exponents extracted using linear regression of measured data in t_{STR} range to 10 s to 1 Ks are shown as a function of hold delay in Fig. 2.12d for different sweep directions. All measurements are done on identical but separate devices, stressed at identical $V_{G\text{-}STR}$ and T. Note that H to L sweep shows largest ΔV_T, smallest n and negligible impact of hold time, while for L to H sweep, ΔV_T reduces and n increases with increase in hold time, larger ΔV_T and smaller n are obtained for hold at −0.6 V compared to 0 V. This is due to negligible

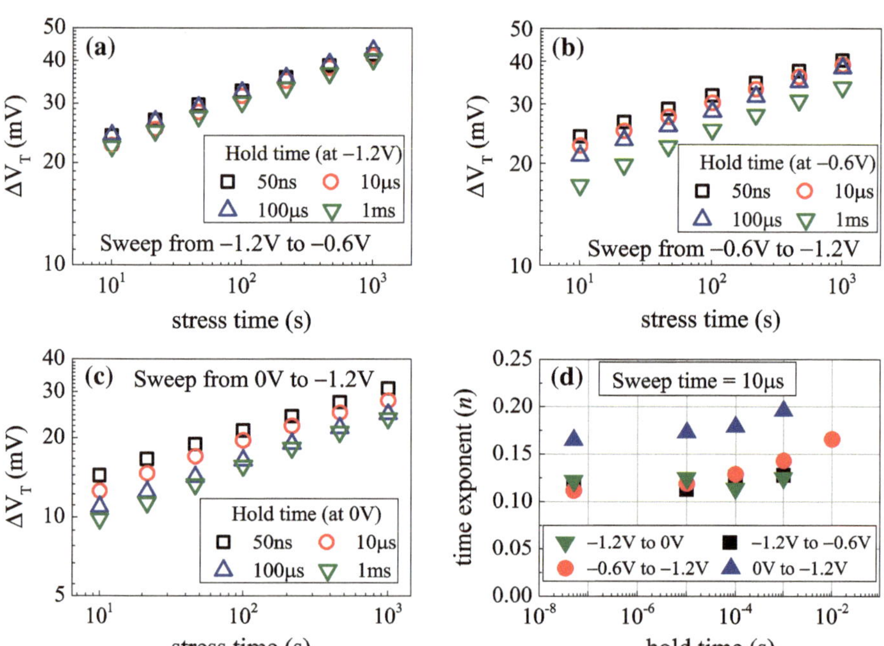

Fig. 2.12 **a–c** Impact of hold time on time evolution of UF-MSM measured ΔV_T for NBTI stress in HKMG p-MOSFETs for different sweep conditions. **d** Measured power-law time exponent n versus hold time for different sweep conditions

ΔV_T recovery at −1.2 V, and much larger recovery at 0 V compared to −0.6 V. Therefore, a non-zero lower limit of V_G sweep reduces the impact of recovery especially for L to H sweep, and the overall measurement time can also be reduced due to reduction in range of sweep V_G.

Besides hold time, the sweep time also impacts time evolution of measured ΔV_T. Higher recovery at lower $|V_G|$ influences both H to L and L to H sweep based *I*–*V* measurements, even for sweeps with relatively higher, non-zero lower limit of V_G. Figure 2.13 plots ΔV_T time evolution for (a) H to L sweep from −1.2 to −0.6 V, as well as L to H sweep from (b) −0.6 to −1.2 V and (c) 0 to −1.2 V. The power-law time exponents obtained using linear regression of measured data in t_{STR} range to 10 s to 1k s are shown as a function of sweep delay in Fig. 2.13d for different sweep directions. Higher sweep time results in lower ΔV_T and higher *n* for both H to L as well as L to H sweep. However contrary to hold time, the sweep time has relatively larger impact on H to L compared to L to H sweep. This is due to larger recovery at larger sweep delay for H to L compared to L to H sweep, as the later part of the sweep goes towards lower $|V_G|$ for H to L and towards higher $|V_G|$ for L to H sweep. Therefore, the sweep direction, hold time and sweep time affect ΔV_T time evolution and need to be carefully chosen for UF-MSM measurements.

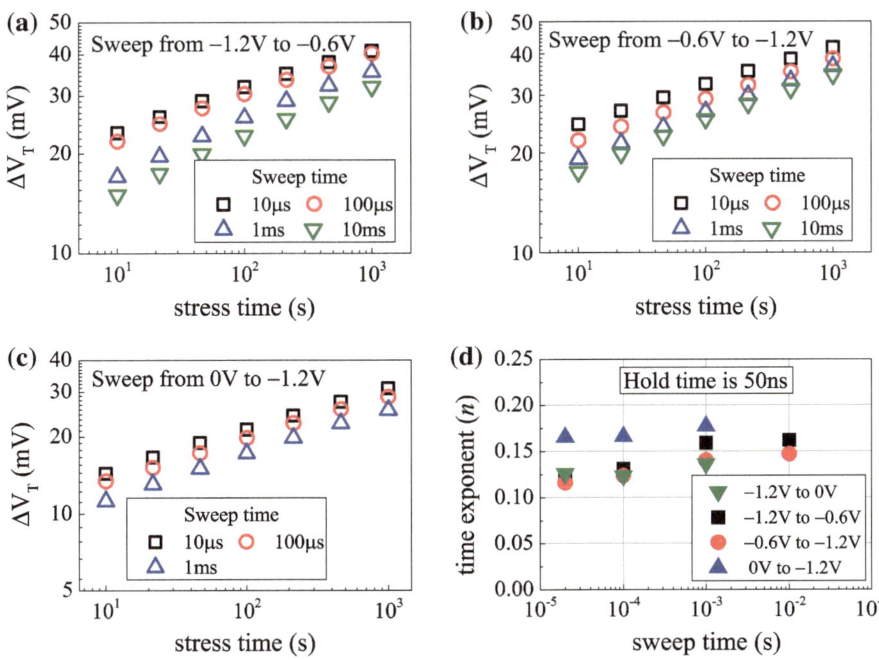

Fig. 2.13 **a–c** Impact of sweep time and sweep direction on UF-MSM measured ΔV_T time evolution for NBTI stress in HKMG p-MOSFETs. **d** Measured power-law time exponent *n* versus sweep time for different sweep conditions

2.4.2 Measurements During AC Stress

Figure 2.14 shows the gate voltage sequence for AC stress and measurement. AC pulse having a particular frequency (*f*) and duty cycle (PDC) is applied to the gate for stress, pulse $V_{G\text{-}HIGH}$ and pulse $V_{G\text{-}LOW}$ being the stress high and low values, respectively. The device gets stressed during pulse high phase, and recovers during pulse low phase. The pulse low is usually kept at 0 V, although non-zero $V_{G\text{-}LOW}$ can also be used to study the impact of V_G on recovery for AC stress. *I–V* sweeps are measured before and during logarithmically spaced interruptions in BTI stress to extract time evolution of ΔV_T. Measure $V_{G\text{-}HIGH}$ and measure $V_{G\text{-}LOW}$ denote the range of V_G sweep for *I–V* measurements. Note that AC stress can be interrupted for *I–V* measurements after the completion of last half cycle or last full cycle, defined respectively as Mode-A and Mode-B AC stress. It is important to note that AC BTI results shown in Chap. 1 are from Mode-B stress. Similar to DC stress, the *I–V* curves can be measured using H to L or L to H directions of V_G sweep for AC stress, although H to L direction is used in this book unless mentioned otherwise. It is important to use identical sweep parameters, such as sweep direction, hold time and sweep time for accurate estimation of the ratio of AC to DC degradation.

Figure 2.15a plots time evolution of ΔV_T for Mode-A and Mode-B AC stress at identical PDC, *f*, $V_{G\text{-}HIGH}$ (=$V_{G\text{-}STR}$) and $V_{G\text{-}LOW}$ (=$V_{G\text{-}REC}$) and measured using H to L gate sweep from −1.2 and −0.6 V. Time evolution of ΔV_T for AC stress also shows power-law dependence at longer t_{STR}. However, Mode-A stress shows higher ΔV_T magnitude and lower time exponent *n* when compared to Mode-B stress, which is due to higher recovery for the later as measurements are done after the pulse low phase. The difference between Mode-A and Mode-B AC stress is analyzed in detail later in this book.

Figure 2.15 also shows the impact of (b) PDC and (c) *f* of the AC gate pulse on measured ΔV_T at fixed t_{STR} for Mode-A and Mode-B stress at fixed $V_{G\text{-}HIGH}$ but

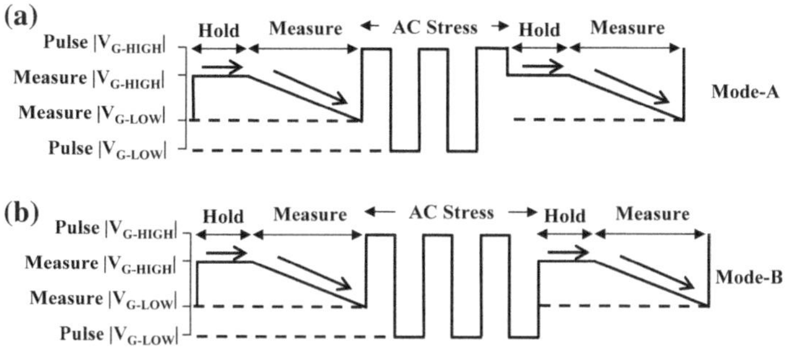

Fig. 2.14 Schematic of UF-MSM method for AC NBTI measurements. Mode-A and Mode-B represents start of *I–V* sweep after last half and full cycle, respectively. Identical sweep directions should be used for both modes, which should also be identical to that used for DC stress

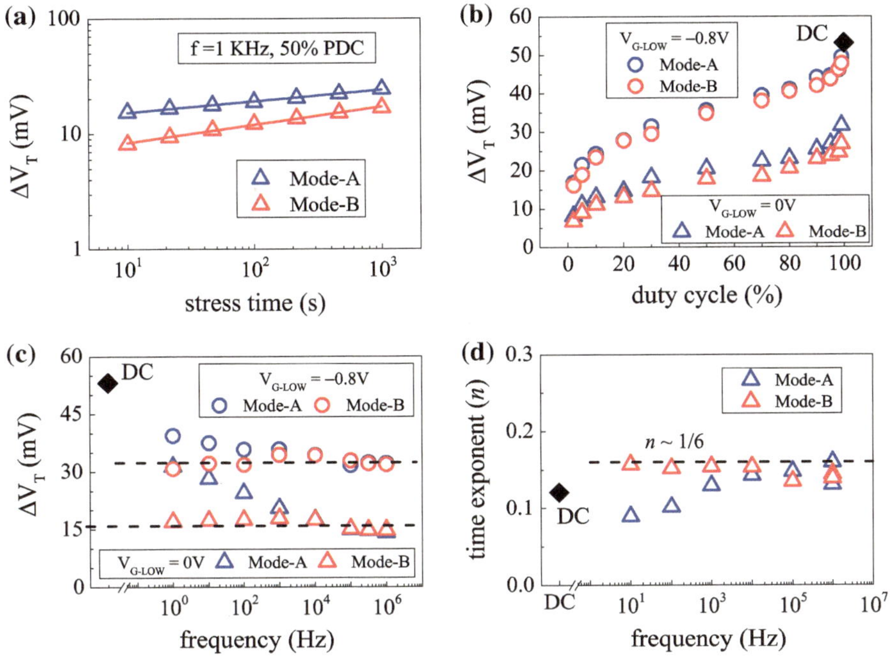

Fig. 2.15 AC NBTI measurements in HKMG p-MOSFETs using UF-MSM method: **a** Time evolution of ΔV_T for Mode-A and Mode-B stress. Fixed time ΔV_T for Mode-A and Mode-B stress using AC pulse having different pulse low bias, as a function of **b** duty cycle and **c** frequency. **d** Power-law time exponent n versus frequency for Mode-A and Mode-B AC stress. DC data are shown as reference

different $V_{G\text{-}LOW}$, the corresponding DC stress value is shown as reference. Measurements are done using H to L V_G sweep from −1.2 to −0.6 V. Note that ΔV_T for DC and AC stress are obtained at fixed t_{STR}, which, for AC stress includes both pulse on and off time, and the net stress or pulse on time for AC stress would depend on PDC of the gate pulse. Higher ΔV_T is obtained at higher $|V_{G\text{-}LOW}|$ for all PDC and f, which is consistent with lower recovery at higher $|V_G|$. ΔV_T increases with increase in PDC due to increase in net pulse high or stress time. The PDC dependence shows a typical "S" shaped characteristics, and a kink or jump is observed between high PDC AC and DC stress. Lower kink in ΔV_T is observed for higher $|V_{G\text{-}LOW}|$ and vice versa, which will be explained later in this book. Measured ΔV_T for Mode-B stress is independent of f for different $V_{G\text{-}LOW}$, although larger AC to DC ratio is observed for higher $|V_{G\text{-}LOW}|$ and vice versa. On the other hand, ΔV_T for Mode-A stress is higher when compared to Mode-B ΔV_T especially at lower f, however, the former reduces with increase in f and merges with the later at higher f and also shows f independence.

The impact of AC pulse f on measured power-law time exponent n for Mode-A and Mode-B stress is shown in Fig. 2.15d. As mentioned before, n is extracted by using linear regression of measured data in t_{STR} range of 10 s to 1k s. Note that Mode-B stress shows universal $n \sim 1/6$ value and f independence of n. On the other hand, Mode-A stress results in lower n compared to Mode-B especially at lower f, however, n for the former increases with increase in f and merges with the later at higher f and shows f independence. It is important to note the f independence of ΔV_T and n for Mode-B stress. Moreover, the observed f dependence of ΔV_T and n for Mode-A stress especially at lower f is also of interest; ΔV_T reduces but n increases with increase in f. These aspects will be explained in Chap. 6.

The impact of hold and sweep delay on AC stress is also of interest. Figure 2.16 shows ΔV_T time evolution measured using L to H V_G sweep from −0.6 to −1.2 V for (a, c) Mode-A and (b, d) Mode-B AC stress, for different (a, b) hold time and (c, d) sweep time. Similar to DC, Mode-A AC stress is also affected by both hold and sweep time; lower ΔV_T magnitude and higher power-law time exponent n are observed for higher measurement delay. However, the sweep and hold delay have negligible impact on Mode-B AC stress as shown. This can be explained by noting that *I–V* sweeps for Mode-B stress are done after the pulse off phase where recovery takes place, and therefore, additional delay has negligible impact on measured ΔV_T time evolution. However, this is not the case for Mode-A stress as *I–V* sweeps are done after pulse high phase, and hence, delay influences measured time evolution of ΔV_T as shown. Therefore, it is necessary to use UF-MSM for accurate estimation of DC to AC ratio for Mode-A and Mode-B AC stress.

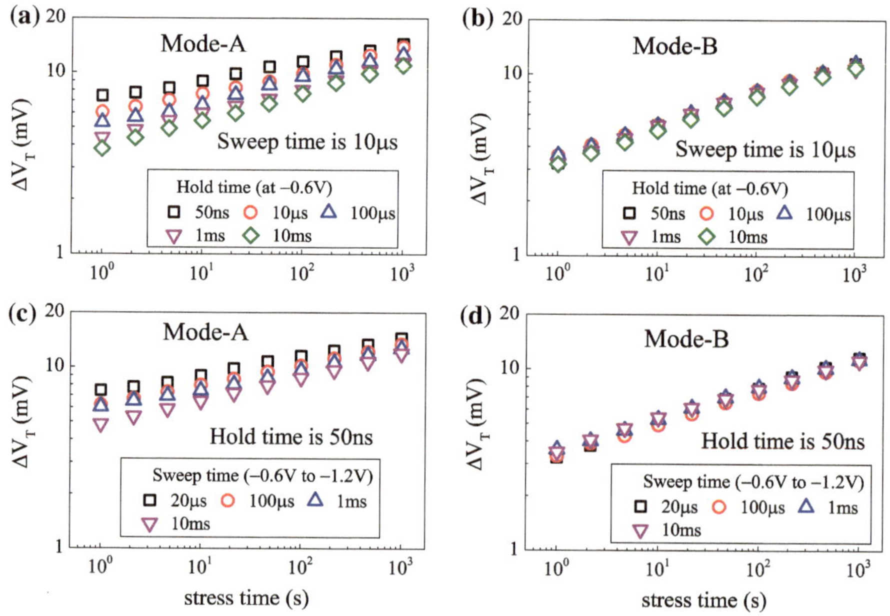

Fig. 2.16 Impact of (**a**, **b**) hold time and (**c**, **d**) sweep time for UF-MSM measured ΔV_T time evolution during AC (**a**, **c**) Mode-A and (**b**, **d**) Mode-B NBTI stress in HKMG p-MOSFETs

2.5 One Spot Drop Down (OSDD) Method

In OSDD technique, illustrated in Fig. 2.17, BTI stress is interrupted and V_G is reduced from $V_{G\text{-}STR}$ to a suitable sense bias ($V_{G\text{-}SNS}$) to measure I_{DLIN} in time t_M. Similar to MSM, the OSDD technique would also suffer from recovery issues, however, it takes much shorter time to measure a single spot I_D than full I_D–V_G sweep and hence recovery can be minimized. Although specialized commercial setups are now available to reliably perform full I_D–V_G sweeps in few microseconds, OSDD has been particularly useful in the past when such specialized instruments were not available. As shown in Fig. 2.17, once post-stress I_{DLIN} is measured at $V_{G\text{-}SNS}$, it can be compared to pre-stress I_D–V_G sweep to determine BTI degradation. In the vertical shift method, $\Delta V = -\Delta I_{DLIN}/I_{DLIN0} * (V_{G\text{-}SNS} - V_{T0})$ can be estimated by noting the difference in I_{DLIN} between pre- and post-stress at $V_{G\text{-}SNS}$, and in the absence of mobility variation, ΔV equals ΔV_T. Note that the absence of mobility degradation implies parallel I_D–V_G curves before and after stress. In the lateral shift method, the voltage corresponding to post-stress I_{DLIN} is noted from the pre-stress I_D–V_G curve, which is denoted as $V_{G\text{-}PST}$ as shown in Fig. 2.17, and the difference between $V_{G\text{-}PST}$ and $V_{G\text{-}SNS}$ is used to calculate ΔV_{LS}, which becomes equal to ΔV_T in the absence of mobility degradation.

In the technique proposed in [20], I_{DLIN} is measured at $V_{G\text{-}SNS} = V_{T0}$, where V_{T0} is pre-stress threshold voltage of the device and lateral shift method is used. Note that the lateral shift method assumes parallel I_D–V_G curves before and after stress and hence no mobility degradation, which is a fair assumption for PBTI stress as discussed in Chap. 1, and also at lower $V_{G\text{-}SNS}$ close to V_{T0} for NBTI stress, which will be shown later in this section. However, as recovery magnitude increases with increase in the difference between $V_{G\text{-}STR}$ and $V_{G\text{-}SNS}$ [16, 21], a low $V_{G\text{-}SNS}$ readout would suffer from recovery artifacts, shown in [22] and discussed later in this section, unless I_{DLIN} is recorded in few microseconds. There exists a tradeoff

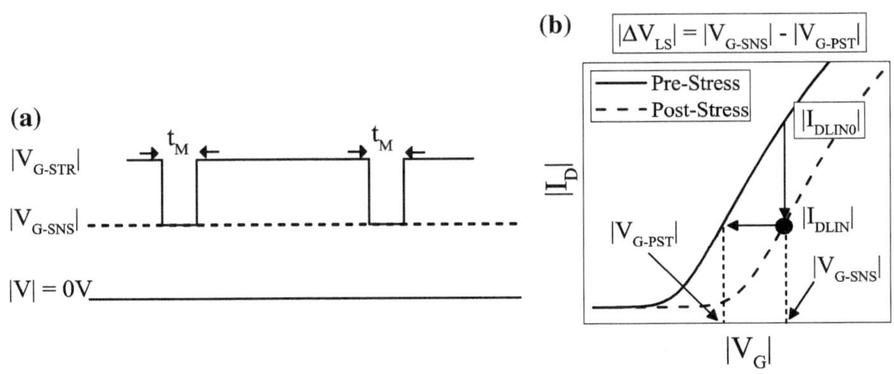

Fig. 2.17 **a** Schematic of OSDD method. **b** Measured I_{DLIN} versus V_G sweep before stress and one spot I_{DLIN} measurement after stress. The lateral and vertical shift methods for ΔV_T extraction are illustrated

between $V_{G\text{-}SNS}$ and drop down delay to minimize the impact of recovery, refer to Chap. 1, Fig. 1.29 for details. Alternatively, especially for relatively slower measurements, the recovery can be reduced by dropping down to $V_{G\text{-}SNS} = V_{DD}$ for I_{DLIN} measurement, where V_{DD} is the operating voltage of the device [23, 24]. ΔV_T can be calculated either using the lateral shift or vertical shift methods. However, measured ΔI_{DLIN} at high $V_{G\text{-}SNS}$ is affected by both ΔV_T and $\Delta \mu_{eff}$, especially for NBTI stress, and the impact of $\Delta \mu_{eff}$ needs to be corrected as described later in this section. Although OSDD method has been used to characterize NBTI in SiON and HKMG p-MOSFETs and PBTI in HKMG n-MOSFETs [20, 24], in this chapter, results are only shown for NBTI in HKMG p-MOSFETs.

Figure 2.18 plots the time evolution of measured ΔV_{LS} during NBTI stress in HKMG p-MOSFETs by using the lateral shift method, for (a) fixed t_M and different $V_{G\text{-}SNS}$ and (b) fixed $V_{G\text{-}SNS}$ and different t_M. The same I_{DLIN} data can be used to calculated ΔV using the vertical shift method using the equation shown above, and obtained time evolution is shown in Fig. 2.18 for (c) fixed t_M and different $V_{G\text{-}SNS}$ and (d) fixed $V_{G\text{-}SNS}$ and different t_M. The impact of $V_{G\text{-}SNS}$ on power-law time exponent n, extracted by linear regression of measured data in t_{STR} range of 10 s to 1 Ks, is plotted in Fig. 2.18 for (e) lateral and (f) vertical shift methods, for different measurement delay t_M. Finally, the time evolution of ΔV_T obtained using 10 μs UF-MSM method is shown in Fig. 2.18a, c, and power-law time exponent n extracted from MSM data for different measurement delay is shown in Fig. 2.18e, f as reference.

Note that when compared to the UF-MSM method for a particular t_M, the magnitude of ΔV_{LS} from the lateral shift and ΔV from the vertical shift OSDD method increases with increase in $V_{G\text{-}SNS}$ as shown respectively in Fig. 2.18a, c, and the corresponding time exponent n reduces as shown in Fig. 2.18e, f. Since recovery is negligible due to the use of small t_M, this is an artifact of mobility degradation at higher $V_{G\text{-}SNS}$ and is discussed below. However, for a particular $V_{G\text{-}SNS}$, the magnitude of ΔV_{LS} and ΔV reduces and corresponding time exponent n increases with increase in t_M, as shown in Fig. 2.18b–f, which can be attributed to recovery artifacts as discussed in the previous section.

To illustrate the impact of mobility degradation, Fig. 2.19a shows I_{DLIN} versus V_G characteristics measured using UF-MSM method in HKMG p-MOSFETs before and after NBTI stress. Pre- and post-stress V_T is determined using the peak g_m method, and the post-stress I–V curve is shifted by ΔV_T to align with the pre-stress curve. Note that although the shifted post-stress curve aligns with the pre-stress curve at lower V_G, the curves do not match at higher V_G, and the post-stress curve is below the pre-stress curve due to additional mobility degradation. The mobility impact for vertical shift method is assessed by $\Delta V^* = -\Delta I_{DLIN}/I_{DLIN0} * (V_{G\text{-}SNS} - V_{T0})$, where ΔI_{DLIN} is difference between pre-stress and ΔV_T shifted post-stress curve. In a similar manner, the mobility impact for lateral shift method is assessed by noting ΔV_{LS}*, the lateral difference between pre-stress and ΔV_T shifted post-stress curves as shown in Fig. 2.19a. Similar analysis can be done at different $V_{G\text{-}SNS}$ and for post-stress I–V measured at different t_{STR}.

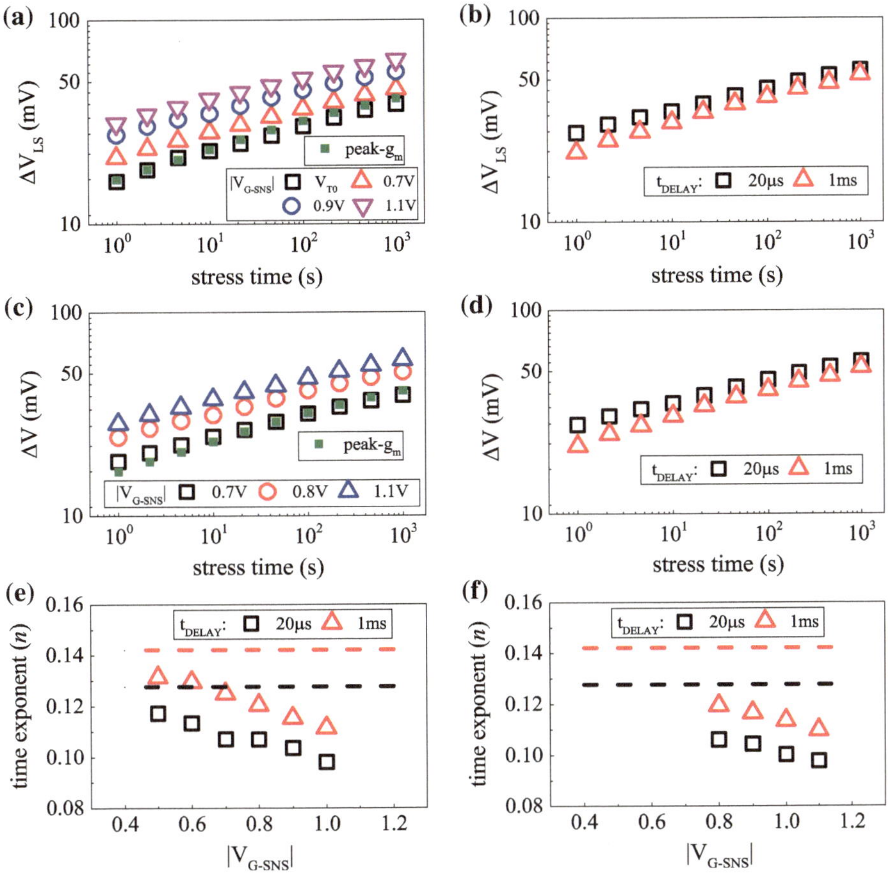

Fig. 2.18 OSDD measurements: Time evolution of (**a**, **b**) ΔV_{LS} using lateral shift and (**c**, **d**) ΔV using vertical shift methods, for different (**a**, **c**) sense bias and (**b**, **d**) measurement delay, for NBTI stress in HKMG p-MOSFETs. Impact of sense V_G on power-law time exponent n for **e** lateral shift and **f** vertical shift method for different measurement delay. UF-MSM data obtained using peak g_m method are shown as reference in (**a**, **c**). *Lines* in (**e**, **f**) represent UF-MSM measured time exponent n for different delay

The resulting time evolution of ΔV_{LS}* and ΔV* at different $V_{G\text{-}SNS}$ is shown respectively in Fig. 2.19b, c. It is evident that post-stress I_{DLIN} is affected by both ΔV_T and $\Delta\mu_{eff}$ especially at higher $V_{G\text{-}SNS}$. The contribution due to $\Delta\mu_{eff}$, manifested as ΔV_{LS}* or ΔV* for lateral or vertical shift methods respectively, increases with increase in $V_{G\text{-}SNS}$. Therefore, ΔV_{LS} and ΔV, extracted from lateral or vertical shift in I_D–V_G curve after stress, increases at higher $V_{G\text{-}SNS}$ due to additional contribution respectively from ΔV_{LS}* and ΔV*. Although time evolution of ΔV_{LS}* and ΔV* shows power-law time dependence, they have lower time exponent n compared to

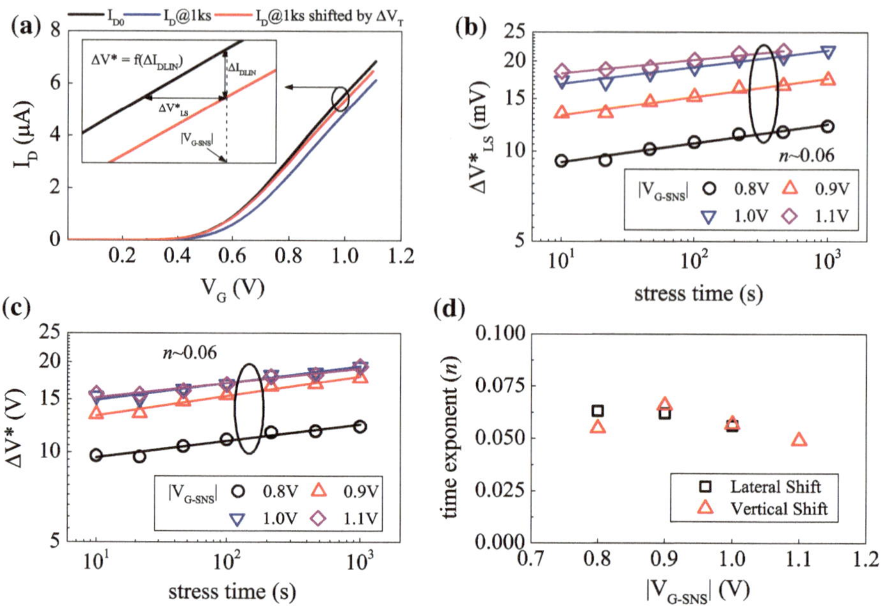

Fig. 2.19 Isolation of mobility impact: **a** Measured I_{DLIN} versus V_G curves before and after stress, and shift of post-stress *I–V* curve by ΔV_T. Calculated degradation using **b** lateral and **c** vertical shift methods, between pre-stress and ΔV_T shifted post-stress data for different sense V_G. **d** Power-law time exponent of mobility degradation induced degradation obtained at different sense V_G from (**b**) and (**c**). Data from NBTI stress in HKMG p-MOSFETs

that for ΔV_T, which is evident from Fig. 2.19b, c. The impact of $V_{G\text{-SNS}}$ on time exponent *n*, extracted using linear regression of time evolution of ΔV_{LS}* and ΔV* data in t_{STR} range of 10 s to 1 Ks is shown in Fig. 2.19d. Due to lower *n* for ΔV_{LS}* and ΔV*, increased ΔV_{LS} and ΔV magnitude is always associated with reduced time exponent *n* when OSDD measurements are done at higher $V_{G\text{-SNS}}$, as shown in Fig. 2.18e, f.

Figure 2.20a, b plot the time evolution of $\Delta V_{LS} - \Delta V_{LS}$* and $\Delta V - \Delta V$* obtained at different $V_{G\text{-SNS}}$ respectively from lateral and vertical shift methods for different measurement delay. It is important to remark that once the voltage shift ΔV_{LS}* and ΔV* corresponding to mobility degradation is subtracted, time evolution of $\Delta V_{LS} - \Delta V_{LS}$* and $\Delta V - \Delta V$* would correspond to that of ΔV_T. Figure 2.20c, d shows the impact of $V_{G\text{-SNS}}$ on corresponding time exponent *n*, obtained using linear regression of the mobility corrected measured data in t_{STR} range of 10 s to 1 Ks, for lateral and vertical shift methods for different measurement delay. UF-MSM measured ΔV_T time evolution for different measurement delay and corresponding time exponents are also shown as reference. It is important to note that once the impact of mobility is corrected, the lateral and vertical shift methods provide identical ΔV_T and *n* values at different $V_{G\text{-SNS}}$, which matches well with UF-MSM measured data, and this holds for different measurement delay as shown.

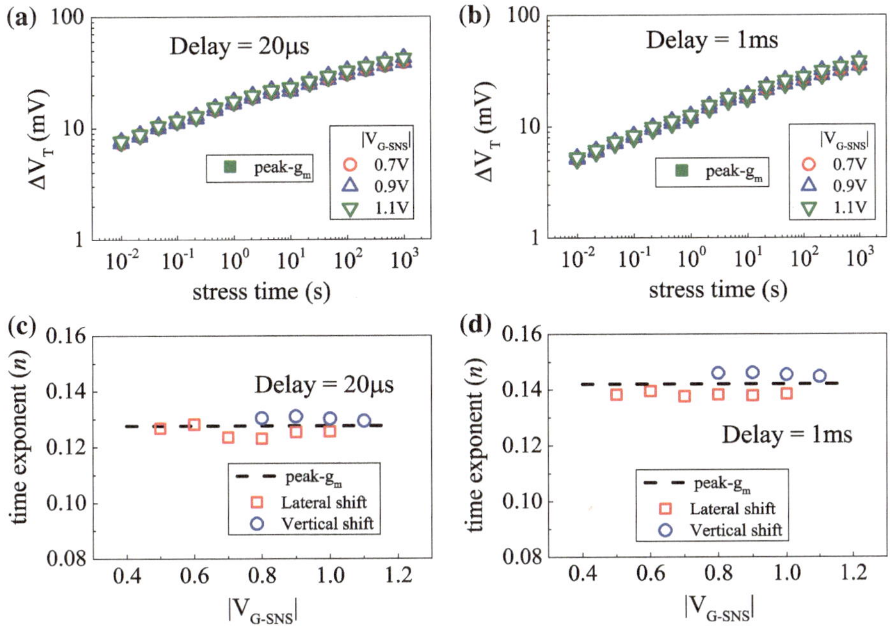

Fig. 2.20 **a**, **b** Time evolution of mobility corrected ΔV_T from lateral and vertical shift OSDD methods at different sense V_G for different measurement delay, for NBTI stress in HKMG p-MOSFETs. **c**, **d** Extracted long-time power-law time exponent n versus sense V_G, obtained from data in (**a**, **b**). UF-MSM measured data obtained using peak g_m method are shown as reference

2.6 Flicker Noise

Figure 2.21 shows the schematic of a flicker noise setup. Gate of the MOSFET is connected to a power supply via a low pass filter; drain is connected to a digital spectrum analyzer via a low noise amplifier. Power spectral density of drain current noise (S_{ID}) is measured in frequency (f) domain for different values of gate overdrive ($V_G - V_{T0}$), and input referred noise (S_{VG}) is obtained from the relation $S_{VG} = S_{ID}/{g_m}^2$, where V_{T0} and g_m are threshold voltage and transconductance respectively of the device under test [25].

Several mechanisms have been proposed in the past to explain flicker noise in MOSFETs. Some reports suggest noise is due to fluctuation in inversion layer carrier density [26–28], while others relate noise to bulk mobility fluctuation [29, 30]. The number fluctuation model [26–28] is based on McWhorter's theory of random trapping and detrapping of inversion layer carriers in the gate insulator traps [31], which in turn causes surface potential fluctuation and hence variation in inversion layer carrier density. The model suggests S_{VG} to be independent of gate overdrive, and strong correlation of S_{VG} to density of gate insulator traps [32, 33]. The mobility fluctuation model on the other hand is based on Hooge's empirical

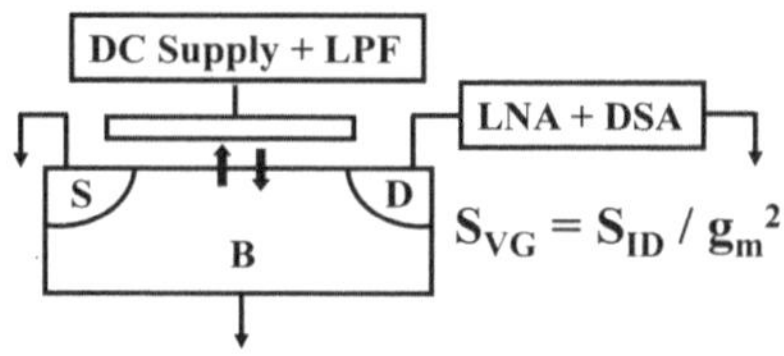

Fig. 2.21 Schematic of flicker noise measurement setup

formulation derived for bulk semiconductors [34], and in MOSFETs it is associated to fluctuations in bulk mobility as a result of fluctuation in phonon population due to phonon scattering [35]. In contrast to the number fluctuation theory, the mobility fluctuation model suggests S_{VG} to be linearly dependent on $(V_G - V_{T0})$.

Flicker noise method has been used in the past to directly estimate gate insulator defects in Silicon Dioxide (SiO_2) MOSFETs [32, 33]. More recently, the method has been used to determine pre-existing trap density in MOSFETs having differently processed SiON and HKMG gate insulators [25, 36–38]. Although flicker noise can also be utilized to determine trap generation after BTI stress [25], accurate flicker noise measurement is a time consuming process and hence the method would suffer from recovery issues and cannot estimate correct magnitude of generated defects.

Figure 2.22 plots S_{VG} as a function of (a) frequency at fixed $(V_G - V_{T0})$ and (b) gate overdrive at fixed f, measured in SiON p-MOSFETs before and after NBTI stress [25]. Note that S_{VG} increases after stress due to generation of new gate insulator traps as discussed in detail later in this chapter, which is consistent with previous reports [32, 33] and number fluctuation theory [26–28].

As discussed in [39] and shown in Fig. 2.23, noise contribution due to trapping and detrapping of inversion carrier for a single gate insulator trap has Lorentzian f dependence of the form $1/[1 + (f/f_0)^2]$, where f_0 is the corner frequency of the trap;

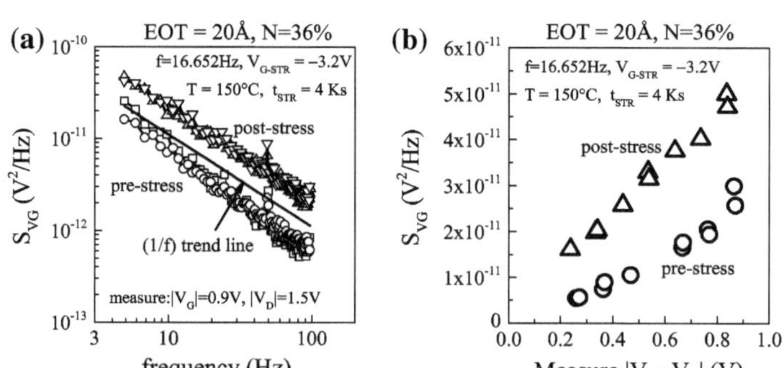

Fig. 2.22 Measured input referred noise versus **a** frequency and **b** gate voltage overdrive, before and after NBTI stress in SiON p-MOSFETs. Multiple measurements are shown in pre- and post-stress, obtained from different devices

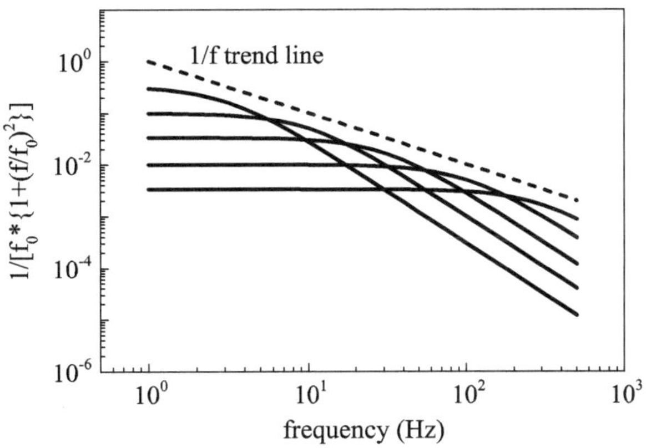

Fig. 2.23 Schematic representation of flicker noise as weighted sum of Lorentzian functions with different corner frequencies, resulting in $1/f$ frequency response

a weighted summation of contribution from uniform spatial distribution of gate insulator traps with different f_0 results in $1/f$ dependence. As evident from Fig. 2.22, measured frequency dependence of S_{VG} shows f^{-k} dependence, with $k \sim 0.7$ before stress, which increases to $k \sim 1$ after stress. A plausible explanation has been provided in [25], which relates noise in SiON devices to Nitrogen (N) related gate insulator traps. Since PNO devices have been used having N concentration that peaks at the SiON/poly-Si interface and exponentially reduces towards the Si/SiON interface, a suitably weighted sum of Lorentzian response from such non-uniform trap distribution results in $k < 1$ for f dependence of S_{VG} before stress. As discussed later in this chapter and in Chap. 3, NBTI stress results in generation of traps at and near the Si/SiON interface and hence the overall spatial distribution of gate insulator traps become relatively more uniform. As a consequence, the f dependence of S_{VG} after stress increases and shows $k \sim 1$.

Although the dependence of S_{VG} on gate insulator traps agrees with the number fluctuation theory, the variation of S_{VG} with gate overdrive shown in Fig. 2.22b cannot be explained in this framework. It is now well known that charges in gate insulator traps not only impacts the inversion carrier density but also influences their mobility by Coulomb scattering. The number fluctuation model has been enhanced to incorporate the effect of surface mobility fluctuation caused by gate insulator charges [40]. The combined number and mobility fluctuation model predicts the following relation for S_{VG} [40]:

$$S_{VG} = \frac{kT\,q^2}{\gamma f W L C_{OX}^2}(1 + \alpha\mu N_C)^2 N_T(E_{FN}) \tag{2.4}$$

In (2.4), q is electronic charge, kT is thermal energy, f is frequency, C_{OX}, W and L are gate oxide capacitance, device width and length respectively; μ is inversion layer mobility, $N_C = C_{OX}/q * (V_G - V_{T0})$ is inversion carrier density; γ is the attenuation factor of electron or hole wave function into the gate insulator and can be calculated using WKB tunneling framework; α is scattering coefficient and N_T is the gate insulator trap density near the Fermi level (E_{FN}). Although missing from the conventional number fluctuation model, the combined number-mobility fluctuation model has linear dependence on gate overdrive via the term N_C and therefore is consistent with measured data.

As discussed later in this book, gate insulator processes impact the density of pre-existing traps and hence the trapping component of BTI degradation. Of particular interest is the impact of Nitrogen, which is discussed in detail in different chapters of this book. Figure 2.24 plots as processed, pre-existing trap density in (a) SiON p-MOSFETs as well as (b) HKMG p-MOSFETs and n-MOSFETs extracted using the flicker noise method [25, 38]. As discussed in Chap. 1, Table 1.1, PNO SiON devices with different N dose have been used. On the other hand, the HKMG devices were fabricated without (D2) and with (D3) nitridation after ALD High-K deposition and with nitrided IL (D4), refer to Chap. 3, Fig. 3.2 for details. Note that flicker noise in n- and p-channel MOSFETs is respectively due to trapping and detrapping of inversion layer electrons and holes in gate insulator traps. Therefore, it can be remarked that presence of N in gate insulator increases pre-existing hole trap density in SiON and HKMG p-MOSFETs and electron trap density in HKMG n-MOSFETs, and hence impacts both NBTI and PBTI degradation as discussed in detail in Chaps. 3 and 4.

In spite of strong experimental evidence in the literature of trap generation during BTI stress in p- and n-channel MOSFETs, shown later in this chapter and also in Chap. 3 , some reports have suggested NBTI and PBTI to be exclusively due to hole and electron trapping respectively, in pre-existing gate insulator traps [41–43]. Such a framework has to associate strong T activation to the charge trapping process to explain T activation of BTI, which can be independently assessed using flicker noise technique. Figure 2.25 plots T dependence of S_{VG} measured at fixed f and ($V_G - V_{T0}$) in SiON p-MOSFETs [44]. Note that contrary to the above proposition, S_{VG} has negligible T dependence and therefore would suggest negligible T activation of the hole trapping and detrapping process in thin gate insulator stacks used in modern MOSFETs. It is important to remark

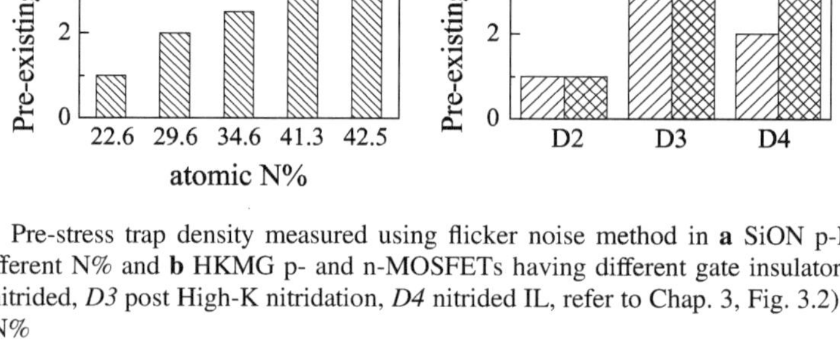

Fig. 2.24 Pre-stress trap density measured using flicker noise method in **a** SiON p-MOSFETs having different N% and **b** HKMG p- and n-MOSFETs having different gate insulator processes (*D2* non-nitrided, *D3* post High-K nitridation, *D4* nitrided IL, refer to Chap. 3, Fig. 3.2) leading to different N%

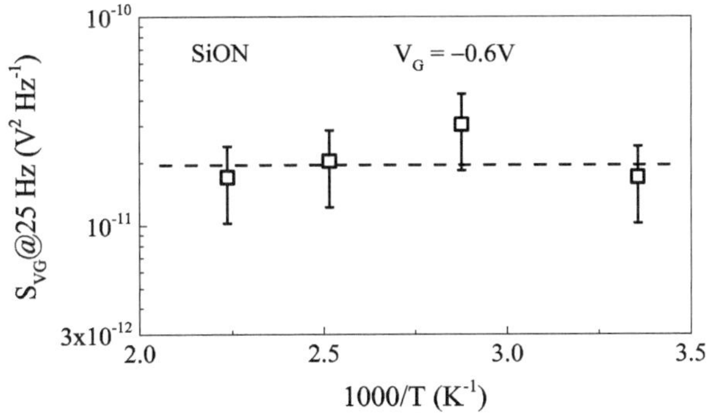

Fig. 2.25 Impact of temperature on measured input referred noise. Data from [44]

that weak *T* activation of the trapping–detrapping process is significant observation, and when combined with relatively stronger *T* activation of the trap generation process, shown in Chap. 3, it can explain *T* activation of BTI for different gate insulator processes as discussed in detail in Chap. 4.

Note that the *T* activation of NBTI measured in SiON and HKMG p-MOSFETs and of PBTI in HKMG n-MOSFETs reduces for devices having N in the gate insulator stack, as discussed in detail in Chaps. 1 and 4. This can be explained by invoking (a) both trap generation and charge trapping processes for BTI that are mutually uncorrelated, (b) relatively lower *T* activation for the charge trapping process, shown in Fig. 2.25, when compared to the trap generation process, shown in Chap. 3, and (c) higher relative contribution of charge trapping in pre-existing gate insulator traps for devices having N in gate insulator, as shown in Fig. 2.24; refer to Chap. 4 for further details. Other manifestations of uncorrelated trap generation and trapping BTI sub-components are discussed in Chaps. 3 and 4 for different gate insulator processes, and flicker noise method has been used for independent verification of pre-existing gate insulator traps.

2.7 Charge Pumping (CP)

CP technique [45] estimates trap density at and near the Silicon (Si) channel and gate insulator interface of a MOSFET by measuring trap assisted electron-hole recombination current. In CP method, illustrated in Fig. 2.26, a large signal gate pulse is applied to drive the MOSFET repetitively between inversion and accumulation, the drain and source terminals are shorted and grounded, and DC current due to recombination of electrons and holes in gate insulator traps is measured at the substrate. Although the method can measure pre-stress gate insulator traps, increase in CP current (ΔI_{CP}) after BTI stress can be used as a direct estimation of generation of new gate insulator traps.

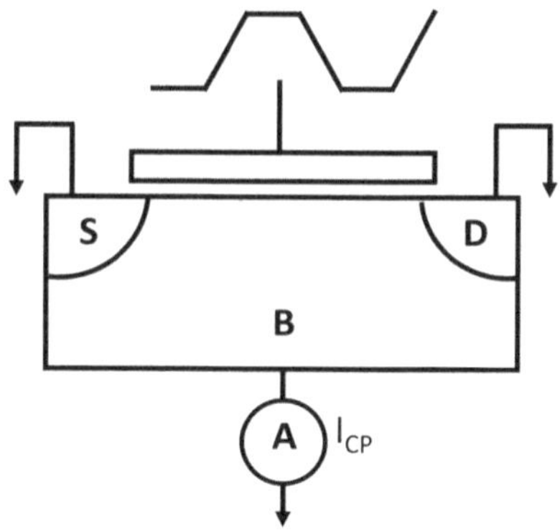

Fig. 2.26 Schematic of charge pumping measurement setup

However, CP is a slow measurement technique and is usually implemented in measure-stress-measure or MSM configuration to estimate trap generation due to BTI stress, and hence suffer from recovery issues. Moreover, CP scans trap generation in an energy range of the Si band gap that is much smaller compared to that scanned by I_{DLIN} and V_T measurement methods. Therefore, measurement delay and band gap correction methods have been proposed [9] for accurate estimation of trap generation contribution to overall BTI degradation and are discussed later in this section. CP technique has been extensively used to characterize trap generation during BTI stress [9, 14, 22, 36, 46–48].

The basic CP method has been modified to estimate the energetic and spatial distribution of traps in the gate insulator [49, 50]. However, such spectroscopic CP techniques are extremely time consuming, and although they can be used to estimate as processed, pre-existing traps, these methods suffer from recovery related issues and hence are not effective to determine the distribution of generated traps. Furthermore, some reports have proposed ultra-fast CP techniques to characterize on-the-fly trap generation without any measurement delay [51, 52]. These methods rely on applying a gate pulse having inversion level equal to the BTI stress bias, $V_{G\text{-}STR}$. The application of such a large gate bias makes these methods prone to gate leakage [53] and unsuitable for thinner gate insulator MOSFETs. More importantly, since CP method involves pulsing the channel from inversion to accumulation, the application of such large bi-polar pulses alters the conventional BTI stress regime, and makes the MOSFET vulnerable to significant bulk trap generation in addition to generation of traps at or near the Si/SiO_2 interface. Owing to the application of large gate pulse, these methods scan traps deep into the gate insulator bulk [49, 50], and therefore, the contribution due to additionally generated bulk traps must be corrected for proper estimation of BTI generated defects. Therefore, although recovery issue can be avoided, the accuracy of these ultra-fast CP methods gets impacted by gate leakage and bulk trap issues, and the methods are of not much use to characterize BTI trap generation.

Figure 2.27 illustrates the dynamics of electron and hole capture-emission processes in gate insulator traps of an n-channel MOSFET under repetitive gate pulses [45]. The channel goes into inversion when V_G goes above V_T, and traps that

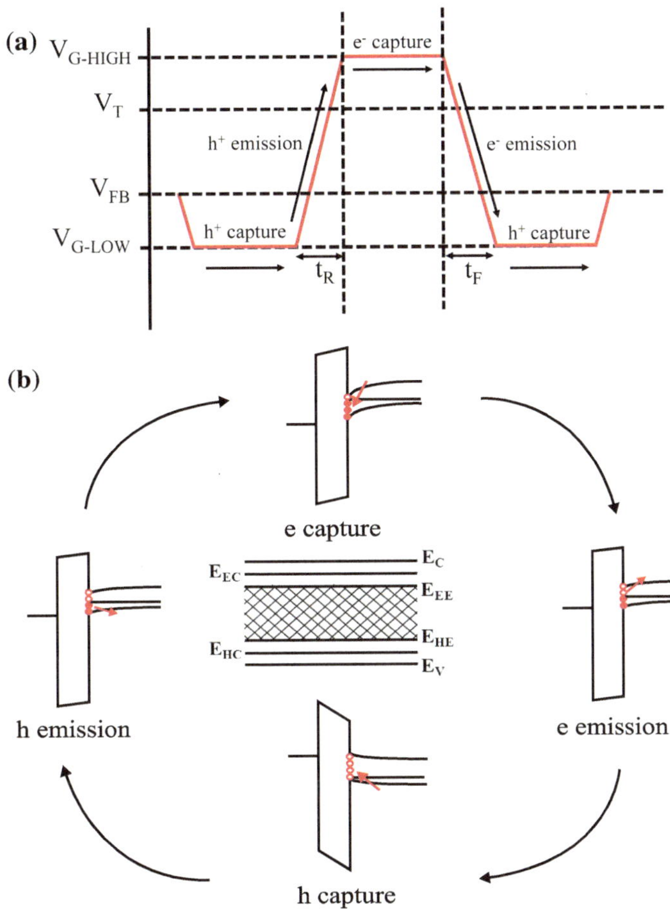

Fig. 2.27 Schematic of charge pumping process: **a** Different regimes of trap assisted capture and emission of electrons and holes as gate is pulsed between inversion and accumulation. **b** Corresponding energy band diagrams showing capture and emission of electrons and holes, and the energy zone for recombination. Example is shown for *n*-channel MOSFET

remain below the Fermi level capture inversion layer electrons from the Si conduction band. The channel transitions from inversion via depletion towards accumulation during the falling edge of the gate pulse, traps start to move above the Fermi level and some traps emit electrons back to the conduction band. Once the channel goes into accumulation as V_G goes below the flatband voltage (V_{FB}), traps above Fermi level capture holes from the valence band of Si. As the channel is pulsed back via depletion to inversion during the rising edge of the gate pulse, traps start to go below the Fermi level and some traps emit holes back to the valence band. Finally, as the channel goes back to inversion as V_G goes above V_T, the electron capture process starts, and subsequent processes get repeated.

Note that time constant of traps for electron emission is smallest near the conduction band edge and increases towards the valence band, while time constant for hole emission is smallest near the valence band edge and increases towards the conduction band. Figure 2.27 also depicts the energy level of traps associated with capture-emission process. During inversion, traps up to the level E_{EC} gets filled with electrons, and electron emission occurs in the levels between E_{EC} and E_{EE} during the falling edge of the gate pulse, as V_G transitions between V_T and V_{FB}. Similarly during accumulation, traps up to the level E_{HC} get filled with holes, and hole emission occurs in the levels between E_{HC} and E_{HE} during the rising edge of the gate pulse, as V_G transitions between V_{FB} and V_T. Therefore, electron-hole recombination occurs in traps between the range E_{EE} and E_{HE} that results in CP current, I_{CP}. The energy zone (ΔE) of traps for electron-hole recombination is given by [45]:

$$\Delta E = -2kT \ln\left[n_i v_{\text{th}} \left(\sigma_n \sigma_p\right)^{0.5} (t_R t_F)^{0.5} \left\{\frac{V_T - V_{FB}}{V_{GH} - V_{GL}}\right\}\right] \tag{2.5}$$

In (2.5), V_{GH} and V_{GL} respectively are pulse high and low levels, t_R and t_F are pulse rise and fall times for transition between V_{GL} to V_{GH} and V_{GH} to V_{GL} respectively, refer to Fig. 2.27; kT and n_i are thermal energy and intrinsic carrier density respectively, v_{th} is thermal velocity, while σ_n and σ_p are trap capture cross sections associated respectively with electron and hole trapping. For gate pulse having frequency f, the CP current I_{CP} is given by [45]:

$$I_{CP} = qfWLN_{IT} \tag{2.6}$$

In (2.6), q is the electronic charge, W and L respectively are width and length of the MOSFET; $N_{IT} = \langle D_{IT}\rangle \, \Delta E$, where $\langle D_{IT}\rangle$ is the average density of traps per energy in the energy zone ΔE scanned by CP, refer to (2.5) and Fig. 2.27, and N_{IT} is the total trap density probed by CP method.

CP measurements are done in slow MSM mode before and during logarithmically spaced interruptions of BTI stress, and by assuming no change in σ_n and σ_p and therefore in ΔE, changes in CP current (ΔI_{CP}) after stress can be attributed to changes in trap density (ΔN_{IT}). Note that to reduce measurement time, CP measurements can be done at fixed gate pulse amplitude, and the V_{GH} and V_{GL} levels of the gate pulse need to be chosen above V_T and below V_{FB} respectively, to drive the MOSFET between strong inversion and accumulation. Pre-stress N_{IT} can be independently estimated from subthreshold slope measurements, and can be used with pre-stress I_{CP} to determine the geometric mean of σ_n and σ_p. Although CP method has been used to quantify trap generation during NBTI stress in SiON and HKMG p-MOSFETs as well as PBTI stress in HKMG n-MOSFETs, in this section, results are shown for NBTI stress in SiON p-MOSFETs.

As mentioned earlier in this chapter and discussed in Chap. 3, NBTI parametric shift is caused by mutually independent contribution from trapping in pre-existing defects and generation of new traps; the later can be independently verified using

CP measurements. However, caution must be applied before the time evolution of ΔN_{IT} obtained from CP measurements is converted to the trap generation component ($\Delta V_{IT} = q \cdot \Delta N_{IT}/C_{OX}$) of overall V_T shift [9]. First, note that CP method scans trap generation in energy range ΔE in the band gap centered on the midgap, refer to (2.5), while V_T shift estimated from inversion I_{DLIN} degradation is impacted by generated traps throughout the band gap. Moreover, while I_{DLIN} is usually measured using ~milliseconds or ~microseconds delay, CP method is implemented in slow MSM mode with measurement delay of ~seconds, and hence, the later would suffer from recovery related artifacts. Therefore, time evolution of ΔN_{IT} from CP measurements must be corrected for band gap and delay differences, which is discussed below.

Figure 2.28a shows ΔN_{IT} time evolution obtained from CP measurements for NBTI stress in SiON p-MOSFETs [9]. Measurements were done using different t_R and t_F values of gate pulse that result in different scanned energy zone ΔE, refer to (2.5) [45]. Note that measured ΔN_{IT} increases with reduction in t_R and t_F as ΔE is increased, and the ΔN_{IT} versus ΔE correlation can be used to determine corrected ΔN_{IT} for traps corresponding to the full band gap, which is also plotted in Fig. 2.28a. Uniform trap generation in the entire band gap is assumed for this correction. Figure 2.28b plots ΔN_{IT} time evolution after band gap correction, obtained from CP measurements with different measurement delay, t_M. Note that NBTI stress induced generated traps do recover after stress is stopped for measurement. Therefore, the magnitude of ΔN_{IT} reduces and the power-law time exponent n increases with increase in t_M as recovery is increased [2, 22].

Time evolution of ΔN_{IT} at different t_M can be modeled using empirical universal recovery expression [54]:

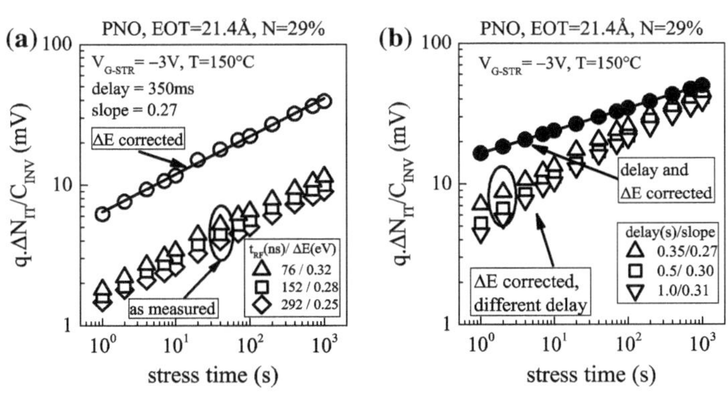

Fig. 2.28 CP measurements: Time evolution of **a** measured ΔN_{IT} for different energy zone of electron-hole recombination determined by different rise and fall time of gate pulse, and estimated ΔN_{IT} after band gap correction. Time evolution of **b** band gap corrected ΔN_{IT} for different measurement delay, and ΔN_{IT} after both band gap and delay correction. Data from NBTI stress in SiON p-MOSFETs

$$\Delta N_{IT}(t_{STR}, t_M) = \frac{\Delta N_{IT}(t_{STR})}{1 + B\left(\frac{t_M}{t_{STR}}\right)^{\beta}} \tag{2.7}$$

The band gap corrected ΔN_{IT} time evolution data can be fitted using (2.7) to determine model parameters, and hence, both band gap and delay corrected ΔN_{IT} can be obtained for any t_M, which is also shown in Fig. 2.28b. It is important to remark that the magnitude of ΔN_{IT} increases and power-law time exponent n reduces after corrections are performed on as-measured CP data. Failure to perform these corrections would severely underestimate the trap generation component and overestimate the trapping component of NBTI, as done in [55]. Since trapping is attributed to the as-processed pre-existing defects, this would severely underestimate quality of the gate insulator; refer to [9, 25] for further details. Once the band gap and delay corrections are performed, time evolution of directly measured ΔN_{IT} from CP would provide correct ΔV_{IT} subcomponent of overall ΔV_T, which is further discussed in later chapters of this book.

Recovery of NBTI stress-generated traps is of interest and is modeled in detail in Chap. 6. An interesting artifact of N_{IT} recovery is shown in Fig. 2.29. CP measurements have been performed in SiON p-MOSFETs before and during logarithmic interruptions of NBTI stress; the stress has been performed at different T, and different measurement delay t_M has been used. Figure 2.29 shows (a) time evolution of ΔN_{IT} for different stress T at fixed t_M, and (b) power-law time exponent n of ΔN_{IT} time evolution data as a function of T for different t_M [22]. The exponent is calculated using linear regression of measured data in t_{STR} range of 10 s to 1 Ks. It is interesting to note that measured n increases at higher T for a given t_M, which has been explained by dispersive Hydrogen (H) transport related trap generation mechanism [20]. However, the above observation is simply a measurement artifact

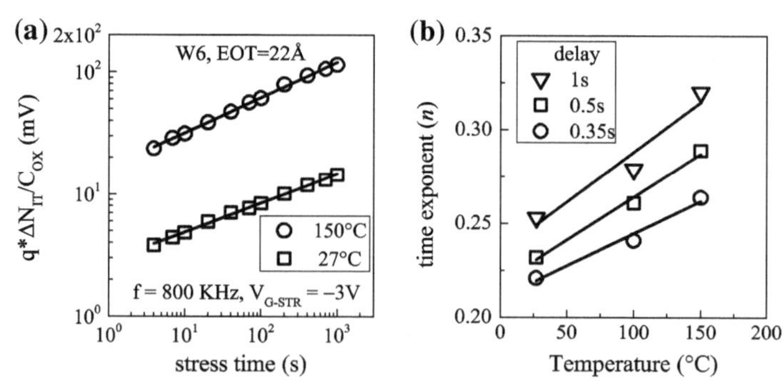

Fig. 2.29 CP measurements: **a** Time evolution of ΔN_{IT} for NBTI stress in SiON p-MOSFETs at different stress T. **b** Extracted power-law time exponent n from ΔN_{IT} time evolution data as a function of stress T for different measurement delay

associated with N_{IT} recovery, since the T dependent increase in n is strongly impacted by t_M as shown in Fig. 2.29b, and larger n is observed at higher t_M for a given T [22]. Therefore, proper corrections to CP measured data must be done before deriving conclusions regarding NBTI physical mechanism.

2.8 Gated Diode (DCIV)

Although CP is a very useful technique to determine trap generation, it requires the application of large pulses to the gate terminal of the MOSFET to drive the channel from strong inversion to accumulation, and hence is not suitable for gate insulators having thin EOT due to gate leakage issues [53]. Alternatively, direct estimation of traps at and near the Si/SiO_2 interface can be obtained using the DCIV method [56, 57], illustrated in Fig. 2.30. The source-drain terminals are shorted and forward biased just below the junction cut-in voltage, the gate bias is swept from accumulation to inversion in the vicinity of $V_G \sim 0$ V, and trap assisted electron hole recombination current I_{DCIV} is measured at substrate terminal. DCIV method has been used to characterize trap generation during NBTI stress in SiON p-MOSFETs [48, 57–59], as well as during NBTI and PBTI stress respectively in HKMG p- and n-MOSFETs [17, 18, 37, 38, 60]; HKMG results are discussed in detail in Chap. 3. In this section, implementation of DCIV method is discussed using the example of NBTI stress in HKMG p-MOSFETs.

Figure 2.31 plots measured I_{DCIV} versus V_G characteristics (a) before and during different intervals of NBTI stress, as well as (b) immediately after and at different intervals following stoppage of NBTI stress in HKMG p-MOSFET. Note that I_{DCIV}

Fig. 2.30 Schematic of gate diode (DCIV) measurement setup

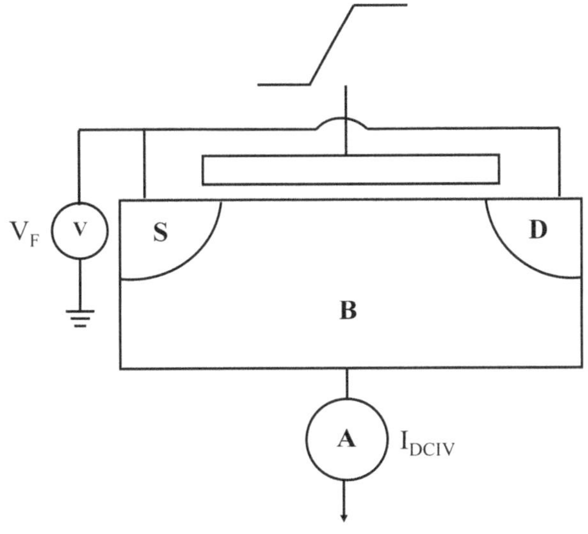

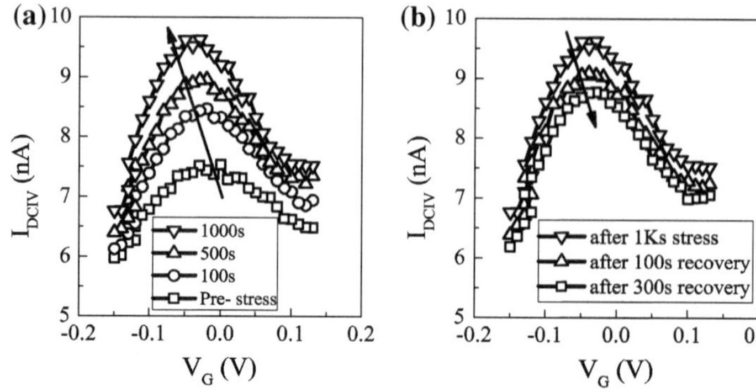

Fig. 2.31 *I–V* characteristics from DCIV measurements **a** before and after different intervals of stress, and **b** immediately after stress and after different recovery intervals, for NBTI stress in HKMG p-MOSFETs

peaks at a particular V_G, and the peak magnitude increases and reduces respectively during and after NBTI stress. The V_G value corresponding to the peak I_{DCIV} also varies during stress and recovery. Since I_{DCIV} is due to electron-hole recombination via the traps at or near the Si/SiO_2 interface, increase and reduction in I_{DCIV} signify generation and recovery of traps during and after NBTI stress. Trap density can be estimated from peak I_{DCIV} using the Shockley–Read–Hall (SRH) formalism of carrier capture and emission [61].

For a single defect situated at energy level E_T having density N_{IT}, the difference between base and peak values of I_{DCIV} versus V_G characteristics is given by the following relation [56, 57, 62]:

$$I_{\text{DCIV,peak}} - I_{\text{DCIV,base}} = \frac{qn_i v_{\text{th}} \left(\sigma_{\text{n}}\sigma_{\text{p}}\right)^{0.5} N_{\text{IT}} \frac{WL}{2} \left[\exp\left(\frac{qV_{\text{F}}}{kT} - 1\right)\right]}{\exp\left(\frac{qV_{\text{F}}}{2kT}\right) + \cos\text{h}(U_{\text{T}})} \tag{2.8}$$

In (2.8), q is electronic charge, W and L are device width and length respectively, kT and n_i are thermal energy and intrinsic carrier density respectively, v_{th} is thermal velocity, σ_n and σ_p are trap capture cross sections associated with electron and hole trapping respectively and V_F is forward bias applied to the shorted source and drain terminals. $U_T = (E_T - E_i)/kT + \ln(\sigma_n/\sigma_p)$, where E_i is intrinsic level. The equation shown is for a single trap situated at a particular energy level, and has to be integrated over different trap energy levels for distribution of traps. The energy zone scanned by DCIV technique is given by $\Delta E = q \cdot |V_F|$, and for relatively larger V_F value, which should be still lower than the junction cut-in voltage, (2.8) can be approximated as follows [59, 62]:

$$I_{\text{DCIV,peak}} - I_{\text{DCIV,base}} = \frac{1}{2} qn_i \left(\sigma_{\text{n}}\sigma_{\text{p}}\right)^{0.5} v_{\text{th}} N_{\text{IT}} (WL) \exp\left(\frac{qV_{\text{F}}}{2kT}\right) \tag{2.9}$$

In (2.9), $N_{IT} = \langle D_{IT} \rangle \, \Delta E$ is average density of traps in energy zone ΔE, and similar values of σ_n and σ_p has been assumed. DCIV measurements are done before and during logarithmically spaced interruptions of BTI stress, and by assuming no change in σ_n and σ_p due to stress, time evolution of ΔN_{IT} can be estimated from time evolution of measured peak ΔI_{DCIV} using (2.9). As before, pre-stress subthreshold slope is used to estimate pre-stress N_{IT} and hence the geometric mean of σ_n and σ_p can be found by using pre-stress I_{DCIV}.

Note that similar to CP, corrections should be done on as-measured DCIV data before directly estimated ΔN_{IT} is compared to the ΔV_{IT} component of ΔV_T. Similar to CP, DCIV is also a slow measurement method and is implemented in the MSM mode. A typical V_G sweep takes ~seconds, and the time evolution of as-measured stress-generated defects should be corrected for measurement delay using the universal recovery expression shown in (2.7). Moreover, DCIV method scans traps in energy zone $\Delta E = q \cdot |V_F|$ in the Si band gap centered on the midgap, while ΔV_T is impacted by traps generated throughout the band gap. Therefore, obtained trap density should also be corrected for band gap difference before comparing to the trap generation subcomponent of overall V_T shift [60]. Failure to do these corrections would result in gross underestimation of the trap generation subcomponent.

DCIV measurements were performed in different HKMG p-MOSFETs, before and immediately after NBTI stress as well as after a certain delay following NBTI stress. Devices having different HKMG gate insulator processes have been used [38]; refer to Chap. 3, Fig. 3.2 for details. Different E_{OX} values have been used for stress, and all devices were stressed and recovered for identical stress time t_{STR} and recovery time t_{REC}, respectively. Figure 2.32a correlates ΔI_{DCIV} measured just after NBTI stress to ΔI_{DCIV} after recovery following NBTI stress. Note that universal correlation between generation and recovery of traps has been observed across different devices. Such correlation can be used to determine the parameters of universal recovery expression shown in (2.7), which can further be used to perform delay correction of as-measured DCIV data. Figure 2.32b plots the time evolution of as-measured ΔN_{IT} using the DCIV method for different measurement delay, t_M, for NBTI stress in HKMG p-MOSFET. Time evolution of ΔN_{IT} after delay correction and both delay as well as band gap correction are also shown. Note that the magnitude of ΔN_{IT} increases as well as power-law time exponent n reduces after delay and band gap corrections.

DCIV measurements have been done before and during logarithmically spaced interruptions in NBTI stress, to determine time evolution of generated traps in different HKMG p-MOSFETs listed in Chap. 3, Fig. 3.2. Measured data are corrected for delay and band gap using the procedure discussed above. Figure 2.32c, d plot power-law time exponent n for these devices obtained from corrected trap generation data, as a function of $V_{G\text{-}STR}$ and T, respectively, obtained using linear regression in t_{STR} range of 10 s to 1 Ks. It is important to remark that universal time exponent of $n \sim 1/6$ is always observed, for different devices and across different $V_{G\text{-}STR}$ and T. The physical mechanism behind this universality will be discussed later in this book. Unless mentioned otherwise, all DCIV data presented in Chap. 3 and elsewhere in this book are delay and band gap corrected.

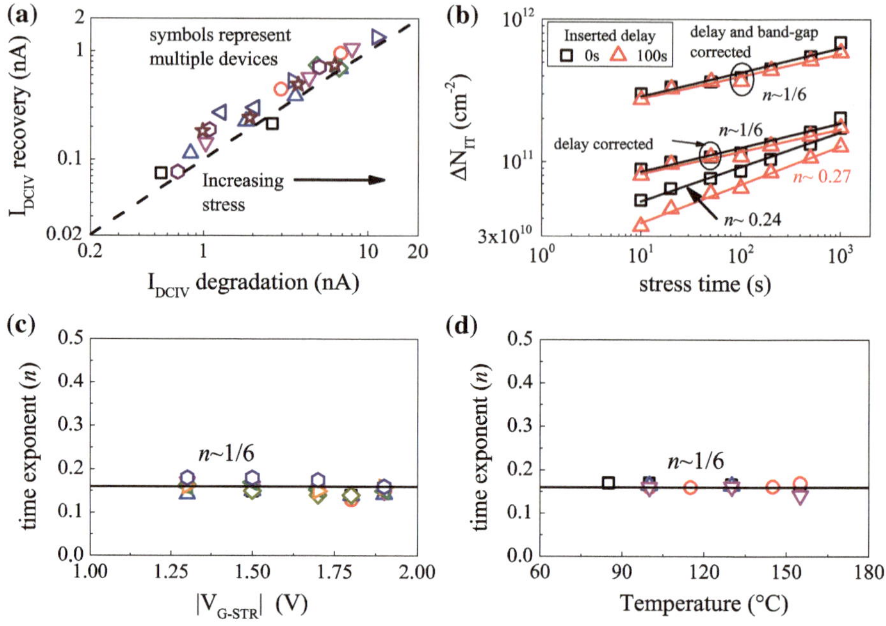

Fig. 2.32 DCIV measurements for NBTI stress in HKMG p-MOSFETs: **a** Correlation of peak DCIV current degradation and recovery for different HKMG devices. **b** Time evolution of measured ΔN_{IT} for different measurement delay, after delay correction and after both delay and band gap correction. Extracted power-law time exponent n versus **c** stress V_G and **d** stress T, obtained from corrected DCIV data for different HKMG devices. Refer to Chap. 3, Fig. 3.2, for details of different HKMG devices

2.9 Stress Induced Leakage Current (SILC)

Increase in gate leakage current (ΔI_G) measured in inversion, refered to as SILC, is a routine characterization technique for Time Dependent Dielectric Breakdown (TDDB) stress in MOSFETs [63–65]. SILC has been attributed to inelastic Trap Assisted Tunneling (TAT) via bulk gate insulator traps generated during TDDB stress, and is illustrated in Fig. 2.33 [63, 66]. Since BTI and TDDB stress regimes are essentially the same, as mentioned in Chap. 1, Fig. 1.2, SILC has been used during NBTI stress in SiON p-MOSFETs [14] and PBTI stress in HKMG n-MOSFETs [24, 37, 38, 67, 68] to access contribution due to generated bulk insulator traps. However, due to band alignment issues, SILC is usually not observed in HKMG p-MOSFETs as discussed in [65]. In this section, the implementation details of SILC measurement are shown using PBTI stress in HKMG n-MOSFETs, and more SILC data are shown later in Chaps. 3 and 4. Independently measured DCIV data are also shown to aid comparison between the two characterization methods.

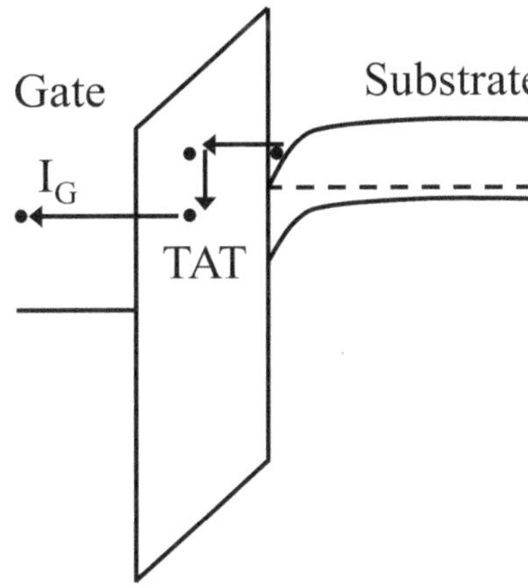

Fig. 2.33 Energy band diagram representing SILC via trap assisted tunneling and structural relaxation of traps. Example shown for *n*-channel MOSFET

Figure 2.34a shows measured I_G versus V_G characteristics before and during logarithmically spaced interruptions of PBTI stress in HKMG n-MOSFET. Time evolution of SILC, manifested in the form of increase in I_G at a particular V_G after stress, can be estimated by noting ΔI_G at a fixed V_G from measured I_G–V_G characteristics. However, note that SILC recovers after stoppage of stress [24] and

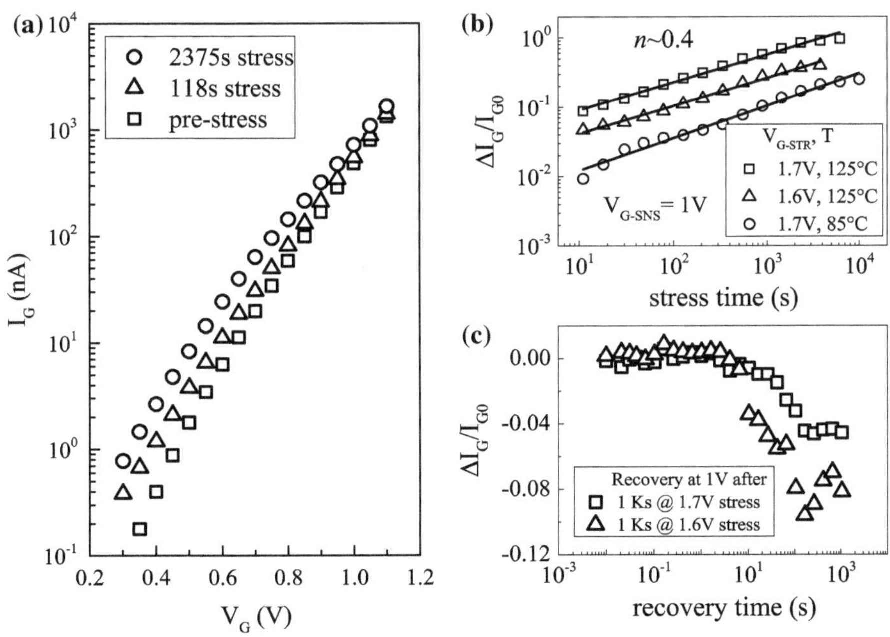

Fig. 2.34 SILC for PBTI stress in HKMG n-MOSFETs: **a** Measured I_G versus V_G characteristics before and after stress. **b** Time evolution ΔI_G obtained at fixed sense V_G for different $V_{G\text{-STR}}$ and T. **c** Time evolution of reduction in ΔI_G sensed at fixed V_G, after removal of stress

I_G–V_G sweep is usually time consuming for accurate I_G measurement. Therefore, it is desirable to avoid full I_G–V_G sweep to minimize recovery, and instead, drop V_G down from $V_{G\text{-STR}}$ to $V_{G\text{-SNS}}$, usually equals to V_{DD}, for one-point I_G measurement. Figure 2.34b shows time evolution of ΔI_G at different $V_{G\text{-STR}}$ and T for PBTI stress in HKMG n-MOSFET. A power-law time dependence is observed with time exponent n ($\sim$0.4), which is much larger than that obtained from DCIV measurements and shown earlier in this chapter. Therefore, trap generation probed by SILC has very different physical origin than that probed by DCIV, which is discussed in detail in Chap. 3. Figure 2.34c shows the reduction of ΔI_G after stoppage of PBTI stress, which signifies the recovery of generated traps as measured using SILC.

Figure 2.35a correlates generation of SILC measured immediately after stress to SILC recovery measured after certain delay following stress in different HKMG n-MOSFETs listed in Chap. 3, Fig. 3.2 [38]. All devices were stressed for fixed t_{STR} duration at different $V_{G\text{-STR}}$, and then allowed to recover for a fixed duration t_{REC}. It is important to note that universal generation to recovery correlation has been observed for different devices, and such data can be used to model SILC recovery using the universal recovery expression of (2.7). As a comparison, Fig. 2.35b correlates trap generation and recovery obtained from measured changes in DCIV current, ΔI_{DCIV}, during and after PBTI stress in these HKMG n-MOSFETs. Identical $V_{G\text{-STR}}$, T, t_{STR} and t_{REC} have been used for SILC and DCIV measurements for fair comparison. Universal correlation has been observed for DCIV measured data across different devices. However, the slope of the correlation line for SILC and DCIV are different, refer to Fig. 2.35a, b. Interestingly, the DCIV measured trap generation to recovery correlation slope is similar between NBTI and PBTI stress, refer to Figs. 2.32a and 2.35b. This aspect is further discussed in Chap. 3.

Figure 2.36 plots the time evolution of as measured and delay corrected (a) ΔI_G and (b) ΔI_{DCIV} respectively from SILC and DCIV measurements, for PBTI stress in

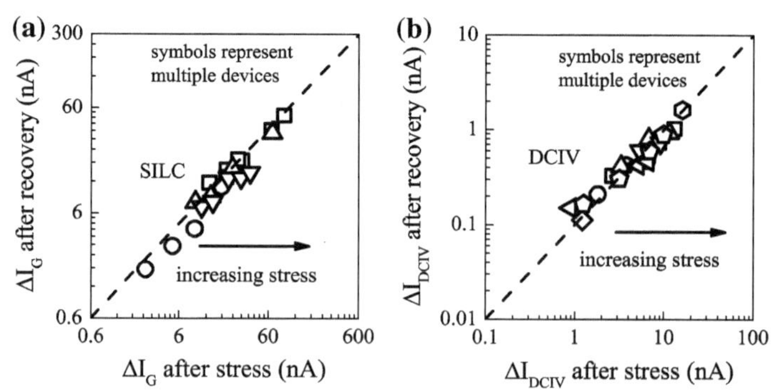

Fig. 2.35 Correlation of trap generation and recovery obtained using **a** SILC and **b** DCIV measurements for PBTI stress in HKMG n-MOSFETs having different gate insulator processes. Refer to Chap. 3, Fig. 3.2, for details of different HKMG devices

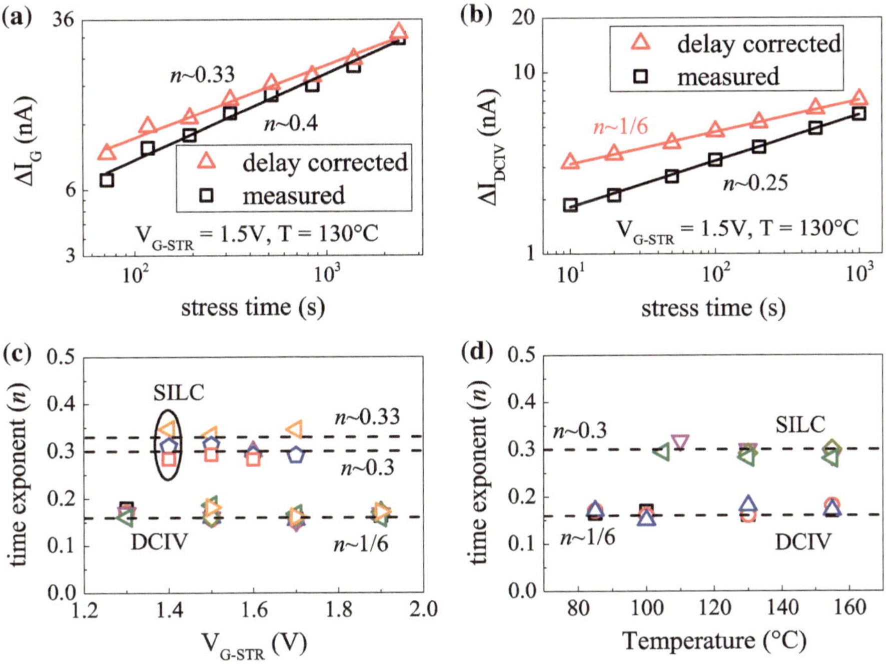

Fig. 2.36 Time evolution of as measured and delay corrected **a** ΔI_G from SILC and **b** ΔI_{DCIV} from DCIV measurements for PBTI stress in HKMG n-MOSFETs. Extracted time exponent *n* from delay corrected SILC and DCIV data as a function of **c** $V_{G\text{-}STR}$ and **d** stress *T*, for PBTI stress in different HKMG devices

HKMG n-MOSFET [38]. Delay correction is done using (2.7), and the parameters of the universal recovery expression are calibrated using SILC and DCIV data shown respectively in Fig. 2.35a, b. Note that power-law time evolution of generated traps is observed for both measurements, although *n* from SILC is much larger than that from DCIV measurements. As expected, magnitude of degradation increases and power-law time exponent *n* reduces after recovery correction. Figure 2.36 also plots delay corrected *n* from SILC and DCIV measurements, as a function of (c) $V_{G\text{-}STR}$ and (d) *T* for PBTI stress in different HKMG n-MOSFETs. The exponent *n* is obtained using linear regression of measured and recovery corrected data in t_{STR} range of 10 s to 1 Ks. Obtained time exponent *n* is independent of $V_{G\text{-}STR}$ and *T* for both SILC and DCIV measurements. SILC measurements show $n \sim 1/3$ after delay correction, which is much larger than $n \sim 1/6$ observed for DCIV measurements. It is important to remark that delay corrected DCIV data show universal $n \sim 1/6$ for both NBTI and PBTI stress, which will be further discussed in Chap. 3.

For a particular bulk trap density N_T (*x*, *E*) at spatial location *x* and energy *E* in the band gap of the gate insulator, increase in inversion gate leakage due to TAT of electrons is given by the following relation [63, 66]:

$$\Delta I_{\mathrm{G}} = \iint q\sigma_{\mathrm{n}} v_{\mathrm{th}} N_{\mathrm{C}} S_{\mathrm{N}} (f_{\mathrm{C}} - f_{\mathrm{A}}) WLN_{\mathrm{T}}(x, E) \frac{T_{\mathrm{C-T}}(E) T_{\mathrm{T-A}}(E - E_{\mathrm{R}})}{T_{\mathrm{C-T}}(E) + T_{\mathrm{T-A}}(E - E_{\mathrm{R}})} \mathrm{d}x\, \mathrm{d}E \tag{2.10}$$

In (2.10), $T_{\text{C-T}}$ and $T_{\text{T-A}}$ are transition probabilities of electron tunneling from cathode, i.e., conduction band of Si substrate to the trap, and from the trap to anode, i.e., gate, respectively, see Fig. 2.33. E_{R} is the energy relaxation of a trap after capture of an electron, q is electronic charge, σ_{n} is electron capture cross section, v_{th} is thermal velocity, N_{C} is density of electrons in the conduction band edge, and S_{N} is energy suppression factor in conduction band, and f_{C} and f_{A} are occupation probabilities of the cathode and anode respectively; W and L respectively are width and length of the MOSFET. The integration is performed over all possible spatial and energy values of traps in the gate insulator. Time evolution of delay corrected ΔI_{G} can be fitted with (2.10) to obtain time evolution of bulk trap density, ΔN_{OT}, and is discussed in detail later in Chap. 4.

2.10 Low Voltage Stress Induced Leakage Current (LV-SILC)

Increase in MOSFET gate leakage current in accumulation or Low Voltage SILC (LV-SILC) has been used as a monitor of interface trap generation for NBTI stress in p-MOSFETs [1, 58, 69] and also TDDB stress in n- and p-MOSFETs [70]. As of now, LV-SILC has been used to characterize trap generation in MOSFETs having SiO_2 and SiON gate insulators, and to the best of our knowledge, this method has not been used in devices having scaled HKMG gate insulators. In this section, LV-SILC method will be briefly reviewed to provide additional evidence of interface trap generation during NBTI stress in SiON p-MOSFETs.

Accumulation I_{G} versus V_{G} sweeps are measured before and during interruptions of NBTI stress, and Fig. 2.37 plots the increase in gate leakage, normalized to pre-stress data, $\Delta I_{\mathrm{G}}/I_{\mathrm{G0}}$, as a function of measurement V_{G} for different t_{STR} [69]. Two different $\Delta I_{\mathrm{G}}/I_{\mathrm{G0}}$ peaks are observed that evolve with increase in t_{STR}, one near $V_{\mathrm{G}} \sim 0$ V and the other near $V_{\mathrm{G}} \sim V_{\mathrm{FB}}$. As also illustrated in Fig. 2.37, LV-SILC is associated with elastic or small energy loss tunneling between traps generated at the Si/SiO_2 and SiO_2/poly-Si interfaces [58, 69, 70]. The peak near $V_{\mathrm{G}} \sim 0$ V is due to electron tunneling from the valence band of poly-Si to traps at Si/SiO_2 interface as well as hole tunneling via traps at Si/SiO_2 interface. On the other hand, the peak near $V_{\mathrm{G}} \sim V_{\mathrm{FB}}$ is due to electron tunneling from Si channel to poly-Si gate via traps at both Si/SiO_2 and SiO_2/poly-Si interfaces. The increase in peak ΔI_{G} magnitude with stress implies generation of new traps at these interfaces during NBTI stress. Although LV-SILC provides crucial evidence of interface trap generation during NBTI stress, this method is not used in this book.

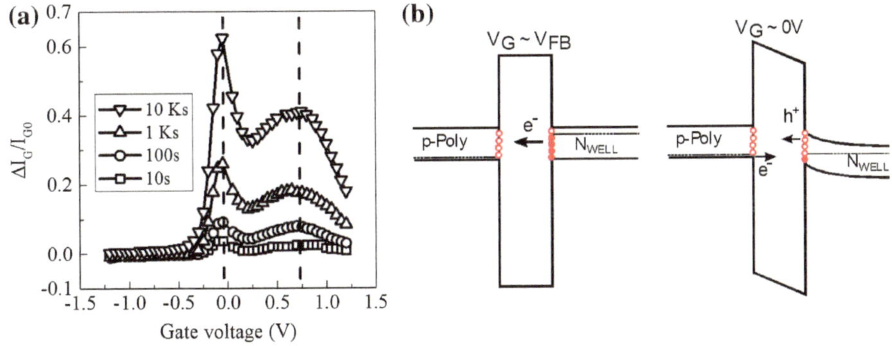

Fig. 2.37 LV-SILC measurements: **a** Increase in I_G after NBTI stress in SiON p-MOSFETs. **b** Energy band diagrams showing trap assisted tunneling corresponding to peaks in *I*–*V* characteristics. Data from [69]

2.11 Experimental Artifacts

During BTI test, MOSFET is stressed at $V_G = V_{G\text{-}STR}$ that is higher than the nominal operating bias, and the resultant degradation is measured either on-the-fly or by using logarithmically spaced interruptions in stress. Time evolution of degradation, measured under accelerated stress condition but for short duration, is then extrapolated to end-of-life under use condition to determine device lifetime. It is important to choose proper stress and measurement conditions so that measured data remain free from different extraneous artifacts discussed in this section.

Measurement speed is a very important parameter and has very different impact on OTF and MSM or OSDD techniques. Since OTF measures I_{DLIN} without reducing stress V_G, refer to Sect. 2.2, it does not suffer from the usual recovery artifacts. However, since the device starts to degrade as soon as $V_{G\text{-}STR}$ is applied, a faster measurement or smaller t_0 delay would capture a less degraded I_{DLIN0}, which is important as I_{DLIN0} is assumed to be undegraded in this method. A larger t_0 delay results in lower than actual degradation magnitude and higher than actual power-law time exponent as discussed earlier in this chapter. For a particular t_0 delay, the accuracy of OTF measurement would depend on how fast the device degrades after the application of $V_{G\text{-}STR}$. Devices with significant fast trapping component are relatively more sensitive to variations in t_0 delay than devices having negligible trapping, as shown using the example of RTNO and PNO SiON p-MOSFETs.

On the other hand, pre-stress measurement is used as a reference for MSM and OSDD methods, while the gate bias is reduced from $V_{G\text{-}STR}$ for measurements during stress, refer to Sects. 2.3 and 2.4. Therefore, these methods suffer from recovery issues, and a lower than actual degradation magnitude and higher than actual power-law time exponent are captured depending on the measurement speed, t_M. For a particular t_M, the accuracy of MSM and OSDD measurements would depend on how fast the device recovers after the removal of stress. Note that devices having significant fast trapping would also recover fast, and would be more sensitive to variations in t_M. Therefore, while OTF is impacted by dynamics of degradation at the initiation of stress, MSM and OSDD methods are impacted by dynamics of recovery after removal of stress. Impact of measurement speed on OTF, MSM, and OSDD measurements has been discussed earlier in this chapter.

Recovery plays an important role when BTI degradation is estimated using different measurement methods and compared against each other. For example, since BTI degradation is due to contribution from trap generation and trapping that are mutually uncorrelated, refer to Chap. 4 for details, trap generation is often estimated directly using CP or DCIV techniques and compared to overall V_T shift to determine the trapping component. Multiple measurements are often performed sequentially after stress is interrupted, and depending on measurement delay and the exact sequence, they would produce different results owing to BTI recovery. As an illustration, Fig. 2.38 plots time evolution of ΔV_T and ΔN_{IT} obtained respectively from I_{DLIN}-V_G and CP measurements during NBTI stress in SiON p-MOSFETs [71]. Two different measurement sequences have been used, i.e., *I–V* first CP last and CP first *I–V* last, and produce different results as shown. Note that CP requires larger measurement time than *I–V* measurements, and since NBTI recovers quickly in time as soon as stress is stopped, ΔV_T obtained from *I–V* first CP last is much larger than that obtained from CP first *I–V* last sequence, as much of the degradation recovers during CP measurements for the latter sequence. Therefore, it is important to avoid sequential measurements and perform different measurements independently on identical but different devices.

Even when independently measured, *I–V* and CP measurements result in different time evolution of degradation under identical measurement time. Time evolution of ΔV_T and ΔN_{IT} due to NBTI stress in SiON p-MOSFETs is measured using OSDD and CP methods, respectively. Experiments were performed under

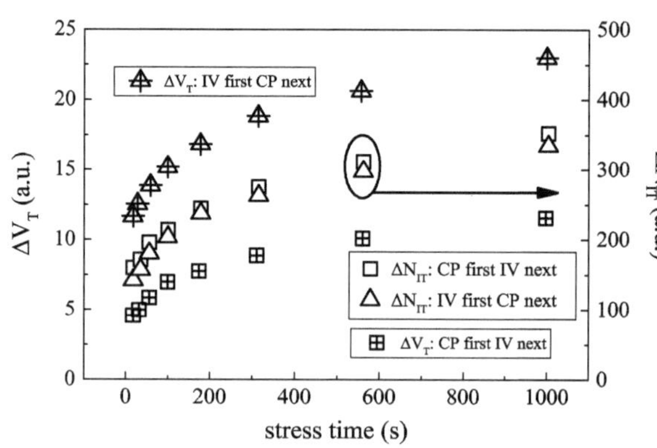

Fig. 2.38 Impact of sequential measurements using multiple methods: Time evolution of ΔV_T and voltage shift corresponding to ΔN_{IT} obtained respectively from *I–V* and CP measurements, for *I–V* first and CP next and CP first and *I–V* next sequence. Data from [71]

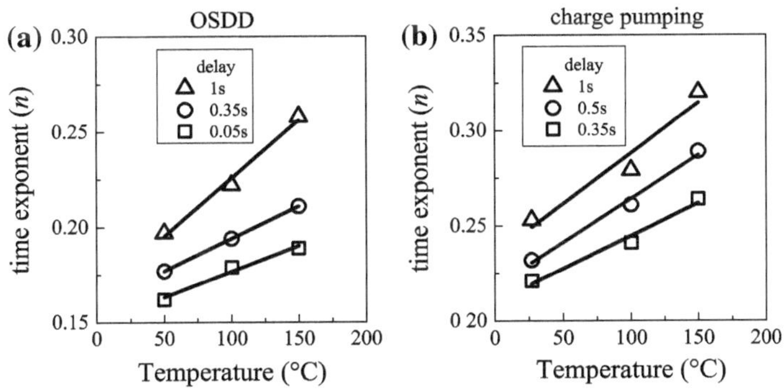

Fig. 2.39 Impact of measurement delay: Measured power-law time exponent n as a function of stress T, extracted from **a** ΔV_T time evolution using OSDD and **b** ΔN_{IT} time evolution using CP methods, for NBTI stress in SiON p-MOSFETs

identical $V_{G\text{-}STR}$ but different stress T, and for different measurement delay t_M. Figure 2.39 plots extracted power-law time exponent n for (a) OSDD and (b) CP measurements as a function of stress T for different t_M, obtained using linear regression of measured data in t_{STR} range of 10 s to 1k s [22]; the CP data are reproduced from Fig. 2.29b. Note that as-measured values of n are different between OSDD and CP measurements at identical stress T and t_M, as these methods suffer from different magnitudes of recovery. For a particular t_M, n increases with increase in stress T, and for both measurement methods, the rate of increase in n is higher for larger t_M due to higher recovery. Therefore, just ensuring identical t_M is not sufficient for error-free comparison of different measurement methods.

As discussed earlier in this chapter, as-measured time evolution of trap generation obtained directly from CP or DCIV must be corrected for measurement delay and band gap differences before comparing with the trap generation component of ΔV_T obtained from ultra-fast OTF, MSM or OSDD I_{DLIN} measurements. It is important to remark that failure to do these corrections would severely underestimate the trap generation and over estimate the trapping component of BTI.

BTI stress is usually performed at different $V_{G\text{-}STR}$, which aids in projecting the accelerated stress data to use condition. BTI and TDDB stress regimes are essentially same, as discussed in Chap. 1, Fig. 1.2, and therefore, it is important to carefully choose $V_{G\text{-}STR}$ so that TDDB effects do not significantly corrupt BTI degradation. As an example, Fig. 2.40a plots the time evolution of ΔI_{DLIN} measured in SiON p-MOSFETs under NBTI stress at different $V_{G\text{-}STR}$ [72]. Note that a power-law time dependence is observed along with the normally observed time exponent of $n \sim 1/6$ for stress at lower $V_{G\text{-}STR}$. However, ΔI_{DLIN} breaks off from simple power-law time dependence and increases at longer t_{STR} for stress at higher $V_{G\text{-}STR}$; the break-off happens earlier in time as $V_{G\text{-}STR}$ is increased. Increased degradation at longer t_{STR} can be modeled using the sum of two independent degradation, having power-law dependence with $n \sim 1/6$ and $n \sim 1/3$ corresponding to NBTI and TDDB respectively, also shown in Fig. 2.40a, refer to

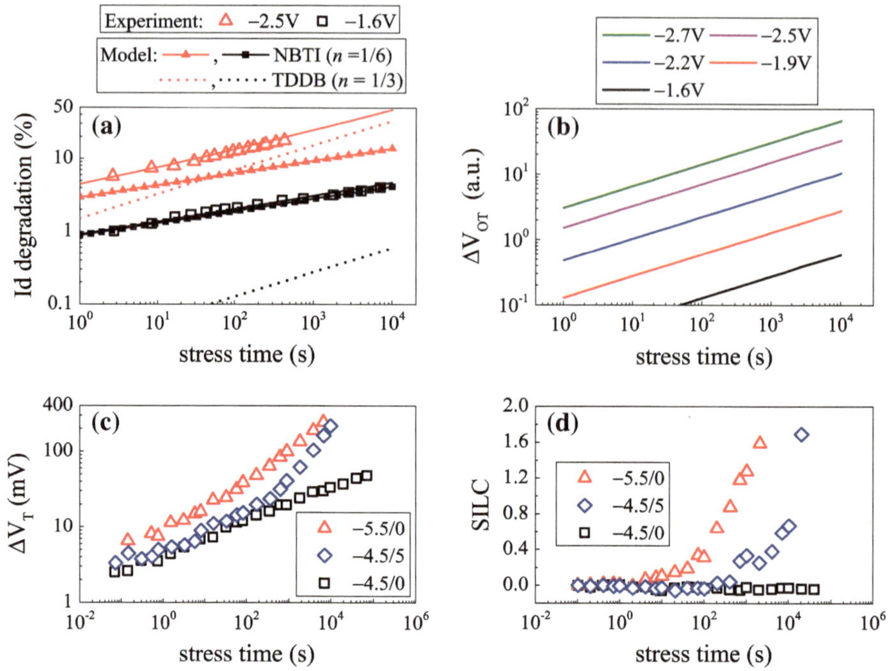

Fig. 2.40 Impact of TDDB defects at high $V_{G\text{-}STR}$: **a** Time evolution of measured ΔI_{DLIN} at different $V_{G\text{-}STR}$, and its decomposition into NBTI (n = 1/6) and TDDB (n = 1/3) components, and **b** time evolution extracted TDDB component for different $V_{G\text{-}STR}$, for NBTI stress in SiON p-MOSFETs. Data obtained from [72]. Time evolution of **c** ΔV_T from I–V and **d** ΔI_G from SILC measurements, for NBTI stress at different $V_{G\text{-}STR}$ and reverse substrate bias V_B in SiO_2 p-MOSFETs

Chap. 4 for further details. Figure 2.40b shows the time evolution of extracted additional degradation for different $V_{G\text{-}STR}$. Identical $n \sim 1/3$ power-law time dependence is observed across all $V_{G\text{-}STR}$, with a V_G acceleration factor of $\Gamma_V \sim 30$ corresponding to time to reach a fixed degradation, which is similar to that observed for TDDB stress [48, 63, 68].

As an additional proof, Fig. 2.40c plots the time evolution of ΔV_T measured in thicker SiO_2 p-MOSFETs for NBTI stress at low and high $V_{G\text{-}STR}$, as well as at low $V_{G\text{-}STR}$ and high reverse substrate bias (V_B) [14]. Note that ΔV_T shows power-law time dependence with a single time exponent of $n \sim 0.25$ for the entire duration of stress, when stress is performed at lower V_G and at $V_B = 0$ V; higher n is observed due to recovery issues as data were measured using slow MSM method. However, ΔV_T breaks off from the $n \sim 0.25$ power-law trend and increases at longer t_{STR} for stress at higher $V_{G\text{-}STR}$ and $V_B = 0$ V or at low $V_{G\text{-}STR}$ but $V_B > 0$ V. To understand the mechanism responsible for enhanced degradation at longer t_{STR}, independent SILC measurements were performed in these devices at identical stress conditions. Figure 2.40d plots time the evolution of $\Delta I_G/I_{G0}$ measured in inversion, showing

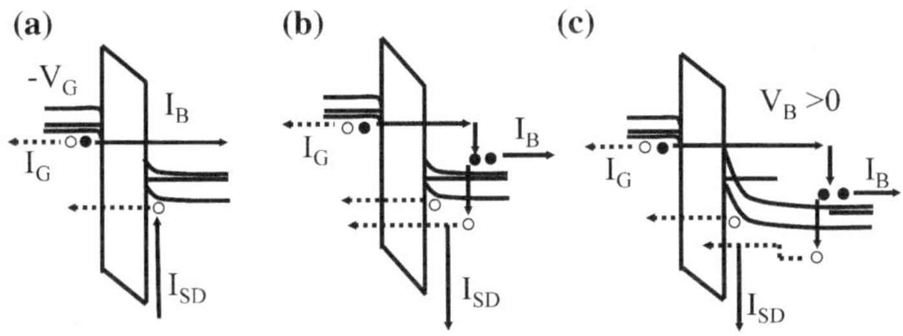

Fig. 2.41 Energy band diagrams for NBTI stress in p-MOSFETs using **a** low V_G and $V_B = 0$, **b** high V_G and $V_B = 0$, and **c** low V_G and high reverse V_B

absence of SILC for low V_G and $V_B = 0$ V stress. However, SILC is observed at longer t_{STR} for stress at higher $V_{G\text{-}STR}$ and $V_B = 0$ V or at lower $V_{G\text{-}STR}$ but $V_B > 0$ V. As SILC is a measure of bulk trap generation, refer to Sect. 2.9, increased ΔV_T at longer t_{STR} can be attributed to the charging of generated TDDB like bulk traps for stress at higher $V_{G\text{-}STR}$ and $V_B = 0$ V or for lower $V_{G\text{-}STR}$ but $V_B > 0$ V.

The energy band diagrams corresponding to NBTI stress in p-MOSFETs, illustrated in Fig. 2.41, can help explain the physical mechanism responsible for TDDB like bulk oxide trap generation under certain stress conditions [14]. For low stress V_G, inversion holes tunnel from Si substrate to gate and electrons tunnel from gate to substrate, refer to Fig. 2.41a, which is normal NBTI condition. At higher V_G, refer to Fig. 2.41b, electrons tunneling from gate to substrate gain enough energy and undergo impact ionization at the substrate and create hot holes, which can get injected back into the oxide by Anode Hole Injection (AHI) mechanism and create oxide defects [63]. Note that high stress V_G results in high E_{OX} in the gate insulator as well as generation of hot holes, and it is difficult to distinguish and identify the main reason behind bulk trap generation. Therefore to further verify the role of hot holes, Fig. 2.41c shows energy bands corresponding to stress at low V_G but high reverse V_B. In spite of low E_{OX}, hot holes can be generated via impacted ionization in the substrate due to high V_B, and can also generate bulk oxide traps as shown. Therefore, TDDB can affect the time evolution of NBTI degradation at long stress time, unless the upper limit of stress gate bias is carefully chosen.

2.12 Summary

To summarize, three different ultra-fast characterization methods are discussed to determine V_T shift of a MOSFET due to BTI stress. The MSM method involves I_D versus V_G sweep measurements before and during periodic interruptions of stress and provides direct estimation of ΔV_T; pre- and post-stress V_T being calculated using peak g_m method. The MSM method is implemented for DC and AC stress and the impact of pulse hold and sweep delay is discussed. The OTF method senses I_D on-the-fly without reducing stress V_G, while the OSDD method measures I_D by reducing V_G from stress to a fixed sense bias. The impact of mobility degradation at different sense bias is discussed, and a post-processing mobility correction procedure is discussed for these one spot methods to properly estimate ΔV_T from measured I_D degradation. The impact of time-zero delay for OTF and drop down delay for OSDD is discussed. It is shown that OTF measurement accuracy is dependent on how fast a device degrades after stress, while that for MSM or OSDD depends on how fast a device recovers after the removal of stress.

Several characterization techniques are discussed that directly estimate the pre-existing and newly generated defects in MOSFET gate insulator and are responsible for BTI degradation. Pre-existing process related traps are estimated in different SiON and HKMG devices using flicker noise method; a correlated number and surface mobility fluctuation model is used to determine trap density from measured drain current noise. The implementation details of CP and gated diode (or DCIV) techniques are discussed to determine density of generated traps during BTI stress. CP and DCIV are relatively slow methods and scan generated traps only in a limited energy range corresponding to the center of the Si band gap. Delay and band gap correction methods are discussed, which can be used to correct as-measured data for proper comparison with the trap generation component obtained from ultra-fast *I–V* methods. Increase in gate leakage in inversion or SILC is used to characterize generation of bulk traps in the gate insulator. The importance of proper choice of stress V_G is discussed, to minimize corruption of BTI degradation by TDDB like bulk trap generation. Finally, the LV-SILC method that also probes interface trap generation during BTI is briefly discussed.

In later chapters, NBTI in SiON p-MOSFETs is characterized using the mobility corrected OTF method, while flicker noise and CP methods are used respectively to characterize the pre-existing and generated traps. NBTI and PBTI in HKMG MOSFETs are characterized using the MSM method; flicker noise and DCIV are used to characterize the pre-existing and generated traps, respectively. When applicable, SILC is used to characterize generated bulk insulator traps.

Acknowledgments The authors would like to acknowledge E. Naresh Kumar and Vrajesh Maheta for OTF measurements, Gautam Kapila and Bijesh Rajamohanan for flicker noise measurements, Dhanoop Varghese for CP measurements and Applied Materials for providing devices used in this work.

References

1. N. Kimizuka, T. Yamamoto, T. Mogami, K. Yamaguchi, K. Imai, T. Horiuchi, The impact of bias temperature instability for direct-tunneling ultra-thin gate oxide on MOSFET scaling, in *Symposium on VLSI Technology: Digest of Technical Papers* (1999), p. 73
2. M. Ershov, S. Saxena, H. Karbasi, S. Winters, S. Minehane, J. Babcock, R. Lindley, P. Clifton, M. Redford, A. Shibkov, Dynamic recovery of negative bias temperature instability in p-type metal–oxide–semiconductor field-effect transistors. Appl. Phys. Lett. **83**, 1647 (2003)

3. S. Rangan, N. Mielke, E.C.C. Yeh, Universal recovery behavior of negative bias temperature instability [PMOSFETs], in *IEEE International Electron Devices Meeting Technical Digest* (2003), p. 14.3.1
4. H. Reisinger, O. Blank, W. Heinrigs, A. Muhlhoff, W. Gustin, C. Schlunder, Analysis of NBTI degradation- and recovery-behavior based on ultra fast VT-measurements, in *IEEE International Reliability Physics Symposium Proceedings* (2006), p. 448
5. E.N. Kumar, V.D. Maheta, S. Purawat, A.E. Islam, C. Olsen, K. Ahmed, M.A. Alam, S. Mahapatra, Material dependence of NBTI physical mechanism in silicon oxynitride (SiON) p-MOSFETs: a comprehensive study by ultra-fast on-the-fly (UF-OTF) IDLIN technique, in *IEEE International Electron Devices Meeting Technical Digest* (2007), p. 809
6. C. Shen, M.-F. Li, C.E. Foo, T. Yang, D.M. Huang, A. Yap, G.S. Samudra, Y.-C. Yeo, Characterization and physical origin of fast Vth transient in NBTI of pMOSFETs with SiON dielectric, in *IEEE International Electron Devices Meeting Technical Digest* (2006). doi:10.1109/IEDM.2006.346776
7. A. Kerber, M. Kerber, Fast wafer level data acquisition for reliability characterization of sub-100 nm CMOS technologies. IEEE International Integrated Reliability Workshop Final Report (2004), p. 41
8. V.D. Maheta, E.N. Kumar, S. Purawat, C. Olsen, K. Ahmed, S. Mahapatra, Development of an ultrafast on-the-fly I_{DLIN} technique to study NBTI in plasma and thermal oxynitride p-MOSFETs. IEEE Trans. Electron Devices **55**, 2614 (2008)
9. S. Mahapatra, K. Ahmed, D. Varghese, A.E. Islam, G. Gupta, L. Madhav, D. Saha, M.A. Alam, On the physical mechanism of NBTI in silicon oxynitride p-MOSFETs: can differences in insulator processing conditions resolve the interface trap generation versus hole trapping controversy?, in *IEEE International Reliability Physics Symposium Proceedings* (2007), p. 1
10. A. Chaudhary, S. Mahapatra, A physical and SPICE mobility degradation analysis for NBTI. IEEE Trans. Electron Devices **60**, 2096 (2013)
11. M. Denais, A. Bravaix, V. Huard, C. Parthasarathy, G. Ribes, F. Perrier, Y. Rey-Tauriac, N. Revil, On-the-fly characterization of NBTI in ultra-thin gate oxide PMOSFET's, in *IEEE International Electron Devices Meeting Technical Digest* (2004), p. 109
12. A.E. Islam, V.D. Maheta, H. Das, S. Mahapatra, M. A. Alam, Mobility degradation due to interface traps in plasma oxynitride PMOS devices, in *IEEE International Reliability Physics Symposium Proceedings* (2008), p. 87
13. V.D. Maheta, C. Olsen, K. Ahmed, S. Mahapatra, The impact of nitrogen engineering in silicon oxynitride gate dielectric on negative-bias temperature instability of p-MOSFETs: a study by ultrafast on-the-fly I_{DLIN} technique. IEEE Trans. Electron Devices **55**, 1630 (2008)
14. S. Mahapatra, P. Bharath Kumar, M.A. Alam, Investigation and modeling of interface and bulk trap generation during negative bias temperature instability of p-MOSFETs". IEEE Trans. Electron Devices **51**, 1371 (2004)
15. Y. Taur, T.H. Ning, *Fundamentals of Modern VLSI Devices* (Cambridge University Press, Cambridge)
16. N. Goel, N. Nanaware, S. Mahapatra, Ultrafast AC–DC NBTI characterization of deep IL scaled HKMG p-MOSFETs. IEEE Electron Device Lett. **34**, 1476 (2013)
17. N. Goel, K. Joshi, S. Mukhopadhyay, N. Nanaware, S. Mahapatra, A comprehensive modeling framework for gate stack process dependence of DC and AC NBTI in SiON and HKMG p-MOSFETs. Microelectron. Reliab. **54**, 491 (2014)
18. N. Goel, S. Mukhopadhyay, N. Nanaware, S. De, R. K. Pandey, K.V.R.M. Murali, S. Mahapatra, A comprehensive DC/AC model for ultra-fast NBTI in deep EOT scaled HKMG p-MOSFETs, in *IEEE International Reliability Physics Symposium Proceedings* (2014), p. 6A.4.1
19. S. Mukhopadhyay, Private communication
20. B. Kaczer, V. Arkbipov, R. Degraeve, N. Collaert, G. Groeseneken, M. Goodwin, Disorder-controlled-kinetics model for negative bias temperature instability and its experimental verification, in *IEEE International Reliability Physics Symposium Proceedings* (2005), p. 381

21. B. Kaczer, T. Grasser, J. Roussel, J. Martin-Martinez, R. O'Connor, B.J. O'Sullivan, G. Groeseneken, Ubiquitous relaxation in BTI stressing—new evaluation and insights, in *IEEE International Reliability Physics Symposium Proceedings* (2008), p. 20
22. D. Varghese, D. Saha, S. Mahapatra, K. Ahmed, F. Nouri, M. Alam, On the dispersive versus Arrhenius temperature activation of NBTI time evolution in plasma nitrided gate oxides: measurements, theory, and implications, in *IEEE International Electron Devices Meeting Technical Digest* (2005), p. 684
23. S. Deora, P. Narayanasetti, M. Thakkar, S. Mahapatra, Development of a novel ultrafast direct threshold voltage (UF-DVT) technique to study NBTI stress and recovery. IEEE Trans. Electron Devices **58**, 3506 (2011)
24. K. Joshi, S. Mukhopadhyay, N. Goel, S. Mahapatra, A consistent physical framework for N and P BTI in HKMG MOSFETs, in *IEEE International Reliability Physics Symposium Proceedings* (2012), p. 5A.3.1
25. G. Kapila, N. Goyal, V.D. Maheta, C. Olsen, K. Ahmed, S. Mahapatra, A comprehensive study of flicker noise in plasma nitrided SiON p-MOSFETs: process dependence of pre-existing and NBTI stress generated trap distribution profiles. , in *IEEE International Electron Devices Meeting Technical Digest* (2008). doi:10.1109/IEDM.2008.4796625
26. S. Christensson, I. Lundström, C. Svensson, Low frequency noise in MOS transistors—I theory. Solid State Electron. **11**, 797 (1968)
27. F. Horng-Sen, C.-T. Sah, Theory and experiments on surface 1/f noise. IEEE Trans. Electron Devices **19**, 273 (1972)
28. Z. Celik, T.Y. Hsiang, Study of 1/f noise in N-MOSFET's: Linear region. IEEE Trans. Electron Devices **32**, 2797 (1985)
29. L.K.J. Vandamme, Model for 1/f; noise in MOS transistors biased in the linear region. Solid State Electron. **23**, 317 (1980)
30. L.K.J. Vandamme, H.M.M. de Werd, 1/f; noise model for MOSTs biased in nonohmic region. Solid State Electron. **23**, 325 (1980)
31. A.L. McWhorter, 1/f noise and germanium surface properties, in *Semiconductor Surface Physics* (University of Pennsylvania Press, Philadelphia, 1957), p. 207
32. G. Abowitz, E. Arnold, E.A. Leventhal, Surface states and 1/f noise in MOS transistors. IEEE Trans. Electron Devices **14**, 775 (1967)
33. H.E. Maes, S.H. Usmani, G. Groeseneken, Correlation between 1/f noise and interface state density at the Fermi level in field-effect transistors. J. Appl. Phys. **57**, 4811 (1985)
34. F.N. Hooge, 1/f noise. Phys. B + C **83**, 14 (1976)
35. R.P. Jindal, Phonon fluctuation model for flicker noise in elemental semiconductors. J. Appl. Phys. **52**, 2884 (1981)
36. V. Huard, Two independent components modeling for negative bias temperature instability, in *IEEE International Reliability Physics Symposium Proceedings* (2010), p. 33
37. K. Joshi, S. Hung, S. Mukhopadhyay, V. Chaudhary, N. Nanaware, B. Rajamohnan, T. Sato, M. Bevan, A. Wei, A. Noori, B. McDougal, C. Ni, G. Saheli, C. Lazik, P. Liu, D. Chu, L. Date, S. Datta, A. Brand, J. Swenberg, S. Mahapatra, HKMG process impact on N, P BTI: Role of thermal IL scaling, IL/HK integration and post HK nitridation, in *IEEE International Reliability Physics Symposium Proceedings* (2013), p. 4C.2.1
38. S. Mukhopadhyay, K. Joshi, V. Chaudhary, N. Goel, S. De, R.K. Pandey, K.V.R.M. Murali, S. Mahapatra, Trap generation in IL and HK layers during BTI / TDDB stress in scaled HKMG N and P MOSFETs, in *IEEE International Reliability Physics Symposium Proceedings* (2014), p. GD.3.1
39. S. Machlup, Noise in semiconductors: spectrum of a two-parameter random signal. J. Appl. Phys. **25**, 341 (1954)
40. K.K. Hung, P.K. Ko, C. Hu, Y.C. Cheng, A unified model for the flicker noise in metal-oxide-semiconductor field-effect transistors. IEEE Trans. Electron Devices **37**, 654 (1990)

41. T. Grasser, B. Kaczer, W. Goes, T. Aichinger, P. Hehenberger, M. Nelhiebel, Understanding negative bias temperature instability in the context of hole trapping (Invited Paper). Microelectron. Eng. **86**, 1876 (2009)
42. H. Reisinger, T. Grasser, W. Gustin, C. Schlunder, The statistical analysis of individual defects constituting NBTI and its implications for modeling DC- and AC-stress, in *IEEE International Reliability Physics Symposium Proceedings* (2010), p. 7
43. S. Zafar, A. Kerber, R. Muralidhar, Physics based PBTI model for accelerated estimation of 10 year lifetime, in *Symposium on VLSI Technology: Digest of Technical Papers* (2014). doi:10.1109/VLSIT.2014.6894388
44. B. Kaczer, T. Grasser, J. Martin-Martinez, E. Simoen, M. Aoulaiche, P.J. Roussel, G. Groeseneken, NBTI from the perspective of defect states with widely distributed time scales, in *IEEE International Reliability Physics Symposium Proceedings* (2009), p. 55
45. G. Groeseneken, H.E. Maes, N. Beltran, R.F. De Keersmaecker, A reliable approach to charge-pumping measurements in MOS transistors. IEEE Trans. Electron Devices **31**, 42 (1984)
46. Y. Mitani, "Influence of nitrogen in ultra-thin SiON on negative bias temperature instability under AC stress, in *IEEE International Electron Devices Meeting Technical Digest* (2004), p. 117
47. S. Mahapatra, V.D. Maheta, A.E. Islam, M.A. Alam, Isolation of NBTI stress generated interface trap and hole-trapping components in PNO p-MOSFETs. IEEE Trans. Electron Devices **56**, 236 (2009)
48. S. Mahapatra, A.E. Islam, S. Deora, V.D. Maheta, K. Joshi, A. Jain, M.A. Alam, A critical re-evaluation of the usefulness of R-D framework in predicting NBTI stress and recovery, in *IEEE International Reliability Physics Symposium Proceedings* (2011), p. 6A.3.1
49. M. Zahid, R. Degraeve, M. Cho, L. Pantisano, D.R. Aguado, J. Van. Houdt, G. Groeseneken, M. Jurczak, Deffect profiling in the SiO_2/Al_2O_3 interface using Variable T charge-T discharge amplitude charge pumping (VT2ACP), in *IEEE International Reliability Physics Symposium Proceedings* (2009), p. 21
50. M. Masuduzzaman, A.E. Islam, M.A. Alam, Exploring the capability of multi frequency charge pumping in resolving location and energy levels of traps within dielectric. IEEE Trans. Electron Devices **55**, 3421 (2008)
51. W.J. Liu, Z.Y. Liu, D. Huang, C.C. Liao, L.F. Zhang, Z.H. Gan, W. Wong, C. Shen, M.-F. Li, On-the-fly interface trap measurement and its impact on the understanding of NBTI mechanism for p-MOSFETs with SiON gate dielectric, in *IEEE International Electron Devices Meeting Technical Digest* (2007), p. 813
52. D.S. Ang, Z.Q. Teo, C.M. Ng, Reassessing NBTI mechanisms by ultrafast charge pumping measurement. ieee international integrated reliability workshop final report (2009), p. 25
53. S.S. Chung, S.-J. Chen, C.-K. Yang, S.-M. Cheng, S.-H. Lin, Y.-C. Sheng, H.-S. Lin, K.-T. Hung, D.-Y. Wu, T.-R. Yew, S.-C. Chien, F.-T. Liou, F. Wen, A novel and direct determination of the interface traps in sub-100 nm CMOS devices with direct tunneling regime (12 $\sim$ 16 A) gate oxide, in *Symposium on VLSI Technology. Digest of Technical Papers* (2002), p. 74
54. T. Grasser, W. Gos, V. Sverdlov, B. Kaczer, The universality of NBTI relaxation and its implications for modeling and characterization, in *IEEE International Reliability Physics Symposium Proceedings* (2007), p. 268
55. J. Franco, B. Kaczer, J. Mitard, M. Toledano-Luque, P.J. Roussel, L. Witters, T. Grasser, G. Groeseneken, NBTI Reliability of SiGe and Ge channel pMOSFETs With SiO_2/HfO_2 dielectric stack. IEEE Trans. Device Mater. Reliab. **13**, 497 (2013)
56. J. Cai, R.-Y. Sah, Monitoring interface traps by DCIV method. IEEE Electron Device Lett. **20**, 60 (1999)
57. A. Neugroschel, G. Bersuker, R. Choi, Applications of DCIV method to NBTI characterization. Microelectron. Reliab. **47**, 1366 (2007)

58. J.H. Stathis, G. LaRosa, A. Chou, Broad energy distribution of NBTI-induced interface states in p-MOSFETs with ultra-thin nitrided oxide, in *IEEE International Reliability Physics Symposium Proceedings* (2004). doi:10.1109/RELPHY.2004.1315292
59. J.P. Campbell, P.M. Lenahan, A.T. Krishnan, S. Krishnan, NBTI: an atomic-scale defect perspective, in *IEEE International Reliability Physics Symposium Proceedings* (2006), p. 447
60. S. Mahapatra, N. Goel, S. Desai, S. Gupta, B. Jose, S. Mukhopadhyay, K. Joshi, A. Jain, A.E. Islam, M.A. Alam, A comparative study of different physics-based NBTI models. IEEE Trans. Electron Devices **60**, 901 (2013)
61. W. Shockley, W. Read, Statistics of the recombinations of holes and electrons. Phys. Rev. **87**, 835 (1952)
62. D.J. Fitzgerald, A.S. Grove, Surface recombination in semiconductors. Surf. Sci. **9**, 347 (1968)
63. M.A. Alam, SILC as a measure of trap generation and predictor of TBD in ultrathin oxides. IEEE Trans. Electron Devices **49**, 226 (2002)
64. S. Pae, T. Ghani, M. Hattendorf, J. Hicks, J. Jopling, J. Maiz, K. Mistry, J. O'Donnell, C. Prasad, J. Wiedemer, J. Xu, Characterization of SILC and its end-of-life reliability assessment on 45nm high-K and metal-gate technology, in *IEEE International Reliability Physics Symposium Proceedings* (2009), p. 499
65. S. Ramey, A. Ashutosh, C. Auth, J. Clifford, M. Hattendorf, J. Hicks, R. James, A. Rahman, V. Sharma, A. St. Amour, C. Wiegand, Intrinsic transistor reliability improvements from 22 nm tri-gate technology, in *IEEE International Reliability Physics Symposium Proceedings* (2013), p. 4C.5.1
66. S. Takagi, M. Takayanagi, A. Toriumi, Experimental examination of physical model for direct tunneling current in unstressed/stressed ultrathin gate oxides, in *IEEE International Electron Devices Meeting Technical Digest* (1999), p. 461
67. E. Cartier, A. Kerber, Stress-induced leakage current and defect generation in nFETs with HfO2/TiN gate stacks during positive-bias temperature stress, in *IEEE International Reliability Physics Symposium Proceedings* (2009), p. 486
68. J. Yang, M. Masuduzzaman, K. Joshi, S. Mukhopadhyay, J. Kang, S. Mahapatra, M.A. Alam, Intrinsic correlation between PBTI and TDDB degradations in nMOS HK/MG dielectrics, in *IEEE International Reliability Physics Symposium Proceedings* (2012), p. 5D.4.1
69. A.T. Krishnan, C. Chancellor, S. Chakravarthi, P.E. Nicollian, V. Reddy, A. Varghese, R.B. Khamankar, S. Krishnan, Material dependence of hydrogen diffusion: implications for NBTI degradation, in *IEEE International Electron Devices Meeting Technical Digest* (2005), p. 688
70. P.E. Nicollian, Insights on trap generation and breakdown in ultra thin SiO_2 and SiON dielectrics from low voltage stress-induced leakage current measurements. Microelectron. Reliab. **48**, 1171 (2008)
71. T. Yang, M.F. Li, C. Shen, C.H. Ang, C. Zhu, Y.C. Yeo, G. Samudra, S.C. Rustagi, M.B. Yu, D.L. Kwong, Fast and slow dynamic NBTI components in p-MOSFET with sion dielectric and their impact on device life-time and circuit application, in *Symposium on VLSI Technology: Digest of Technical Papers* (2005), p. 92
72. Y.M. Lin, C.J. Wang, K. Wu, A new finding on NBTI lifetime model and an investigation on NBTI degradation characteristic for 1.2 nm ultra thin oxide, in *IEEE International Reliability Physics Symposium Proceedings* (2005), p. 704

Chapter 3
Physical Mechanism of BTI Degradation—Direct Estimation of Trap Generation and Trapping

Subhadeep Mukhopadhyay and Souvik Mahapatra

Abstract In this chapter, direct characterization techniques have been used to access the trap generation and trapping subcomponents of BTI degradation in HKMG MOSFETs having different gate stack processes. Generation of new traps is estimated using DCIV for NBTI stress and both DCIV and SILC for PBTI stress respectively in p- and n-channel MOSFETs. Flicker noise is used to estimate the density of process related pre-existing gate insulator traps responsible for hole and electron trapping respectively during NBTI and PBTI stress. The spatial and energetic locations of generated traps for NBTI and PBTI stress are identified. The time, bias, and temperature dependencies of trap generation obtained using the DCIV technique are compared between NBTI and PBTI stress, while these parameters obtained using DCIV and SILC techniques are compared for PBTI stress. The relative dominance of trap generation and trapping on NBTI and PBTI threshold voltage degradation is estimated for different gate insulator processes.

3.1 Introduction

From a practical point of view of technology qualification, it is important to know the magnitude of Bias Temperature Instability (BTI) at end-of-life of devices and hence of circuits and products under normal use condition. Estimation of Negative BTI (NBTI) and Positive BTI (PBTI) respectively in p- and n-channel Metal Oxide Semiconductor Field Effect Transistors (MOSFETs) is usually done by stressing the devices at higher than normal gate bias ($V_G = V_{G\text{-}STR}$) and measuring the resulting device parametric degradation with minimal impact of recovery artifacts. Different "recovery-free" measurement techniques have been discussed in Chap. 2. As mentioned before, stress tests are usually performed for few hours or

S. Mukhopadhyay · S. Mahapatra (✉)
Department of Electrical Engineering, Indian Institute of Technology Bombay, Mumbai 400076, India
e-mail: souvik@ee.iitb.ac.in; mahapatra.souvik@gmail.com

S. Mahapatra (ed.), *Fundamentals of Bias Temperature Instability in MOS Transistors*, Springer Series in Advanced Microelectronics 139,
DOI 10.1007/978-81-322-2508-9_3

days in wafer level setup, although sometimes the test can go on for few months in package level setup. Measured data at short time and at accelerated aging condition are then extrapolated to end-of-life and normal operating condition by using suitable models. Proper understanding of BTI degradation mechanism is necessary to develop reliable extrapolation models. It also helps in understanding the impact of gate insulator processes on BTI, which in turn helps in process optimization for keeping BTI under acceptable limits for technology qualification.

The physical mechanism of BTI has remained as a topic of great debate. It was discussed in Chap. 1 that NBTI has emerged as a crucial p-MOSFET reliability issue since 90 nm technology node, when Silicon Oxynitride (SiON) replaced Silicon Dioxide (SiO_2) as the gate insulator [1, 2]. NBTI remains as an important issue even today for planar MOSFETs and FinFETs having state-of-the-art High-K Metal Gate (HKMG) gate stacks [3, 4]. Over the years, different physical mechanisms have been proposed to explain buildup of positive charges in the gate insulator during NBTI stress and were recently reviewed [5]. Figure 3.1 shows schematic of p- and n-channel HKMG MOSFETs having bi-layer gate insulator stack with SiO_2 or SiON Interlayer (IL) and Hafnium Oxide (HfO_2) High-K dielectrics. Positive charge buildup during NBTI stress can be due to either one or both of the two processes, i.e., generation of new traps at or near the interface between Silicon (Si) channel and SiO_2 (or SiON) IL and/or charging of pre-existing, as-processed traps in IL bulk. It is believed that HfO_2 High-K layer presumably acts primarily as a voltage divider.

Different characterization techniques have been discussed in Chap. 2 for accessing the density of process related and generated traps respectively before and after BTI stress. Note that direct characterization techniques such as Charge Pumping (CP) [6] and Gated Diode (or DCIV) [7] have been used in several reports to provide irrefutable proof of interface trap generation (ΔN_{IT}) for NBTI stress in SiON [5, 7–15] and HKMG [5, 16–18] p-MOSFETs. In spite of these experimental evidences, some reports have suggested hole trapping in pre-existing traps (ΔN_{HT}) as the only NBTI mechanism [19–23], which is definitely not correct. Similarly, reports suggesting ΔN_{IT} as the exclusive NBTI mechanism [10, 11] are also not correct, as they cannot explain ultra-fast threshold voltage shift (ΔV_T) measurements [20, 24, 25] and gate insulator process dependence of NBTI [13–15]. As of

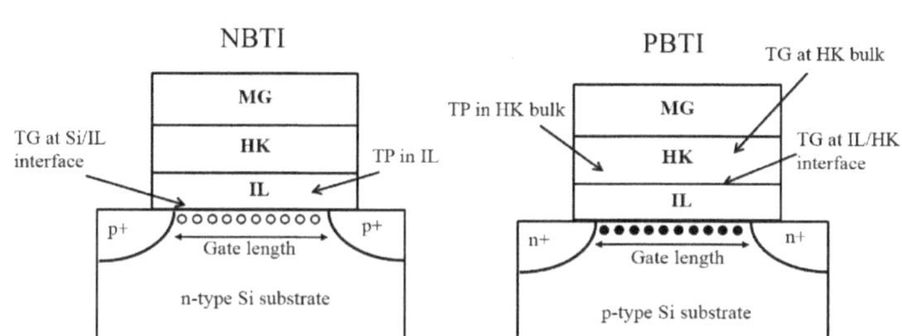

Fig. 3.1 Schematic of p- and n-channel HKMG MOSFETs showing different trap generation (TG) and trapping (TP) processes under NBTI and PBTI stress

today, the prevailing notion of NBTI mechanism involves contribution from both ΔN_{IT} and ΔN_{HT}; although some have suggested strong coupling or correlation between the two processes [26], most reports suggest that they are independent and mutually uncorrelated [5, 9, 13–18, 27–35]. Furthermore, trap generation (ΔN_{OT}) in bulk IL also contributes for situations involving high stress gate bias ($V_{G\text{-}STR}$). Mutually uncoupled ΔN_{IT} and ΔN_{HT} (and also ΔN_{OT} for certain situations) mechanisms can explain different gate insulator process dependent NBTI data [5, 33–35], and is discussed later in Chap. 4.

As discussed in Chap. 1, PBTI remained negligible for SiON n-MOSFETs and became important with the introduction of HKMG technology [36]. As mentioned before, NBTI in HKMG MOSFETs results in positive charge buildup in the SiO_2 or SiON IL due to trap generation at Si/IL interface and hole trapping in IL bulk [16, 17, 34]. In contrast, PBTI results in negative charge buildup in HfO_2 High-K layer as shown in Fig. 3.1 [36]. Note that initial HKMG MOSFETs had thick and not fully optimized HfO_2 High-K layer and showed very large PBTI degradation primarily due to significant electron trapping in pre-existing traps (ΔN_{ET}) [37, 38]. However, PBTI magnitude reduces with reduction in High-K layer thickness [36] and with optimization of HKMG process [39] as mentioned in Chap. 1. For well-optimized gate insulator stacks, various reports have suggested trap generation in the High-K layer measured directly using DCIV [16, 17, 40] and Stress Induced Leakage Current (SILC) [40, 41] methods. Moreover, a recent report has suggested two different and mutually uncoupled PBTI trap generation processes, presumably at the IL/High-K interface ($\Delta N_{IT\text{-}HK}$) and High-K bulk ($\Delta N_{OT\text{-}HK}$), respectively, probed by DCIV and SILC techniques [17]. In spite of such direct experimental evidences of trap generation, some report still suggests PBTI to be solely due to electron trapping in pre-existing process related traps in the High-K layer even for state-of-the-art technology nodes [42], which is obviously not correct. However, many reports have suggested mutually uncorrelated trap generation and trapping in High-K as the physical mechanism of PBTI [16, 17, 32, 40, 43]. PBTI model is discussed later in Chap. 4.

Furthermore, transconductance degradation (Δg_m) has been reported for NBTI stress in HKMG p-MOSFETs since generated IL charges are closer to the channel and results in mobility degradation due to Coulomb scattering. However, Δg_m is negligible for PBTI stress in well-optimized HKMG n-MOSFETs due to negligible IL degradation [39], refer to Chap. 1 for details. While the magnitude of PBTI degradation reduces with HfO_2 thickness scaling, it increases with reduction in IL thickness since generated High-K layer charges come closer to the channel, and also due to possible modification of the High-K layer quality introduced by the IL thickness scaling process [17, 32, 44, 45]. Therefore, NBTI and PBTI charges are shown to have very different physical location in the HKMG gate stack.

As mentioned in Chap. 1, the relative magnitude of NBTI and PBTI degradation respectively in p- and n-channel HKMG MOSFETs stressed under identical oxide field (E_{OX}) and temperature (T) depends on the gate insulator process and is industry specific [3, 4]. Although NBTI results in trap generation and trapping in the IL layer while PBTI degrades the High-K layer, they demonstrate very similar behavior listed as follows, when measured by using Ultra-Fast Measure-Stress-Measure (UF-MSM) technique, refer to Chap. 1 for details:

(a) Time evolution of ΔV_T shows rapid increase at the beginning of stress and power-law time dependence with time exponent n for longer stress time. The longer time power-law dependence is observed for both DC and AC stress, refer to Figs. 1.25 and 1.30.
(b) The longer time power-law time exponent n is independent of stress E_{OX} and T when stress and measurements are performed without any extraneous artifacts, such as influence of recovery or stress saturation, and shows similar values for NBTI and PBTI stress. The exponent is also independent of AC pulse duty cycle (PDC) and frequency (f), although n for AC stress is higher compared to DC stress. Refer to Figs. 1.26 and 1.31.
(c) Measured ΔV_T increases with stress E_{OX} and T. The power-law E_{OX} acceleration factor (Γ_E) and Arrhenius T activation energy (E_A) extracted using ΔV_T measured at fixed stress time (t_{STR}) are found to be independent of stress T and E_{OX} respectively, when stress and measurements remain free from certain extraneous artifacts mentioned above, refer to Fig. 1.27.
(d) BTI recovery results in lower ΔV_T for AC when compared to DC stress. AC ΔV_T is independent of f, and shows a typical "S" shaped characteristic with variation in PDC, with a large "kink" or "jump" in ΔV_T observed between high PDC AC and DC stress, refer to Fig. 1.32.
(e) Although the magnitude of ΔV_T increases for NBTI but reduces for PBTI when Nitrogen (N) is incorporated in the gate insulator stack, the parameters n, E_A and Γ_E show a reduction for both NBTI and PBTI stress, refer to Fig. 1.34.
(f) Equivalent Oxide Thickness (EOT) scaling achieved either by reduction in IL thickness or post High-K nitridation results in higher ΔV_T but reduction in parameters n, E_A and Γ_E, refer to Fig. 1.35.

In this chapter, the underlying trap generation and trapping subcomponents of NBTI and PBTI degradation are independently assessed respectively in p- and n-channel HKMG MOSFETs. Devices having different HKMG gate insulator processes are used. The contribution of trap generation and trapping on ΔV_T measured using ultra-fast MSM method is also assessed to determine physical mechanism of BTI degradation. The concept developed in this chapter will be used in Chap. 4 to develop quantitative NBTI and PBTI models.

3.2 Description of HKMG Devices

Figure 3.2 shows the gate insulator process flow of different HKMG MOSFETs studied in this chapter. A Gate First integration scheme has been used.

The devices have different Rapid Thermal Process (RTP) based thermal IL layers [46] but identical HfO_2 High-K layer obtained using the Atomic Layer Deposition (ALD) method [47]. Different pre-clean surface treatments have been

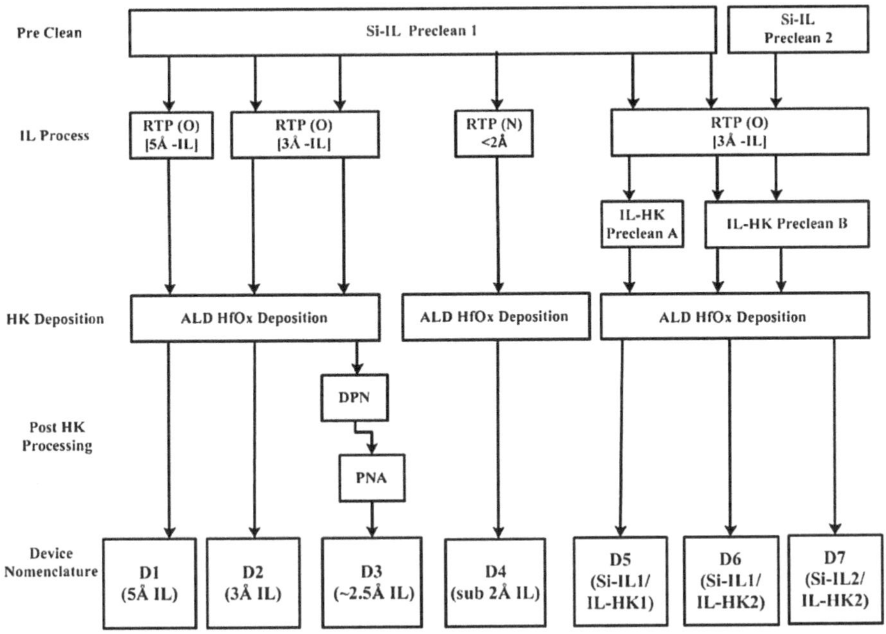

Fig. 3.2 Schematic process flow of different HKMG gate insulators studied in this chapter

used before thermal IL growth and before ALD High-K deposition. The Si/IL pre-clean is done before the IL growth and therefore affect the IL quality, and for thinner IL, it can also affect the High-K quality. The IL/HK pre-clean is done after IL growth but before High-K deposition, and impacts the High-K layer quality. EOT scaling has been achieved (a) by using RTP based thermal IL having thickness of 5 Å (D1) and 3 Å (D2), (b) by introducing Nitrogen (N) in the gate stack after High-K deposition, using Decoupled Plasma Nitridation (DPN) [48] with proper Post Nitridation Anneal (PNA) [49] (D3) and (c) by using N based surface passivation before RTP IL growth (D4). Devices D1 through D4 have identical Si/IL and IL/HK pre-clean processes, different IL but identical High-K thickness. Devices D5 through D7 have identical IL and High-K layer thickness, however, D5 and D6 have similar Si/IL but different IL/HK pre-clean processes, while D6 and D7 have different Si/IL but similar IL/HK pre-clean processes. All thicknesses are measured using X-ray Photoelectron Spectroscopy; refer to [32] for further details.

3.3 Trap Generation During NBTI

In this section, trap generation during NBTI stress in HKMG p-MOSFETs is studied using the Gated Diode or DCIV technique [7], which has been discussed in detail in Chap. 2. In this technique, the source and drain terminals of a MOSFET are shorted together and forward biased below the junction cut-in voltage, the gate is swept from accumulation to inversion, and the current due to electron-hole recombination in traps at and near the Si channel and gate insulator interface is measured at the substrate. DCIV current (I_{DCIV}) is proportional to the density of these traps, and therefore, increase in I_{DCIV} after NBTI stress is a direct measure of generation of new traps. In HKMG MOSFETs, DCIV can measure trap generation at Si/IL interface, IL bulk, as well as at IL/High-K interface especially for gate stacks having thin IL layers, refer to Fig. 3.1. However, it is unlikely that DCIV would probe much deeper inside High-K bulk. Therefore, in addition to generated traps associated with NBTI, DCIV technique can also measure generated traps in the IL due to Time Dependent Dielectric Breakdown (TDDB) process as both mechanisms get triggered at identical stress conditions, refer to Chap. 1, Fig. 1.2. However, as discussed in Chaps. 2 and 4, TDDB has much larger V_G acceleration factor compared to NBTI [15], and for HKMG MOSFETs, the stress V_G gets divided between the IL and High-K layers. Therefore, for moderate values of stress V_G, trap generation in HKMG devices measured by DCIV can be largely associated to the NBTI process.

Furthermore, DCIV is a slow measurement technique and it takes approximately few seconds to perform the V_G sweep and measure I_{DCIV} using conventional instrumentation. Note that DCIV characterization is performed in MSM mode, where measurements are performed before and during logarithmically spaced intervals of NBTI stress. Since generated traps start to reduce when NBTI stress is stopped for measurement, measured DCIV data get corrupted by recovery artifacts and therefore needs to be corrected for measurement delay, as discussed in Chap. 2. Finally, note that DCIV scans trap generation located energetically in $\sim$0.3 eV around the Si band gap [7]. Since ΔV_T gets affected by trap generation in the entire band gap, measured DCIV data must also be corrected for such band gap difference before compared to ΔV_T obtained from I–V measurements. DCIV data in this and following sections are plotted after correction for measurement delay and band gap difference using the procedure discussed in Chap. 2 [17].

3.3.1 DCIV Measurements in DC Stress

Figure 3.3 plots time evolution of generated interface traps (ΔN_{IT}) in HKMG p-MOSFETs having (a) non-nitrided (D2) and (b) N surface passivated (D4) IL for NBTI stress at different $V_{G\text{-}STR}$ and T. As mentioned before, ΔN_{IT} is extracted from DCIV measurements after delay and band gap correction. Only longer time data are plotted, and ΔN_{IT} increases with increase in $V_{G\text{-}STR}$ and T as expected, and the nitrided device D4 shows slightly higher degradation compared to its non-nitrided counterpart due to larger stress E_{OX} and different IL quality. Note that E_{OX} is calculated using ($V_{G\text{-}STR} - V_{T0}$)/EOT; where V_{T0} is pre-stress V_T and EOT is the SiO_2 equivalent gate insulator thickness. The time evolution of ΔN_{IT} shows power-law dependence with identical time exponent ($n \sim 1/6$) for both devices and

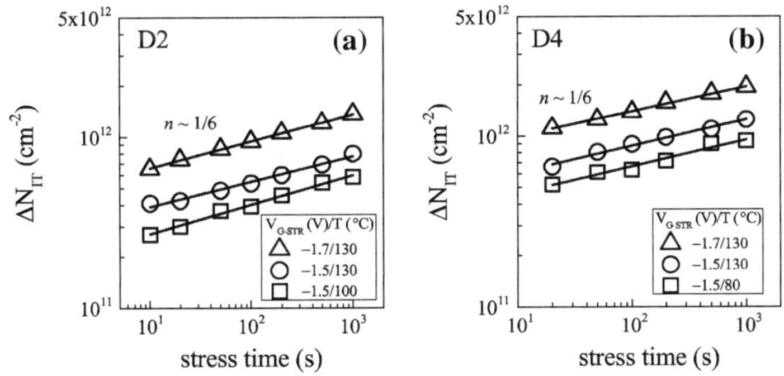

Fig. 3.3 DCIV measured time evolution of ΔN_{IT} for NBTI stress at different $V_{G\text{-}STR}$ and T in HKMG p-MOSFETs having different IL processes

for different $V_{G\text{-}STR}$ and T. The exponent n is extracted by linear regression of measured ΔN_{IT} time evolution data in t_{STR} range of 10 s to 1 Ks. Interestingly, the HKMG process dependence of time exponent n for ΔN_{IT} is different from that for ΔV_T shown earlier in Chap. 1, Figs. 1.25, 1.33 and 1.35. Note that although ΔV_T shows power-law time dependence with exponent n that is independent of $V_{G\text{-}STR}$ and T for a particular HKMG process, its value reduces for nitrided devices unlike that shown here for ΔN_{IT} data. Moreover, the value of n for ΔN_{IT} is always higher than that for ΔV_T when extracted in the same range of t_{STR} as shown.

As a further proof of the universality of power-law time exponent n, time evolution of ΔN_{IT} is measured using DCIV method in different HKMG devices shown in Fig. 3.2. Figure 3.4 plots extracted n versus (a) $V_{G\text{-}STR}$ and (b) T after correction for measurement delay. Note that within measurement error, a universal power-law time dependence with $n \sim 1/6$ is obtained for devices having different HKMG gate insulator processes and also across different $V_{G\text{-}STR}$ and T. Once again, while a universal exponent is obtained for time evolution of ΔN_{IT}, this is in contrast

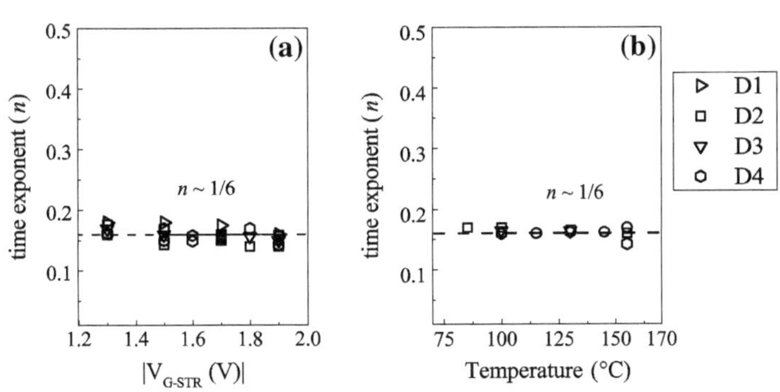

Fig. 3.4 Extracted long-time power-law time exponent n for different HKMG devices as a function of **a** stress V_G and **b** stress T, obtained using DCIV measurements during NBTI stress

to the exponent obtained for time evolution of ΔV_T from UF-MSM technique; the latter has strong gate insulator process dependence as discussed in Chap. 1, Sect. 1. 6. This universality of ΔN_{IT} time evolution is a very significant result and underlines the robustness of physical mechanism governing interface trap generation, which will be discussed later in this book.

Figure 3.5 plots (a) stress E_{OX} and (b) stress T dependence of ΔN_{IT} at fixed t_{STR}, obtained from DCIV measurements after delay and band gap corrections. Experiments have been performed at three sets of stress T, and for each T, three different values of E_{OX} have been used. This facilitates extraction of power-law E_{OX} acceleration factor Γ_E and Arrhenius T activation energy E_A of ΔN_{IT} at different T and E_{OX}, respectively. Since ΔN_{IT} time evolution has power-law dependence with similar n for different stress E_{OX} and T as shown above, extracted Γ_E and E_A would be independent of t_{STR}. It is interesting to note that similar to E_{OX} and T dependence of ΔV_T shown in Chap. 1, Fig. 1.27, measured values of Γ_E and E_A from E_{OX} and T dependence of ΔN_{IT} are also independent of T and E_{OX}, respectively. Note that mutually independent E_{OX} and T dependencies are obtained when stress and measurement remain free from extraneous artifacts mentioned in Chap. 1, Sect. 1.3.

Figure 3.6 shows the impact of IL thickness on (a) power-law field acceleration factor Γ_E and (b) Arrhenius T activation energy E_A extracted from DCIV measured ΔN_{IT} data after delay and band gap correction, for NBTI stress in different HKMG devices shown in Fig. 3.2. As discussed before, IL thickness scaling is achieved by using different RTP based thermal IL (D1, D2), post High-K nitridation (D3), and RTP based IL on N passivated Si substrate (D4). Both Γ_E and E_A reduce with EOT scaling as shown, which is similar to EOT dependence of Γ_E and E_A for ΔV_T shown in Fig. 1.35. For a particular device, the magnitude of Γ_E for ΔN_{IT} is similar to that

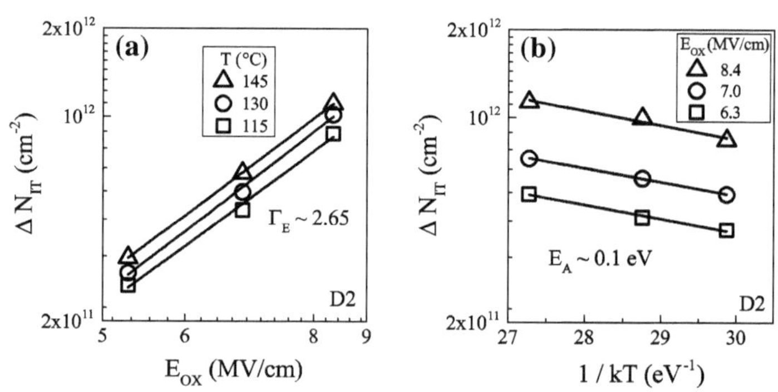

Fig. 3.5 Fixed time DCIV measured ΔN_{IT} versus **a** stress E_{OX} and **b** stress T for NBTI stress in HKMG p-MOSFETs. E_{OX} dependence is shown for different T and T dependence shown for different E_{OX}. E_{OX} dependence is plotted in a log–log scale and T dependence is plotted in a semi-log scale

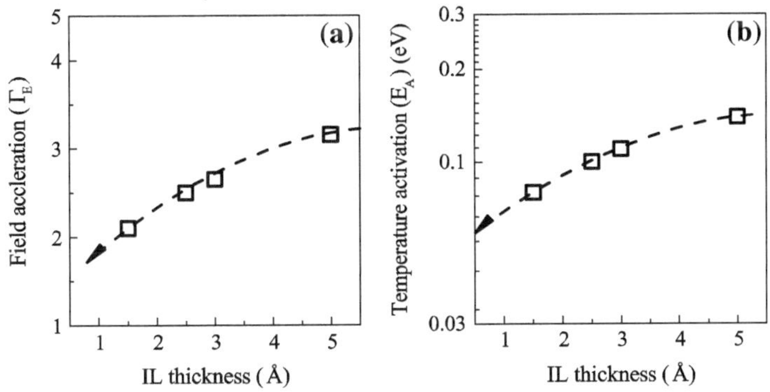

Fig. 3.6 Power-law field acceleration factor (Γ_E) and Arrhenius T activation energy (E_A) of DCIV measured ΔN_{IT} for NBTI stress, as a function of IL thickness of different HKMG stacks

for ΔV_T, while E_A for ΔN_{IT} is always higher compared to the corresponding value for ΔV_T. This aspect is discussed later in this chapter and also in Chap. 4.

3.3.2 DCIV Measurements in AC Stress

Similar to DC stress, DCIV assessment of trap generation during AC NBTI stress is also done in MSM mode, where measurements are performed before and during logarithmically spaced interruptions of stress. Figure 3.7 plots time evolution of ΔN_{IT} during AC stress at different (a) PDC and (b) f but identical stress E_{OX} and T, obtained from DCIV measurements after delay and band gap corrections.

Only the long t_{STR} data are plotted, and the corresponding DC data are also shown. Note that DC stress bias and AC stress pulses have been applied for identical t_{STR} duration, and therefore the actual duration of AC stress, i.e., the pulse on time would depend on PDC of the applied AC pulse. ΔN_{IT} time evolution shows power-law dependence at longer t_{STR} for both DC and AC stress, and the exponent n extracted in t_{STR} range of 10 s to 1 Ks is also plotted in Fig. 3.7 as a function of (c) PDC and (d) f of the gate pulse; DC value is shown as reference (100 % PDC). Identical n ($\sim 1/6$) is observed for DC and AC stress at different PDC and f. Note that identical n of ΔN_{IT} time evolution for DC and AC stress is in contrast to that observed for time evolution of ΔV_T, measured using the UF-MSM technique and shown in Chap. 1, Fig. 1.31; ΔV_T shows lower n for DC compared to AC stress. Moreover when extracted in same t_{STR} range, ΔN_{IT} shows higher n compared to ΔV_T for DC stress, however, both ΔN_{IT} and ΔV_T show identical n ($\sim 1/6$) for AC stress. Once again, the universality of $n \sim 1/6$ for ΔN_{IT} time evolution during DC and AC stress suggest the robustness of the underlying physical mechanism of interface trap generation and is discussed in Chap. 5.

Figure 3.7 also plots measured ΔN_{IT} at fixed t_{STR} for AC NBTI stress as a function of (e) PDC and (f) f, the PDC dependence is measured using AC pulses

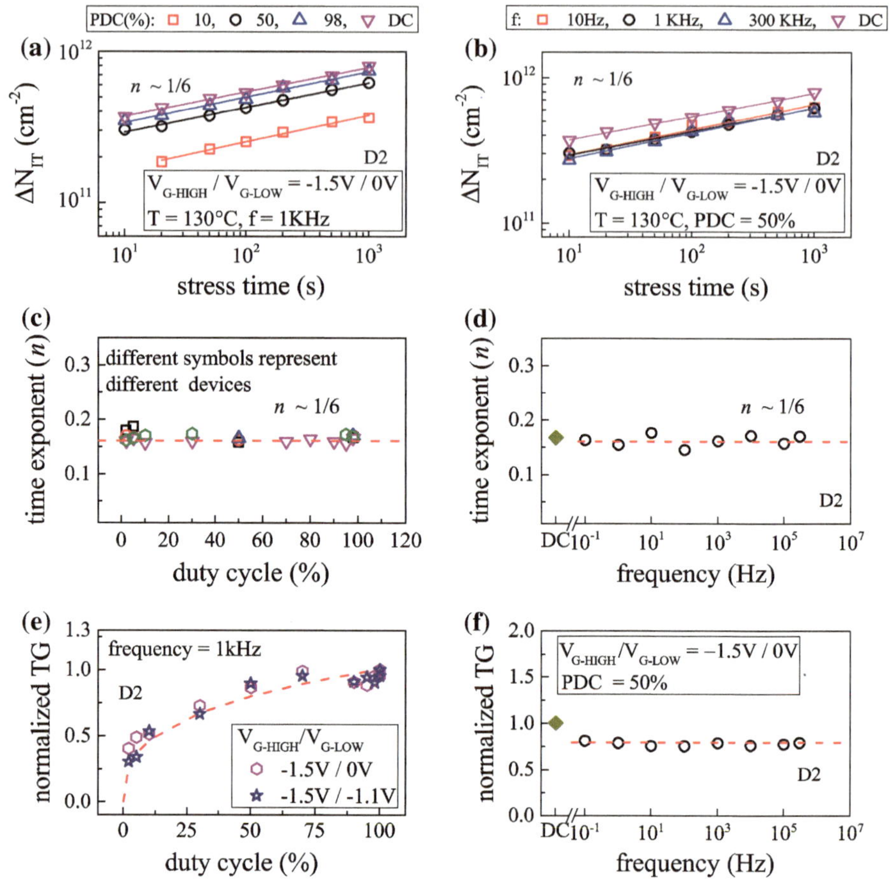

Fig. 3.7 DCIV measured time evolution of ΔN_{IT} for AC NBTI stress at different **a** PDC and **b** frequency; the PDC and f values used in panel **a** and **b** respectively are mentioned at the *top*. Measured long-time power-law time exponent n as a function of **c** PDC for different devices and **d** frequency. Fixed time measured ΔN_{IT} versus **e** PDC for different $V_{G\text{-}LOW}$ and **f** frequency

having identical pulse high but different pulse low values. AC data are normalized to the corresponding DC value at identical t_{STR}. Note that ΔN_{IT} magnitude increases with increase in PDC, but remains independent of the pulse low value and f. Although f independence of ΔN_{IT} is qualitatively similar to that observed for ΔV_T as shown in Chap. 1, Fig. 1.32, ΔN_{IT} and ΔV_T have very different AC to DC ratio. On the other hand, the PDC dependence of ΔN_{IT} is both qualitatively and quantitatively different from the PDC dependence of ΔV_T. Note, ΔN_{IT} does not show the "S" shaped PDC dependence as in ΔV_T, and moreover, no "kink" or "jump" is seen for ΔN_{IT} data between high PDC AC and DC stress. Furthermore, not only ΔN_{IT} has very different AC to DC ratio compared to ΔV_T as mentioned above, unlike ΔV_T, the AC to DC ratio for ΔN_{IT} does not depend on the pulse low value. Time evolution of ΔN_{IT} and ΔV_T for AC stress will be explained later in this book.

3.4 Hole Trapping During NBTI

As discussed earlier in this chapter, although interface trap generation plays a crucial role, it alone cannot explain UF-MSM measured ΔV_T during NBTI stress. As an evidence of additional contribution from the hole-trapping component, Fig. 3.8 plots time evolution of measured ΔV_T and ΔV_{IT} ($=q/C_{OX} * \Delta N_{IT}$) obtained using UF-MSM and DCIV techniques, respectively. Experiments were performed on HKMG p-MOSFETs having (a) non-nitrided (D2) and (b) N surface passivated (D4) IL, refer to Fig. 3.2; C_{OX} is gate insulator capacitance and q is electronic charge. Identical stress E_{OX} and T have been used for both devices, and experimental DCIV data were corrected for measurement delay and band gap differences as mentioned earlier. Since ΔV_{IT} signifies the component of ΔV_T that is contributed by generated interface traps (ΔN_{IT}), the difference between ΔV_T and ΔV_{IT} signifies contribution due to hole trapping in pre-existing traps, ΔV_{HT} ($=q/C_{OX} * \Delta N_{HT}$). Time evolution of ΔV_{HT} is also plotted in Fig. 3.8. Only longer t_{STR} data are shown.

Time evolution of ΔV_{IT} has power-law dependence with exponent $n \sim 1/6$ as discussed above, while ΔV_{HT} saturates at longer t_{STR} as shown. Therefore, time evolution of ΔV_T ($=\Delta V_{IT} + \Delta V_{HT}$) shows power-law dependence with lower n compared to that for ΔV_{IT}. This explains the reason behind lower time exponent n observed for ΔV_T compared to ΔN_{IT} data across different $V_{G\text{-}STR}$ and T as mentioned before. Although the ΔV_{IT} contribution increases slightly for the N containing device D4 compared to the non-nitrided device D2, a significantly large increase is observed for the ΔV_{HT} component. Therefore, ΔV_T magnitude increases while time exponent n reduces for the D4 device as shown. Figure 3.8 clearly indicates that the underlying ΔN_{IT} and ΔN_{HT} components of NBTI are uncorrelated; a relatively larger increase in ΔN_{HT} is observed for the D4 device having N in

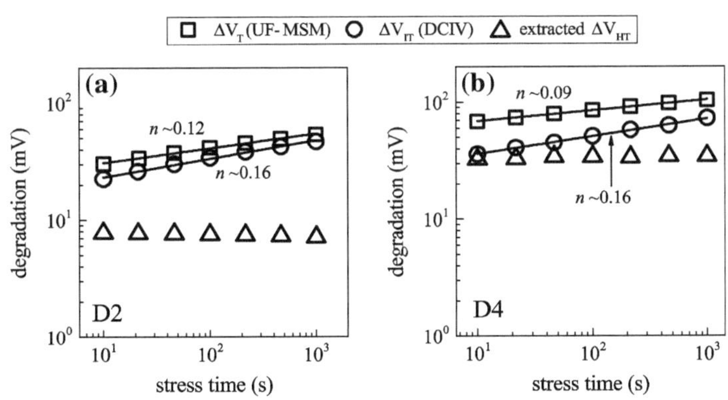

Fig. 3.8 Time evolution of UF-MSM measured ΔV_T and DCIV measured ΔN_{IT} contribution after measurement delay and band gap correction (ΔV_{IT}), for NBTI stress in HKMG p-MOSFETs having different IL processes. Extracted difference (ΔV_{HT}) is also shown

the gate stack, which can explain the measured reduction in the time exponent n. However, ΔV_{IT} component dominates ΔV_T for both devices as shown.

As a further proof, Fig. 3.9a shows the correlation of measured ΔV_T and ΔV_{IT} for HKMG devices D2 and D4. As mentioned before, ΔV_T is obtained using UF-MSM and ΔV_{IT} using DCIV after delay and band gap correction. The 1:1 correlation line is also shown, which signifies zero hole-trapping contribution. Note that for a particular ΔV_{IT}, D2 device shows slightly higher ΔV_T than the 1:1 correlation line, while a somewhat larger ΔV_T is observed for device D4, which is consistent with relatively larger ΔN_{HT} contribution for device D4 having N in the gate insulator stack. Larger magnitude of hole trapping during NBTI stress in gate insulators containing N is a well-known result and reported by various groups [13–15, 50].

Hole trapping occurs in pre-existing, process related gate insulator traps, and as mentioned in Chap. 2, flicker noise technique can be used to access the density of these traps. In flicker noise method, the gate of the MOSFET is biased in weak inversion and the power spectral density of drain current noise (S_{ID}) is measured using a spectrum analyzer. The inversion layer carrier density in the channel remains low in weak inversion, and drain current noise arises due to trapping and detrapping of carriers in gate insulator traps. Higher trap density results in larger S_{ID} and vice versa. Figure 3.9b shows measured pre-existing trap density in devices D2 and D4. Note that the nitrided device D4 shows higher trap density, which is consistent with higher ΔN_{HT} contribution shown in Figs. 3.8 and 3.9a. The impact of N on pre-existing hole traps has been studied in detail in SiON [50] and HKMG [51] p-MOSFETs and also verified by Density Functional Theory (DFT) calculations as discussed in detail in [17, 51].

Time evolution of ΔV_T and ΔN_{IT} during NBTI stress has been measured respectively using UF-MSM and DCIV methods for three sets of stress T, and for each T, three different E_{OX} values have been used for stress. The E_{OX} and T dependencies of ΔV_T at fixed t_{STR} are shown in Chap. 1, Fig. 1.27 and that for

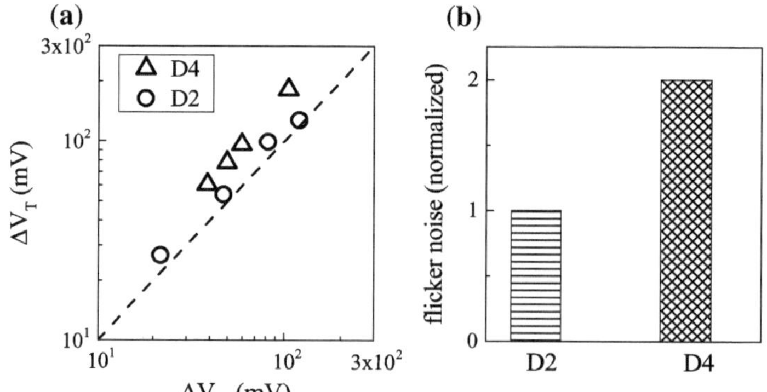

Fig. 3.9 **a** Correlation of UF-MSM measured ΔV_T and DCIV measured ΔV_{IT} for NBTI stress, and **b** pre-stress trap density measured by using flicker noise method, in HKMG p-MOSFETs having different HKMG gate insulator processes

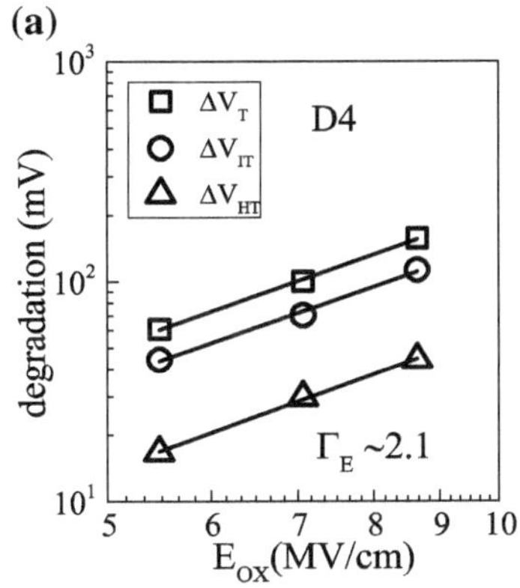

Fig. 3.10 **a** Fixed time UF-MSM measured ΔV_T, DCIV measured ΔV_{IT} and their difference (ΔV_{HT}) versus NBTI stress E_{OX} plotted in a log–log scale. **b** Extracted power-law field acceleration factor for ΔV_T, ΔV_{IT} and ΔV_{HT} versus stress T

ΔN_{IT} are shown before in Fig. 3.5. Measured ΔV_T and ΔV_{IT} as well as the extracted difference ΔV_{HT} obtained at a fixed t_{STR} for the HKMG device D4 are plotted in Fig. 3.10a versus stress E_{OX} for a particular stress T, and plotted in Fig. 3.11a versus stress T for a particular stress E_{OX}. Note that the ΔV_{IT} subcomponent dominates overall ΔV_T for all E_{OX} and T, even for D4 device having N in the gate insulator stack. Moreover, identical power-law E_{OX} dependence Γ_E is obtained for ΔV_T, ΔV_{IT} and therefore for ΔV_{HT}. However, extracted ΔV_{HT} has much lower Arrhenius T activation energy E_A compared to measured ΔV_{IT}, which explains lower E_A for ΔV_T when compared to E_A for ΔV_{IT} as shown.

As discussed before, ΔV_T and ΔN_{IT} have power-law time dependence with identical n across different E_{OX} and T, although n for ΔN_{IT} is higher than that for ΔV_T. Therefore, extracted Γ_E and E_A for ΔV_T, ΔV_{IT} and hence for ΔV_{HT} would be

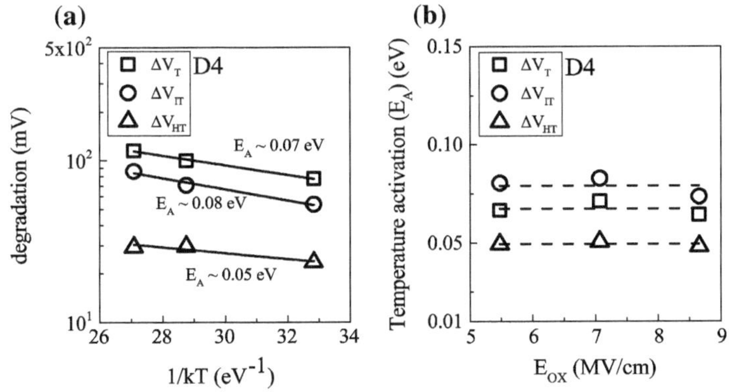

Fig. 3.11 **a** Fixed time UF-MSM measured ΔV_T, DCIV measured ΔV_{IT} and their difference (ΔV_{HT}) versus NBTI stress T plotted in a semi-log scale. **b** Extracted T activation energy for ΔV_T, ΔV_{IT} and ΔV_{HT} versus stress E_{OX}

independent of the value of t_{STR} used for extracting E_{OX} and T dependence, otherwise these terms would not have much meaning. Extracted Γ_E versus T for ΔV_T and its ΔV_{IT} and ΔV_{HT} subcomponents is shown in Fig. 3.10b, while the corresponding E_A versus E_{OX} relation is shown in Fig. 3.11b. Note that in the absence of different extraneous artifacts mentioned in Chap. 1, Sect. 1.3, Γ_E and E_A for both ΔV_T and ΔN_{IT} (or ΔV_{IT}) are independent of T and E_{OX} respectively as shown. Therefore, Γ_E and E_A of extracted ΔN_{HT} (or ΔV_{HT}) also has the same behavior as shown. Mutually uncoupled Γ_E and E_A is observed for other devices, not plotted here for brevity.

3.5 Trap Generation During PBTI

In this section, trap generation during PBTI stress in HKMG n-MOSFETs is studied using the DCIV [7] and SILC [41] techniques. DCIV scans trap generation located energetically in ~0.3 eV around the Si band gap, and in HKMG MOSFETs, it can measure generated traps at Si/IL interface, IL bulk, as well as at IL/High-K interface especially for gate stacks having thin IL layers. As mentioned before, negligible g_m degradation during PBTI stress suggests negligible IL degradation and therefore, PBTI is believed to cause trap generation and trapping predominately in the HfO_2 High-K layer [36, 39]. DCIV measurements can probe trap generation during PBTI stress at and near the IL/High-K interface ($\Delta N_{IT\text{-}HK}$); refer to Fig. 1. As mentioned before, it is unlikely that DCIV method would probe much deeper into the High-K bulk.

As discussed in Chap. 2, SILC is estimated from measured gate current (I_G) before and during logarithmically spaced intervals of BTI stress. Increase in I_G after stress is due to trap assisted tunneling via newly generated traps, and therefore, the magnitude of increased gate current (ΔI_G) can be used to estimate density of generated traps during PBTI stress in HKMG n-MOSFETs [40]. Note that SILC has been used in the past to estimate gate insulator trap generation associated with the TDDB process [52–54], and also bulk trap generation during NBTI stress [55], in SiON MOSFETs. SILC is negligible during NBTI stress in HKMG p-MOSFETs due to band alignment issues mentioned in Chap. 2, while non-negligible SILC is observed in HKMG n-MOSFETs during PBTI stress [3, 16, 17, 40, 41]. Note that unlike DCIV that scans traps that are energetically aligned with Si mid gap, SILC scans traps close to the conduction band edge of the High-K layer [41]. However, there is a debate regarding the exact physical location of generated traps during PBTI stress in HKMG n-MOSFETs as probed by SILC; some report suggests it is at the IL/High-K interface [56], while other suggests it is deeper inside the High-K bulk [41]. Although the exact location of traps is an important aspect especially from the viewpoint of TDDB process, the type of generated traps probed by SILC has been found to have much smaller impact on PBTI degradation when compared to the impact of generated traps probed by the DCIV method [17], and will be discussed later in Chap. 4.

3.5.1 DCIV Measurements in DC Stress

Figure 3.12 plots the time evolution of generated interface traps ($\Delta N_{IT\text{-}HK}$) during PBTI stress at different $V_{G\text{-}STR}$ and stress T in HKMG n-MOSFETs with (a) non-nitrided (D2) and (b) N surface passivated (D4) IL. As mentioned before, $\Delta N_{IT\text{-}HK}$ is extracted from DCIV measurements after delay and band gap correction. The delay correction of DCIV is straightforward and the method used for NBTI can be used. However, the band gap correction is not obvious, as DCIV probes traps at or near the IL/High-K interface for PBTI stress, and the exact energetic extent of trap generation is yet unknown. For simplicity, a similar correction factor as used for NBTI stress is also assumed for PBTI stress in different HKMG devices.

Only longer time data are plotted, and measured $\Delta N_{IT\text{-}HK}$ increases with increase in $V_{G\text{-}STR}$ and T as expected. However unlike NBTI, the $\Delta N_{IT\text{-}HK}$ magnitude reduces significantly for the nitrided device D4 when compared to the non-nitrided device D2 at identical $V_{G\text{-}STR}$ and T. This happens in spite of higher E_{OX} for device D4 due to lower EOT, E_{OX} being calculated as $(V_{G\text{-}STR} - V_{T0})$/EOT, and hence is attributed to presence of N in the gate insulator stack. Similar to NBTI, time evolution of $\Delta N_{IT\text{-}HK}$ for PBTI stress shows power-law dependence with identical time exponent ($n \sim 1/6$) for both devices and for different $V_{G\text{-}STR}$ and T; the exponent n is extracted by linear regression of measured $\Delta N_{IT\text{-}HK}$ time evolution in t_{STR} range of 10 s to 1 Ks. Remarkably, identical n has been obtained for both NBTI and PBTI stress as shown. Note that similar to NBTI stress data shown earlier, the process dependence of $\Delta N_{IT\text{-}HK}$ time exponent for PBTI stress is in contrast with process dependence of ΔV_T time exponent shown in Chap. 1, Figs. 1.25 and 1.33. Although ΔV_T has power-law dependence for PBTI stress with time exponent n that is independent of stress E_{OX} and T for a particular HKMG process, the value of n reduces for nitrided devices, unlike that seen here for $\Delta N_{IT\text{-}HK}$ data. Moreover, similar to that observed for NBTI stress, the value of n for $\Delta N_{IT\text{-}HK}$ is always higher than n for ΔV_T for PBTI stress, when extracted in the same range of t_{STR} as shown.

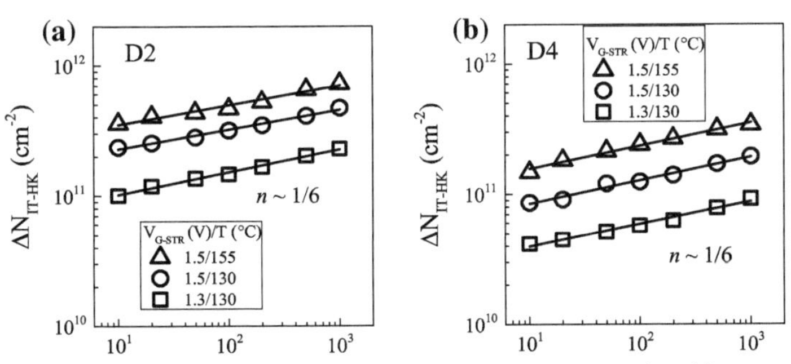

Fig. 3.12 DCIV measured time evolution of $\Delta N_{IT\text{-}HK}$ for PBTI stress at different $V_{G\text{-}STR}$ and T in HKMG n-MOSFETs having different IL processes

As a further proof of the universality of time exponent ($n \sim 1/6$) for PBTI stress, time evolution of $\Delta N_{IT\text{-}HK}$ is measured by using DCIV in different HKMG devices listed in Figs. 3.2, and 3.13 plots extracted n as a function of (a) $V_{G\text{-}STR}$ and (b) T after correction for measurement delay. Similar to NBTI, power-law time dependence with universal $n \sim 1/6$ is also obtained for $\Delta N_{IT\text{-}HK}$ during PBTI stress at different $V_{G\text{-}STR}$ and T in devices having different HKMG gate insulator processes. Once again, similar to NBTI results shown earlier, while a universal $n \sim 1/6$ exponent is obtained for time evolution of $\Delta N_{IT\text{-}HK}$ for PBTI stress in different HKMG devices, this is in contrast to the exponent obtained for time evolution of ΔV_T from UF-MSM technique; the latter shows strong gate insulator process dependence as discussed in Chap. 1, Sect. 1.6. The similarity of time evolution of ΔN_{IT} for NBTI and $\Delta N_{IT\text{-}HK}$ for PBTI stress is a very remarkable result and underlines the similarity of physical mechanism governing trap generation at Si/IL and IL/High-K interfaces, and will be discussed later in this book.

Similar to NBTI results shown earlier, $\Delta N_{IT\text{-}HK}$ time evolution for PBTI stress also has power-law dependence with similar n across different stress E_{OX} and T as shown. Therefore, the power-law E_{OX} acceleration factor Γ_E and Arrhenius T activation energy E_A for $\Delta N_{IT\text{-}HK}$ extracted at a fixed t_{STR} would remain independent of t_{STR}. PBTI experiments were done at three sets of stress T, and at each T, three different E_{OX} values have been used for stress. Figure 3.14 plots (a) stress E_{OX} and (b) stress T dependence of $\Delta N_{IT\text{-}HK}$ obtained from DCIV measurements after delay and band gap corrections. Measured values of Γ_E and E_A extracted from E_{OX} and T dependence of $\Delta N_{IT\text{-}HK}$ for PBTI stress are independent of T and E_{OX}, respectively. This behavior is similar to the E_{OX} and T dependent parameters of ΔV_T shown in Chap. 1, Fig. 1.27. Moreover, this observation is identical to the NBTI results discussed before in this chapter. As mentioned before, mutually independent E_{OX} and T dependencies are observed when stress and measurements remain free from artifacts mentioned earlier in Chap. 1, Sect. 1.3.

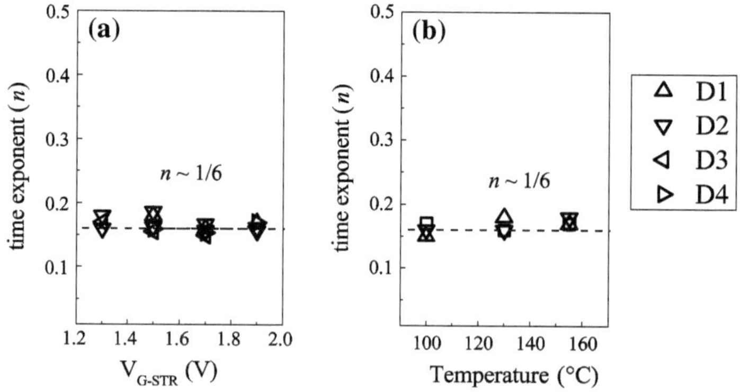

Fig. 3.13 Extracted long-time power-law time exponent n for different HKMG devices as a function of **a** stress V_G and **b** stress T, obtained using DCIV measurements during PBTI stress

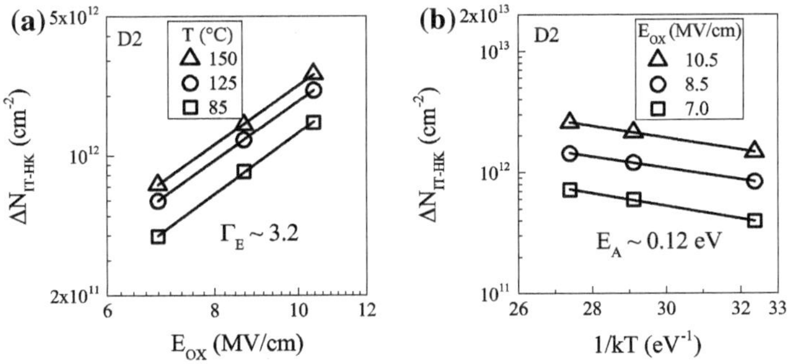

Fig. 3.14 Fixed time DCIV measured $\Delta N_{IT\text{-}HK}$ versus **a** stress E_{OX} and **b** stress T for PBTI stress in HKMG n-MOSFETs. E_{OX} dependence is shown for different T and T dependence shown for different E_{OX}. E_{OX} dependence is plotted in a log–log scale and T dependence is plotted in a semi-log scale

Figure 3.15 plots the impact of IL thickness on (a) power-law field acceleration factor Γ_E and (b) Arrhenius T activation energy E_A extracted from DCIV measured $\Delta N_{IT\text{-}HK}$ data after delay and band gap corrections, obtained for PBTI stress in different HKMG devices shown in Fig. 3.2. Reduction in IL thickness is achieved using different RTP based thermal IL (D1, D2), post High-K nitridation (D3), and RTP based IL on N passivated Si substrate (D4), refer to Fig. 3.2. Similar to NBTI, Γ_E for PBTI reduces with EOT scaling. However unlike NBTI, only negligible reduction in E_A has been observed with EOT scaling. Similar to NBTI results shown before, the magnitude of Γ_E for $\Delta N_{IT\text{-}HK}$ is similar to Γ_E for ΔV_T for a particular device also for PBTI stress, while E_A for $\Delta N_{IT\text{-}HK}$ is always higher compared to the corresponding value for ΔV_T; refer to Chap. 1, Fig. 1.35 for dependence of ΔV_T parameters on IL thickness. This aspect will be discussed later in this chapter.

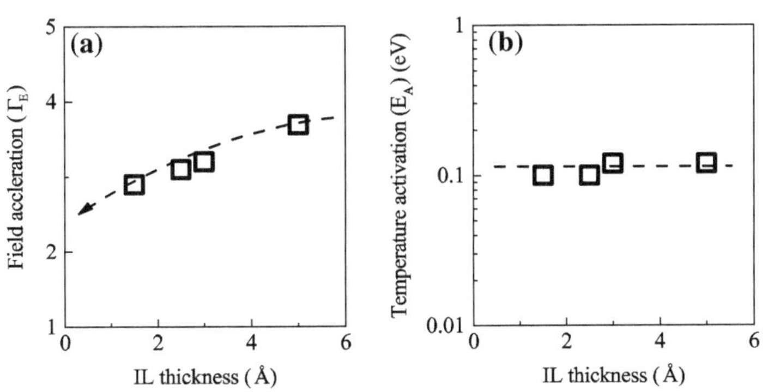

Fig. 3.15 Power-law field acceleration factor (Γ_E) and Arrhenius T activation energy (E_A) of DCIV measured $\Delta N_{IT\text{-}HK}$ for PBTI stress, as a function of IL thickness of different HKMG stacks

3.5.2 DCIV Measurements in AC Stress

DCIV assessment of trap generation during AC stress is done in MSM mode similar to DC stress, and measurements were performed before and during logarithmically spaced stress intervals. Figure 3.16 plots time evolution of $\Delta N_{IT\text{-}HK}$ during AC stress at different (a) PDC and (b) *f* but identical stress E_{OX} and *T*, obtained from DCIV measurements after delay and band gap corrections. Only long t_{STR} data are plotted, and the corresponding DC data are also shown. Similar to NBTI stress as discussed before, the DC stress bias and AC stress pulses for PBTI stress were applied for identical t_{STR} duration, and actual duration of AC stress depends on PDC of the applied AC pulse. Time evolution of $\Delta N_{IT\text{-}HK}$ shows power-law dependence at longer t_{STR}; the exponent *n* extracted in t_{STR} range of 10 s to 1 Ks is

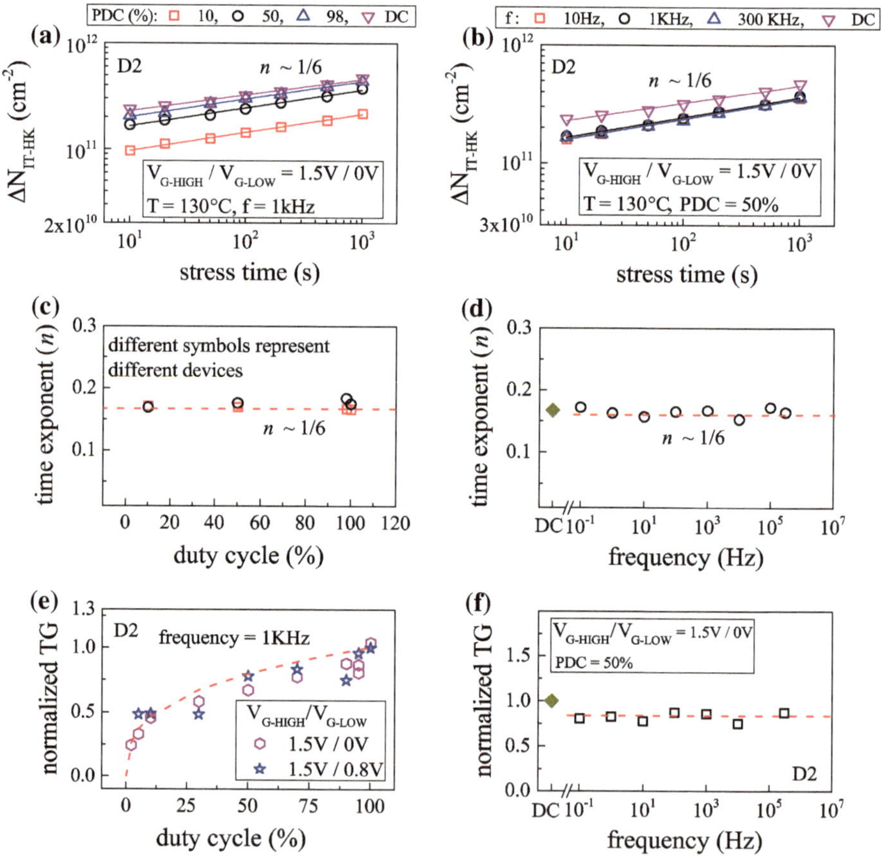

Fig. 3.16 DCIV measured time evolution of $\Delta N_{IT\text{-}HK}$ for AC PBTI stress at different **a** PDC and **b** frequency; the PDC and *f* values used in panel **a** and **b** respectively are mentioned at the *top*. Measured long-time power-law time exponent *n* as a function of **c** PDC for different devices and **d** frequency. Fixed time measured $\Delta N_{IT\text{-}HK}$ versus **e** PDC for different $V_{G\text{-}LOW}$ and **f** frequency

also plotted in Fig. 3.16 as a function of (c) PDC and (d) *f* of the gate pulse. The DC value is shown as reference.

Similar to NBTI stress results shown earlier, identical *n* ($\sim$ 1/6) is observed for DC and AC PBTI stress at different PDC and *f*. Once again, similar to NBTI, identical *n* of $\Delta N_{IT\text{-}HK}$ time evolution for DC and AC PBTI stress is in contrast to the observed time evolution of ΔV_T measured using UF-MSM technique and shown in Chap. 1, Fig. 1.31; ΔV_T time evolution has lower *n* for DC compared to AC stress. Moreover when extracted in the same t_{STR} range, $\Delta N_{IT\text{-}HK}$ shows higher *n* compared to ΔV_T for DC stress, but both $\Delta N_{IT\text{-}HK}$ and ΔV_T show identical *n* ($\sim$ 1/6) for AC stress also for PBTI stress, which is again similar to NBTI stress results shown before. Once again, the universality of $n \sim 1/6$ for $\Delta N_{IT\text{-}HK}$ time evolution during DC and AC PBTI stress suggest the robustness of the underlying physical mechanism of interface trap generation and will be discussed later in the book.

Figure 3.16 also plots the measured $\Delta N_{IT\text{-}HK}$ at fixed t_{STR} for AC PBTI stress as a function of (e) PDC and (f) *f* of the gate pulse. Once again, AC pulses having identical pulse high but different pulse low values were used to study the PDC dependence. AC data are normalized to the corresponding DC value at identical t_{STR}. Similar to that shown earlier for NBTI, the magnitude of $\Delta N_{IT\text{-}HK}$ for PBTI stress increases with increase in PDC but remains independent of the pulse low value and *f* of the gate pulse. Remarkably similar AC to DC ratio and PDC dependent shape has been observed for NBTI and PBTI stress. Moreover, although the *f* independence of $\Delta N_{IT\text{-}HK}$ for PBTI stress is qualitatively similar to that observed for ΔV_T shown in Chap. 1, Fig. 1.32, $\Delta N_{IT\text{-}HK}$ and ΔV_T show very different AC to DC ratio. This observation is similar to that reported for NBTI stress. Once again, similar to NBTI, the PDC dependence of $\Delta N_{IT\text{-}HK}$ for PBTI stress is both qualitatively and quantitatively different from the PDC dependence of ΔV_T shown in Chap. 1. Similar to NBTI results, $\Delta N_{IT\text{-}HK}$ for PBTI also does not have the "S" shaped PDC dependence as observed for ΔV_T, and unlike ΔV_T, no "kink" or "jump" is observed for $\Delta N_{IT\text{-}HK}$ data between high PDC AC and DC stress. Moreover as mentioned earlier, $\Delta N_{IT\text{-}HK}$ has very different AC to DC ratio compared to ΔV_T, and unlike ΔV_T, the AC to DC ratio for $\Delta N_{IT\text{-}HK}$ does not depend on the pulse low value. The PDC and *f* dependencies of $\Delta N_{IT\text{-}HK}$ and ΔV_T for AC PBTI stress have been found to be remarkably similar to AC NBTI stress as shown earlier in this chapter and also in Chap. 1.

3.5.3 SILC Measurements in DC Stress

As mentioned before, increase in gate leakage current (I_G) after stress is also used to calculate trap generation during PBTI stress [40, 41]. While DCIV scans generated traps aligned energetically with the Si [7] mid gap, SILC scans traps closer to the conduction band edge of High-K layer, although the exact physical location of these traps are debated [41, 56]. I_G versus V_G sweeps were taken before and during logarithmic intervals of stress, and generated trap density ($\Delta N_{OT\text{-}HK}$) is calculated as discussed in Chap. 2 [40].

Figure 3.17 plots the time evolution of $\Delta N_{OT\text{-}HK}$ calculated from SILC for different $V_{G\text{-}STR}$ and T in (a) D2 and (b) D4 HKMG n-MOSFETs; refer to Fig. 3.2 for device details. Note that SILC in HKMG devices recovers after removal of stress, and therefore, measured data must be corrected for measurement delay following the methodology shown in Chap. 2. Only long t_{STR} data are plotted; magnitude of $\Delta N_{OT\text{-}HK}$ increases with $V_{G\text{-}STR}$ and T as expected, but reduces drastically for the device D4 having N in the gate stack. Note that D4 device have thinner EOT and would have higher E_{OX} compared to device D2 for a given $V_{G\text{-}STR}$. Therefore, the reduction can be attributed to the presence of N in the gate insulator stack. It is interesting to remark that both DCIV and SILC measurements show reduced trap generation in nitrided devices during PBTI stress. Although the time evolution of $\Delta N_{OT\text{-}HK}$ from SILC shows power-law dependence, the time exponent n is much larger when compared to that for DCIV measured $\Delta N_{IT\text{-}HK}$ and UF-MSM measured ΔV_T time evolution data shown earlier in the book.

As a further proof, Fig. 3.18 plots extracted time exponent n from $\Delta N_{OT\text{-}HK}$ time evolution measured using SILC after delay correction, versus (a) $V_{G\text{-}STR}$ and (b) T. As done earlier, the exponent n is extracted by linear regression of measured time evolution of $\Delta N_{OT\text{-}HK}$ data in t_{STR} range of 10 s to 1 Ks. Data from different HKMG devices shown in Fig. 3.2 are plotted, and the trend line corresponding to DCIV data is also shown as a reference.

Note that SILC data show similar n ($\sim 1/3$) for different $V_{G\text{-}STR}$ and T and for different devices, which signifies the robustness of the underlying physical mechanism. However, the magnitude of time exponent n from SILC is considerably higher than n ($\sim 1/6$) obtained from DCIV, which unequivocally suggests different physical mechanism of generated traps that are probed by these methods. On the other hand, a remarkable similarity of extracted n has been observed during NBTI and PBTI stress

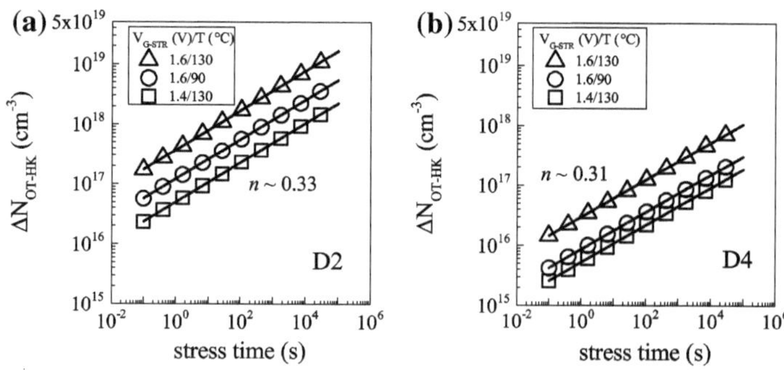

Fig. 3.17 SILC measured time evolution of $\Delta N_{OT\text{-}HK}$ for PBTI stress at different $V_{G\text{-}STR}$ and T in HKMG n-MOSFETs having different IL processes

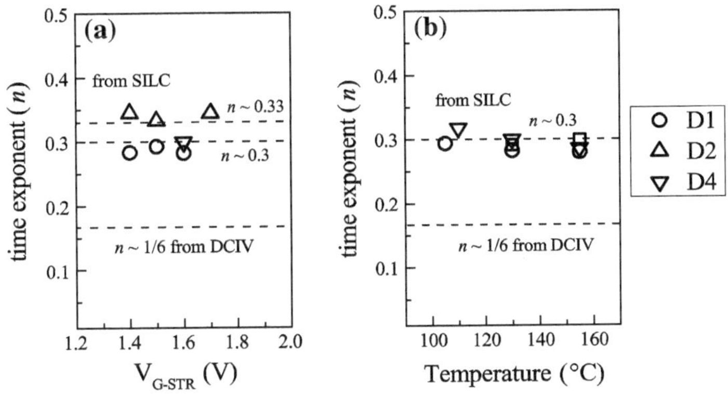

Fig. 3.18 Extracted long-time power-law time exponent n for different HKMG devices as a function of **a** stress V_G and **b** stress T using SILC measurements during PBTI stress. The reference line for DCIV measurement is also shown

for DCIV measurements, refer to Figs. 3.4 and 3.13, which suggests similar generation mechanism for these traps. This will be discussed later in this book.

Figure 3.19 plots measured $\Delta N_{OT\text{-}HK}$ at fixed t_{STR} from delay corrected SILC data versus (a) stress E_{OX} and (b) stress T, for D2 and D4 HKMG n-MOSFETs. The magnitude of $\Delta N_{OT\text{-}HK}$ reduces for the nitrided device D4, although the power-law E_{OX} acceleration factor Γ_E and Arrhenius T activation energy E_A remain independent of HKMG processes. As $\Delta N_{OT\text{-}HK}$ shows power-law time dependence and identical n across stress E_{OX} and T, extracted Γ_E and E_A would be independent of t_{STR}. Note that Γ_E and E_A for $\Delta N_{OT\text{-}HK}$ obtained from SILC has much larger value when compared to $\Delta N_{IT\text{-}HK}$ from DCIV and ΔV_T from UF-MSM measurements

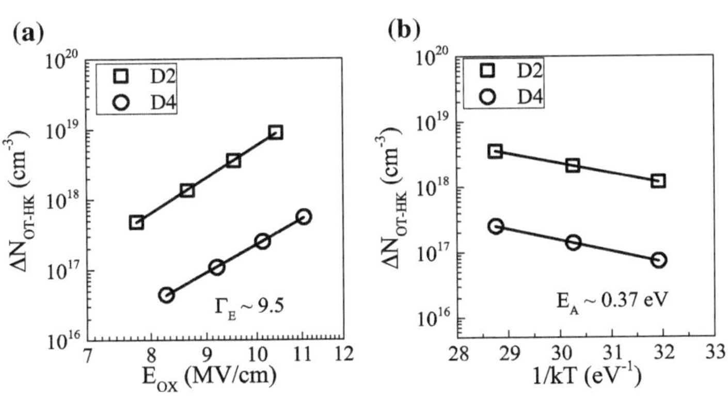

Fig. 3.19 Fixed time SILC measured $\Delta N_{OT\text{-}HK}$ versus **a** stress E_{OX} and **b** stress T for PBTI stress in HKMG n-MOSFETs. E_{OX} dependence is plotted in a log–log scale and T dependence is plotted in a semi-log scale

shown earlier in the book. Different Γ_E and E_A also suggest different physical mechanism of generated traps as probed by DCIV and SILC techniques during PBTI stress.

To understand the relative importance of different trap generation processes on PBTI degradation, Fig. 3.20 compares extracted (a) n, (b) Γ_E and (c) E_A for DCIV, SILC and UF-MSM measurements, as a function of IL thickness for D1 through D4 HKMG devices shown in Fig. 3.2. Note that $\Delta N_{OT\text{-}HK}$ parameters obtained from SILC measurements remain almost constant across IL thickness and show significantly higher values when compared to the ΔV_T and $\Delta N_{IT\text{-}HK}$ parameters obtained respectively from UF-MSM and DCIV measurements. Between DCIV and UF-MSM measurements, power-law time exponent n is somewhat higher for $\Delta N_{IT\text{-}HK}$ compared to ΔV_T for all devices and n reduces with IL scaling for both ΔV_T and $\Delta N_{IT\text{-}HK}$; however, much larger reduction is observed for ΔV_T for deeply scaled IL. The power-law E_{OX} acceleration Γ_E shows similar values for ΔV_T and $\Delta N_{IT\text{-}HK}$ and reduces slightly with IL scaling. Arrhenius T activation energy E_A for $\Delta N_{OT\text{-}HK}$ remains almost constant with IL, but that for ΔV_T and $\Delta N_{IT\text{-}HK}$ reduces slightly with IL scaling. It will be shown in Chap. 4 that the voltage shift corresponding to generated bulk traps from SILC is negligible compared to that for generated interface traps from DCIV, and it is the later that dominates PBTI degradation. Therefore, the parameters n, Γ_E and E_A for ΔV_T are closer to the parameters for $\Delta N_{IT\text{-}HK}$ rather than that for $\Delta N_{OT\text{-}HK}$. Of course, additional contribution from electron trapping in High-K bulk must be considered to compute ΔV_T, as discussed in the following section.

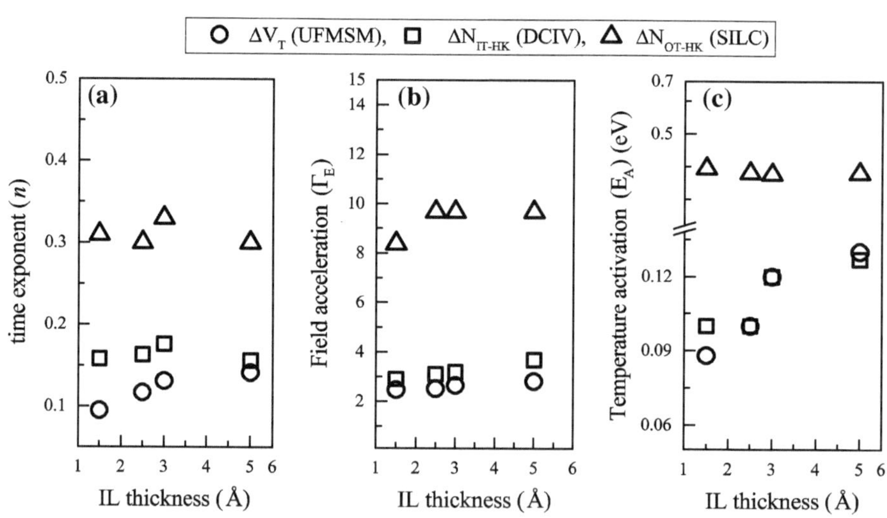

Fig. 3.20 **a** Power-law time dependence n, **b** power-law field acceleration factor Γ_E and **c** Arrhenius T activation energy E_A obtained using UF-MSM, DCIV and SILC methods for PBTI stress, versus IL thickness of different HKMG stacks

3.6 Electron Trapping During PBTI

Although trap generation in the IL is negligible for PBTI stress and the generated traps at or near the IL/High-K interface ($\Delta N_{IT\text{-}HK}$) plays a crucial role, it alone cannot explain ΔV_T measured using UF-MSM method. As an evidence of additional contribution from electron trapping in pre-existing, process related High-K bulk traps (ΔN_{ET}), Fig. 3.21 plots time evolution of ΔV_T and $\Delta V_{IT\text{-}HK}$ ($=q/C_{OX} * \Delta N_{IT\text{-}HK}$) respectively measured using UF-MSM and DCIV techniques. Experiments were performed in HKMG n-MOSFETs having (a) non-nitrided (D2) and (b) N surface passivated (D4) IL, refer to Fig. 3.2; C_{OX} is gate insulator capacitance and q is the electronic charge. Identical $V_{G\text{-}STR}$ and T have been used for both devices, and experimental DCIV data were corrected for measurement delay and band gap differences as discussed before. Since $\Delta V_{IT\text{-}HK}$ signifies the component of ΔV_T contributed by generated interface traps at IL/High-K interface, the difference between ΔV_T and $\Delta V_{IT\text{-}HK}$ signifies the contribution due to electron trapping in pre-existing traps, ΔV_{ET} ($=q/C_{OX} * \Delta N_{ET}$). Time evolution of ΔV_{ET} is also plotted in Fig. 3.21. Only longer t_{STR} data are plotted, and contribution from $\Delta V_{IT\text{-}HK}$ dominates ΔV_T for both devices as shown.

Similar to NBTI data, time evolution of $\Delta V_{IT\text{-}HK}$ for PBTI stress shows power-law dependence with exponent $n \sim 1/6$ and ΔV_{ET} saturates at longer t_{STR}. Therefore, time evolution of ΔV_T ($=\Delta V_{IT\text{-}HK} + \Delta V_{ET}$) shows power-law dependence with lower n compared to that for $\Delta V_{IT\text{-}HK}$. However contrary to NBTI, the $\Delta V_{IT\text{-}HK}$ component of ΔV_T reduces drastically for the N containing device D4 compared to the non-nitrided device D2, while an increase is observed for the ΔV_{ET} component. Note that device D4 has lower EOT and therefore higher E_{OX} when compared to device D2 at identical $V_{G\text{-}STR}$. As $\Delta V_{IT\text{-}HK}$ dominates overall ΔV_T for both D2 and D4 devices, the magnitude of ΔV_T and its power-law time exponent n reduce for

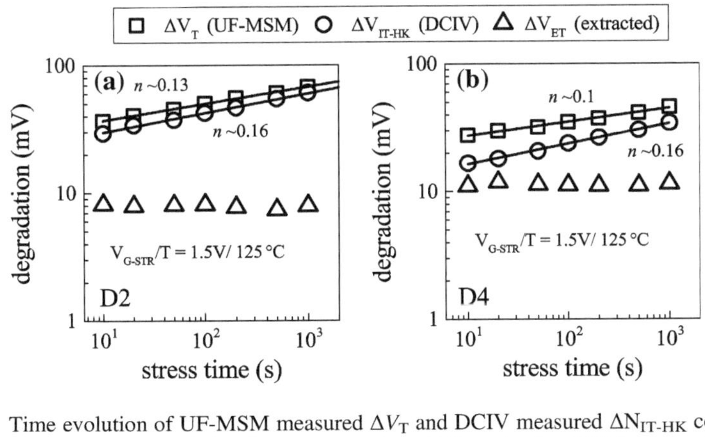

Fig. 3.21 Time evolution of UF-MSM measured ΔV_T and DCIV measured $\Delta N_{IT\text{-}HK}$ contribution after measurement delay and band gap correction ($\Delta V_{IT\text{-}HK}$), for PBTI stress in HKMG n-MOSFETs having different IL processes. Extracted difference (ΔV_{ET}) is also shown

device D4 as shown. Therefore, Fig. 3.21 clearly shows that underlying $\Delta N_{IT\text{-}HK}$ and ΔN_{ET} components of PBTI are clearly uncorrelated; i.e., while $\Delta N_{IT\text{-}HK}$ reduces, ΔN_{ET} increases for the D4 device, which can explain the measured reduction in ΔV_T and n. This aspect is further discussed in detail in Chap. 4.

As an additional proof, Fig. 3.22a plots the correlation of measured ΔV_T and $\Delta V_{IT\text{-}HK}$ for HKMG devices D2 and D4. As mentioned before, ΔV_T is obtained using UF-MSM method and $\Delta V_{IT\text{-}HK}$ using DCIV method after delay and band gap corrections. The 1:1 correlation line corresponding to zero electron trapping contribution is also plotted. Note that for a particular $\Delta V_{IT\text{-}HK}$, D2 device shows slightly higher ΔV_T than the 1:1 correlation line. However, much larger ΔV_T is observed for device D4, which is consistent with relatively larger ΔN_{ET} contribution for device D4 having N in the gate insulator stack. Larger magnitude of electron trapping during NBTI stress in gate insulators containing N has been analyzed using DFT simulations; refer to [17] for additional details.

Similar to hole trapping in NBTI, electron trapping during PBTI also occurs in pre-existing, process related gate insulator traps; while the hole traps are located in IL, electron traps are located in the High-K layer. As done in p-MOSFETs, flicker noise method can also be used to access the density of High-K electron traps in n-MOSFETs. Figure 3.22b shows measured pre-existing trap density in devices D2 and D4. Note that the nitrided device D4 shows higher trap density, which is consistent with higher ΔN_{ET} contribution shown in Figs. 3.21 and 3.22a, and is also consistent with DFT simulation results [17].

Similar to NBTI stress, time evolution of ΔV_T and $\Delta N_{IT\text{-}HK}$ during PBTI stress has been measured respectively using UF-MSM and DCIV methods for three sets of stress T, and for each T, three different stress E_{OX} have been used. The E_{OX} and T dependencies of ΔV_T at fixed t_{STR} are shown in Chap. 1, Fig. 1.27, and the same for ΔN_{IT} are shown earlier in Fig. 3.13. Measured ΔV_T, $\Delta V_{IT\text{-}HK}$ and their difference ΔV_{ET} at a particular t_{STR} for the HKMG device D4 is plotted versus stress E_{OX} at a

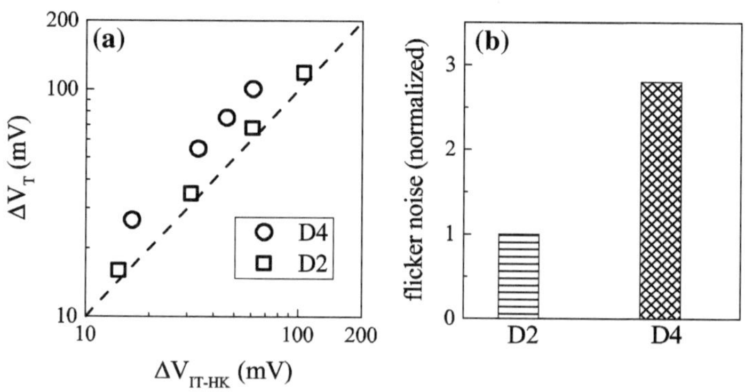

Fig. 3.22 **a** Correlation of UF-MSM measured ΔV_T and DCIV measured $\Delta V_{IT\text{-}HK}$ for PBTI stress, and **b** pre-stress trap density measured by using flicker noise method, in HKMG n-MOSFETs having different HKMG gate insulator processes

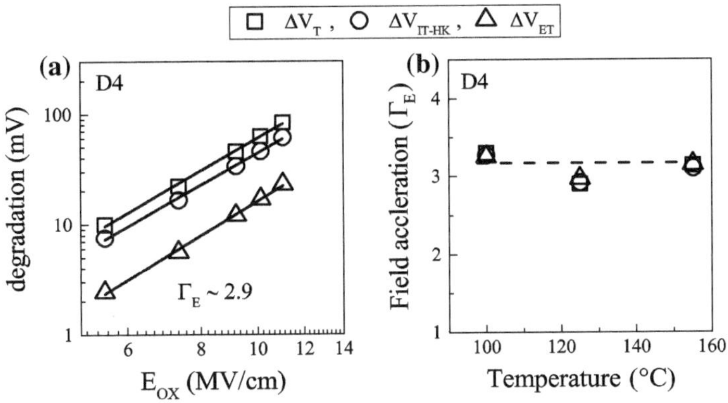

Fig. 3.23 **a** Fixed time UF-MSM measured ΔV_T, DCIV measured $\Delta V_{IT\text{-}HK}$ and their difference (ΔV_{ET}) versus PBTI stress E_{OX} plotted in a log–log scale. **b** Extracted power-law field acceleration factor for ΔV_T, ΔV_{IT} and ΔV_{ET} as a function of stress T

particular stress T in Fig. 3.23a, and also plotted versus stress T for a fixed stress E_{OX} in Fig. 3.24a. Note that $\Delta V_{IT\text{-}HK}$ subcomponent dominates overall ΔV_T for all E_{OX} and T, even for this particular device having N in the gate insulator stack. Moreover, identical power-law E_{OX} dependence Γ_E is obtained for ΔV_T, $\Delta V_{IT\text{-}HK}$ and therefore for ΔV_{ET}. However, ΔV_{ET} shows much lower Arrhenius T activation energy E_A compared to ΔV_T and $\Delta V_{IT\text{-}HK}$. Note that the relative values of Γ_E and E_A for ΔV_T, $\Delta V_{IT\text{-}HK}$ and ΔV_{ET} for PBTI stress are similar to that observed for NBTI stress data shown earlier in this chapter.

As shown before, ΔV_T and $\Delta N_{IT\text{-}HK}$ have power-law time dependence with identical n across different E_{OX} and T, although higher n has been observed for

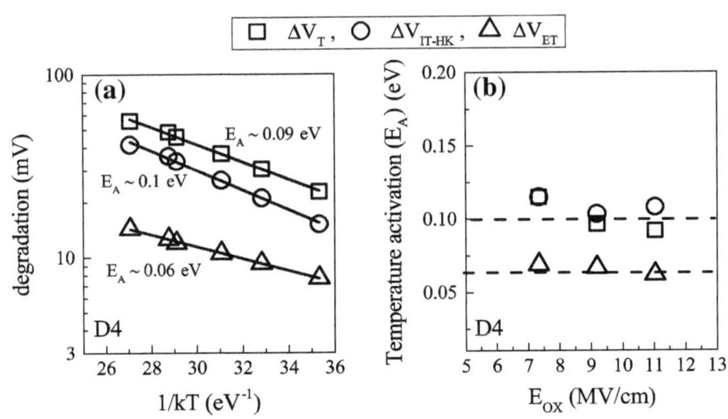

Fig. 3.24 **a** Fixed time UF-MSM measured ΔV_T, DCIV measured $\Delta V_{IT\text{-}HK}$ and their difference (ΔV_{ET}) versus PBTI stress T plotted in a semi-log scale. **b** Extracted T activation energy for ΔV_T, ΔV_{IT} and ΔV_{ET} as a function of stress E_{OX}

$\Delta N_{IT\text{-}HK}$ compared to ΔV_T data. Therefore, extracted Γ_E and E_A for ΔV_T, $\Delta V_{IT\text{-}HK}$ and therefore for ΔV_{ET} would be independent of the value of t_{STR} used for extracting E_{OX} and T dependence. Extracted Γ_E for ΔV_T and underlying $\Delta V_{IT\text{-}HK}$ and ΔV_{ET} subcomponents are plotted versus T in Fig. 3.23b, while their corresponding E_A values are plotted versus E_{OX} in Fig. 3.24b. Note that in the absence of different extraneous artifacts mentioned in Chap. 1, Sect. 1.3, Γ_E and E_A for ΔV_T and $\Delta N_{IT\text{-}HK}$ (or $\Delta V_{IT\text{-}HK}$) are shown to be independent of T and E_{OX}, respectively. Therefore, Γ_E and E_A of extracted ΔN_{ET} (or ΔV_{ET}) also show the same behavior as shown. These features of PBTI stress are exactly identical to NBTI, and similar mutually uncoupled Γ_E and E_A has been observed for other devices, not plotted here for brevity.

3.7 Location of Generated Traps (DCIV Measurements)

As discussed earlier in this chapter, although hole and electron trapping cannot be ignored respectively for NBTI and PBTI stress, trap generation plays the dominant role. DCIV technique has been used to access trap generation during BTI stress, and the resulting data show very similar time, bias and temperature dependence for NBTI and PBTI stress in different HKMG devices. This suggests very similar underlying physical mechanism of DCIV accessed trap generation for NBTI and PBTI. SILC is not observed for NBTI due to band alignment issues but is present for PBTI stress; although the magnitude of generated bulk traps that are responsible for SILC has been found to be smaller compared to that probed by DCIV, refer to Chap. 4 for details. DCIV accessed trap generation during NBTI and PBTI has been attributed to Si/IL and IL/High-K interfaces respectively by noting the presence or absence of degradation in g_m. Trap generation in the IL in any significant quantity is ruled out for PBTI stress due to the absence of Δg_m, while the presence of Δg_m suggests trap generation in IL for NBTI stress. The similarities of DCIV measured trap generation during NBTI and PBTI stress have been shown before, which suggests similarity of underlying physical mechanism. In this section, experimental proof is provided to bring out their differences, which will ascertain the difference in physical location. All DCIV data shown in this section are corrected for delay and band gap using the procedure described in Chap. 2.

Figure 3.25 plots the E_{OX} dependence of generated traps obtained at fixed t_{STR} for (a) NBTI and (b) PBTI stress in HKMG devices having relatively thicker (D1) and thinner (D2) IL, refer to Fig. 3.2. IL thickness has no impact for NBTI stress since generated traps are at the Si/IL interface. However, magnitude of generated traps increases for thinner IL device for PBTI stress. As discussed in Chap. 5, trap generation for NBTI stress in p-MOSFETs is caused due to tunneling of inversion layer holes into the Si–H bonds at Si/IL interface, and therefore, the magnitude is independent of IL thickness. However, trap generation at or near the IL/High-K interface for PBTI stress in n-MOSFETs occurs due to tunneling of inversion layer electrons, which increases with reduction in IL thickness, and therefore causes higher trap generation magnitude for thinner IL devices as shown.

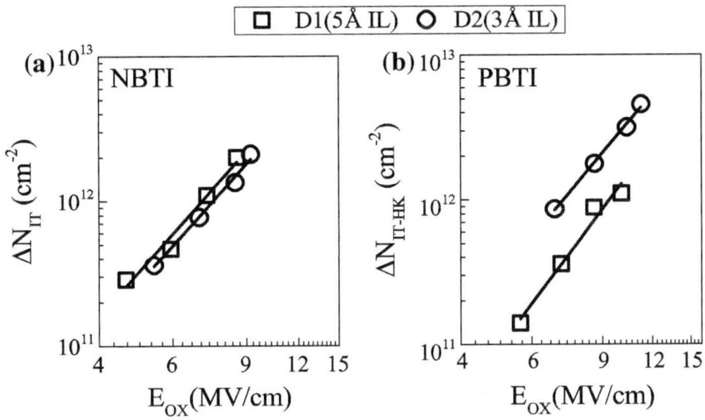

Fig. 3.25 DCIV measured trap generation versus E_{OX} for **a** NBTI and **b** PBTI stress in HKMG MOSFETs having different IL thickness

Figure 3.26 plots the E_{OX} dependence of generated traps obtained at fixed t_{STR} for (a) NBTI and (b) PBTI stress in HKMG devices without (D2) and with (D3) post High-K nitridation using the DPN process, refer to Fig. 3.2. Nitridation impact is negligible for NBTI stress, but trap generation reduces significantly for device D3 for PBTI stress. Since N incorporation is done after High-K deposition, higher N density is expected in the High-K compared to the IL layer, especially for denser thermal IL used in this study that show much lower N penetration, refer to [51] for additional evidence from Angle Resolved X-ray Photoelectron Spectroscopy measurements. Therefore, post High-K nitridation has larger impact on generated

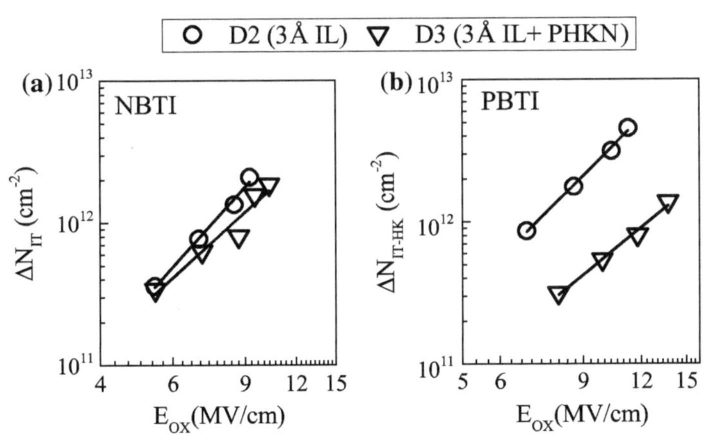

Fig. 3.26 DCIV measured trap generation versus E_{OX} for **a** NBTI and **b** PBTI stress in HKMG MOSFETs without and with nitridation after High-K deposition

traps at or near the IL/High-K interface for PBTI stress compared to that at Si/IL interface for NBTI stress. The exact physical mechanism responsible for reduction in trap generation for nitrided gate stacks is not known and need further study.

Figure 3.27 plots the E_{OX} dependence of generated traps obtained at fixed t_{STR} for (a) NBTI and (b) PBTI stress in HKMG devices D5 through D7 having different pre-cleans before IL growth and High-K deposition, refer to Fig. 3.2. Devices D5 and D6 have similar pre-clean before IL but different pre-clean before High-K and hence similar IL but different High-K quality. On the other hand, devices D6 and D7 have different pre-clean before IL but similar pre-clean before High-K and hence different IL but similar High-K quality. Therefore, trap generation at Si/IL interface for NBTI stress in p-MOSFETs is similar for D5 and D6 but different for D6 and D7 devices, while trap generation at or near the IL/High-K interface for PBTI stress in n-MOSFETs is different for D5 and D6 and similar for D6 and D7 devices as shown.

Although DCIV assessed generated traps for NBTI and PBTI have similar time dynamics and E_{OX} and T dependence, they have very different HKMG process dependence as shown in Figs. 3.25, 3.26 and 3.27 that clearly verifies the difference in physical location of these generated traps. As a further evidence, Fig. 3.28 correlates g_m degradation to generated trap density measured using DCIV for NBTI and PBTI stress. Note that NBTI shows higher Δg_m for a particular trap density when compared to PBTI stress. Since Δg_m is due to mobility degradation resulting from Coulomb scattering by BTI charges and as trap generation dominates both NBTI and PBTI degradation, higher Δg_m is caused by trap generation at Si/IL interface for NBTI than that at IL/High-K interface for PBTI.

Therefore, although DCIV measured traps during NBTI and PBTI stress respectively in p- and n-channel HKMG MOSFETs show very similar kinetics, they have very different physical location as discussed above.

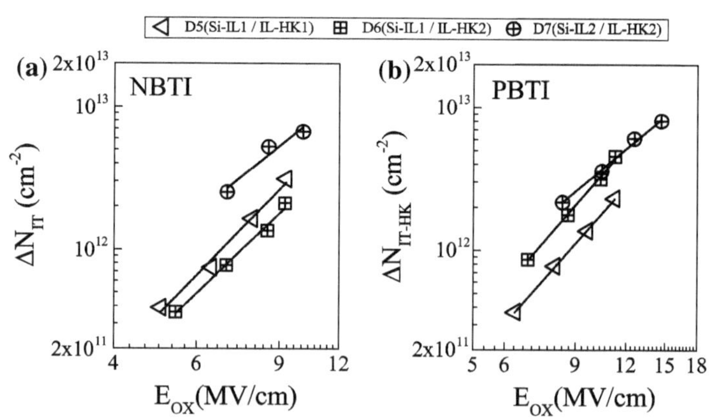

Fig. 3.27 DCIV measured trap generation versus E_{OX} for **a** NBTI and **b** PBTI stress in HKMG MOSFETs having different pre-clean processes before IL growth and High-K deposition. Refer to Fig. 3.2 for details

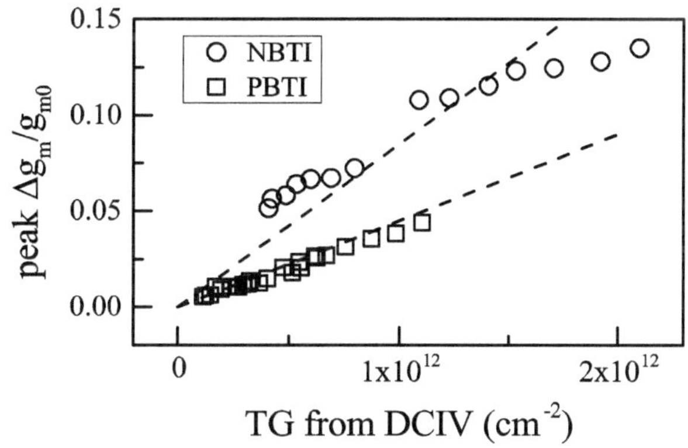

Fig. 3.28 Correlation of peak transconductance degradation to DCIV measured generation of traps for NBTI and PBTI stress

3.8 Summary

To summarize, the underlying subcomponents of NBTI and PBTI degradation in p- and n-channel HKMG MOSFETs respectively have been analyzed using different direct characterization techniques. MOSFETs having different HKMG gate insulator processes have been studied. Process related pre-existing hole and electron traps respectively in the IL and High-K layers is accessed using flicker noise. The generation of traps during NBTI is studied using DCIV method, while that during PBTI stress using DCIV and SILC methods. The relative contribution of trapping and trap generation on BTI degradation has been accessed. Although hole trapping in IL for NBTI and electron trapping in High-K for PBTI contribute, it is observed that trap generation dominates NBTI and PBTI degradation for different HKMG gate insulator processes.

DCIV measures trap generation at Si/IL interface during NBTI and at or near IL/High-K interface during PBTI, refer to Fig. 3.1. Energetically, these traps are aligned to ~0.3 eV around the Si band gap as shown in Chap. 2. Although locations of DCIV accessed generated traps are different, at Si/IL interface for NBTI and at or near IL/High-K interface for PBTI, they show very similar time, bias, temperature, AC duty cycle, and frequency dependence between NBTI and PBTI stress, suggesting strong universality of the underlying physical process. SILC measures trap generation energetically located closer to the HfO_2 conduction band, however, the exact physical location of these traps are still debated. Nevertheless for PBTI stress, SILC accessed generated traps show very different time, temperature, and bias dependence compared to that measured by DCIV and suggests very different underlying physical mechanism. SILC accessed traps have been found to have negligible role in PBTI degradation.

The following observations can be made between trap generation measured using DCIV, hole or electron trapping, and UF-MSM measured threshold voltage degradation for both NBTI and PBTI stress:

(a) Contribution from generated interface traps, ΔV_{IT} for NBTI or $\Delta V_{IT\text{-}HK}$ for PBTI, shows power-law time dependence at long stress time and universal exponent $n \sim 1/6$ for DC stress at different E_{OX} and T, for AC stress at different f and PDC, and for different HKMG processes. Since the trapping component, ΔV_{HT} for NBTI or ΔV_{ET} for PBTI, saturates at long time ($n \sim 0$), the relative magnitude of trap generation and trapping determines the time exponent ($n < 1/6$) of overall ΔV_T. Trap generation and trapping are completely uncorrelated to each other.

(b) Incorporation of N in the gate insulator has minimal impact on ΔV_{IT} for NBTI but significantly reduces $\Delta V_{IT\text{-}HK}$ for PBTI stress, while N increases both ΔV_{HT} and ΔV_{ET} for NBTI and PBTI stress, respectively. This unequivocally indicates that the trap generation and trapping subcomponents are mutually uncorrelated. Although ΔV_{IT} and $\Delta V_{IT\text{-}HK}$ dominate NBTI and PBTI respectively, a relatively larger ΔV_{HT} or ΔV_{ET} contribution results in reduction in n for overall ΔV_T in HKMG devices containing N in the gate stack.

(c) The power-law E_{OX} acceleration factor Γ_E has been found to be identical for ΔV_{IT} and ΔV_{HT} (or for $\Delta V_{IT\text{-}HK}$ and ΔV_{ET}) and therefore for ΔV_T for all HKMG processes, although Γ_E reduces for devices containing N in the gate stack or with reduction in IL thickness.

(d) The Arrhenius T activation energy E_A is larger for ΔV_{IT} (or $\Delta V_{IT\text{-}HK}$) compared to ΔV_{HT} (or ΔV_{ET}). Therefore, the relative magnitude of trap generation and trapping determines E_A for overall ΔV_T. HKMG processes having relatively larger trapping subcomponent show lower E_A for ΔV_T and vice versa.

(e) For AC stress, ΔV_{IT} (or $\Delta V_{IT\text{-}HK}$) remains independent of f but increases with increase in PDC of the gate pulse. However, the AC/DC ratio and PDC dependent shape for ΔV_{IT} (or $\Delta V_{IT\text{-}HK}$) is very different from that for ΔV_T, and will be explained later in the book.

Acknowledgment The authors would like to acknowledge Nilesh Goel for ultrafast NBTI measurements, Bijesh Rajamohanan for flicker noise measurements, Applied Materials for providing devices used for experiments and Ankush Chaudhary for editorial support.

References

1. N. Kimizuka, K. Yamaguchi, K. Imai, T. Iizuka, C.T. Liu, R.C. Keller, T. Horiuchi, NBTI enhancement by nitrogen incorporation into ultrathin gate oxide for 0.10-/spl mu/m gate CMOS generation, in *Symposium on VLSI Technology: Digest of Technical Papers* (2000), p. 92
2. Y. Mitani, M. Nagamine, H. Satake, A. Toriumi, NBTI mechanism in ultra-thin gate dielectric-nitrogen-originated mechanism in SiON, in *IEEE International Electron Devices Meeting Technical Digest* (2002), p. 509
3. S. Ramey, A. Ashutosh, C. Auth, J. Clifford, M. Hattendorf, J. Hicks, R. James, A. Rahman, V. Sharma, A. St.Amour, C. Wiegand, Intrinsic transistor reliability improvements from 22 nm tri-gate technology, in *IEEE International Reliability Physics Symposium Proceedings* (2013), p. 4C.5.1

4. W. McMahon, C. Tian, S. Uppal, H. Kothari, M. Jin, G. LaRosa, T. Nigam, A. Kerber, B. P. Linder, E. Cartier, W.L. Lai, Y. Liu, R. Ramachandran, U. Kwon, B. Parameshwaran, S. Krishnan, V. Narayanan, Intrinsic dielectric stack reliability of a high performance bulk planar 20 nm replacement gate high-k metal gate technology and comparison to 28 nm gate first high-k metal gate process, in *IEEE International Reliability Physics Symposium Proceedings* (2013), p. 4C.4.1
5. S. Mahapatra, N. Goel, S. Desai, S. Gupta, B. Jose, S. Mukhopadhyay, K. Joshi, A. Jain, A.E. Islam, M.A. Alam, A comparative study of different physics-based NBTI models. IEEE Trans. Electr. Dev. **60** (2013), 901
6. G. Groeseneken, H. Maes, N. Beltran, R. De Keersmaecker, A reliable approach to charge-pumping measurements in MOS transistors. IEEE Trans. Electr. Dev. **31** (1984), 42
7. A. Neugroschel, G. Bersuker, R. Choi, Applications of DCIV method to NBTI characterization. Microelectron. Reliab. **47**, 1366 (2007)
8. J.H. Stathis, G. LaRosa, A. Chou, Broad energy distribution of NBTI-induced interface states in p-MOSFETs with ultra-thin nitrided oxide, in *IEEE International Reliability Physics Symposium Proceedings* (2004), p. 25
9. Y. Mitani, Influence of nitrogen in ultra-thin SiON on negative bias temperature instability under AC stress, in *IEEE International Electron Devices Meeting Technical Digest* (2004), p. 117
10. D. Varghese, D. Saha, S. Mahapatra, K. Ahmed, F. Nouri, M.A. Alam, On the dispersive versus Arrhenius temperature activation of NBTI time evolution in plasma nitrided gate oxides: measurements, theory, and implications, in *IEEE International Electron Devices Meeting Technical Digest* (2005), p. 684
11. A.T. Krishnan, C. Chancellor, S. Chakravarthi, P.E. Nicollian, V. Reddy, A. Varghese, R.B. Khamankar, S. Krishnan, Material dependence of hydrogen diffusion: implications for NBTI degradation, in *IEEE International Electron Devices Meeting Technical Digest* (2005), p. 688
12. J.P. Campbell, P.M. Lenahan, A.T. Krishnan, S. Krishnan, NBTI: an atomic-scale defect perspective, in *IEEE International Reliability Physics Symposium Proceedings* (2006), p. 442
13. V. Huard, Two independent components modeling for negative bias temperature instability, in *IEEE International Reliability Physics Symposium Proceedings* (2010), p. 33
14. S. Mahapatra, K. Ahmed, D. Varghese, A.E. Islam, G. Gupta, L. Madhav, D. Saha, M.A. Alam, On the physical mechanism of NBTI in silicon oxynitride p-MOSFETs: can differences in insulator processing conditions resolve the interface trap generation versus hole trapping controversy?, in *IEEE International Reliability Physics Symposium Proceedings* (2007), p. 1
15. S. Mahapatra, A. Islam, S. Deora, V. Maheta, K. Joshi, A. Jain, M. Alam, A critical re-evaluation of the usefulness of R-D framework in predicting NBTI stress and recovery, in *IEEE International Reliability Physics Symposium Proceedings* (2011), p. 6A.3.1
16. K. Joshi, S. Mukhopadhyay, N. Goel, S. Mahapatra, A consistent physical framework for N and P BTI in HKMG MOSFETs, in *IEEE International Reliability Physics Symposium Proceedings* (2012), p. 5A.3.1
17. S. Mukhopadhyay, K. Joshi, V. Chaudhary, N. Goel, S. De, R.K. Pandey, K.V. Murali, S. Mahapatra, Trap generation in IL and HK layers during BTI/TDDB stress in scaled HKMG N and P MOSFETs, in *IEEE International Reliability Physics Symposium Proceedings* (2014), p. GD.3.1
18. S. Mahapatra, V. Huard, A. Kerber, V. Reddy, S. Kalpat, A. Haggag, Universality of NBTI-from devices to circuits and products, in *IEEE International Reliability Physics Symposium Proceedings* (2014), p. 3B.1.1
19. D.S. Ang, S. Wang, Recovery of the NBTI-stressed ultrathin gate p-MOSFET: the role of deep-level hole traps. IEEE Electr. Dev. Lett. **27**, 914 (2006)

20. C. Shen, M.F. Li, C.E. Foo, T. Yang, D.M. Huang, A. Yap, G.S. Samudra, Y.C. Yeo, Characterization and physical origin of fast Vth transient in NBTI of pMOSFETs with SiON dielectric, in *IEEE International Electron Devices Meeting Technical Digest* (2006), p. 12.5.1
21. T. Grasser, B. Kaczer, W. Goes, An energy-level perspective of bias temperature instability, in *IEEE International Reliability Physics Symposium Proceedings* (2008), p. 28
22. D. Ielmini, M. Manigrasso, F. Gattel, G. Valentini, A unified model for permanent and recoverable NBTI based on hole trapping and structure relaxation, in *IEEE International Reliability Physics Symposium Proceedings* (2009), p. 26
23. H. Reisinger, T. Grasser, K. Ermisch, H. Nielen, W. Gustin, C. Schlunder, Understanding and modeling AC BTI, in *IEEE International Reliability Physics Symposium Proceedings* (2011), p. 6A.1.1
24. E.N. Kumar, V.D. Maheta, S. Purawat, A.E. Islam, C. Olsen, K. Ahmed, M. Alam, S. Mahapatra, Material dependence of NBTI physical mechanism in silicon oxynitride (SiON) p-MOSFETs: a comprehensive study by ultrafast on-the-fly (UF-OTF) I_{DLIN} technique, in *IEEE International Electron Devices Meeting Technical Digest* (2007), p. 809
25. V.D. Maheta, C. Olsen, K. Ahmed, S. Mahapatra, The impact of nitrogen engineering in silicon oxynitride gate dielectric on negative-bias temperature instability of p-MOSFETs: a study by ultrafast on-the-fly technique. IEEE Trans. Electr. Dev. **55**, 1630 (2008)
26. T. Grasser, B. Kaczer, W. Goes, T. Aichinger, P. Hehenberger, M. Nelhiebel, A two stage model for negative bias temperature instability, in *IEEE International Reliability Physics Symposium Proceedings* (2009), p. 33
27. H. Reisinger, O. Blank, W. Heinrigs, A. Muhlhoff, W. Gustin, C. Schlunder, Analysis of NBTI degradation-and recovery-behavior based on ultra fast VT-measurements, in *IEEE International Reliability Physics Symposium Proceedings* (2006), p. 448
28. K. Sakuma, D. Matsushita, K. Muraoka, Y. Mitani, Investigation of nitrogen-originated NBTI mechanism in SiON with high-nitrogen concentration, in *IEEE International Reliability Physics Symposium Proceedings* (2006), p. 454
29. Y. Mitani, H. Satake, A. Toriumi, Influence of nitrogen on negative bias temperature instability in ultrathin SiON. IEEE Trans. Dev. Mater. Reliab. **8**, 6 (2008)
30. S. Mahapatra, V.D. Maheta, A.E. Islam, M.A. Alam, Isolation of NBTI stress generated interface trap and hole-trapping components in PNO p-MOSFETs. IEEE Trans. Electr. Dev. **56**, 236 (2009)
31. S. Desai, S. Mukhopadhyay, N. Goel, N. Nanaware, B. Jose, K. Joshi, S. Mahapatra, A comprehensive AC/DC NBTI model: stress, recovery, frequency, duty cycle and process dependence, in *IEEE International Reliability Physics Symposium Proceedings* (2013), p. XT.2.1
32. K. Joshi, S. Hung, S. Mukhopadhyay, V. Chaudhary, N. Nanaware, B. Rajamohanan, T. Sato, M. Bevan, A. Wei, A. Noori, B. McDougall, C. Ni, C. Lazik, G. Saheli, P. Liu, D. Chu, L. Date, S. Datta, A. Brand, J. Swenberg, S. Mahapatra, HKMG process impact on N, P BTI: role of thermal IL scaling, IL/HK integration and post HK nitridation, in *IEEE International Reliability Physics Symposium Proceedings* (2013), p. 4C.2.1
33. N. Goel, K. Joshi, S. Mukhopadhyay, N. Nanaware, S. Mahapatra, A comprehensive modeling framework for gate stack process dependence of DC and AC NBTI in SiON and HKMG p-MOSFETs. Microelectron. Reliab. **54**, 491 (2014)
34. N. Goel, S. Mukhopadhyay, N. Nanaware, S. De, R.K. Pandey, K.V.R.M. Murali, S. Mahapatra, A comprehensive DC/AC model for ultra-fast NBTI in deep EOT scaled HKMG p-MOSFETs, in *IEEE International Reliability Physics Symposium Proceedings* (2014), p. 6A.4.1
35. N. Goel, T. Naphade, S. Mahapatra, Combined trap generation and transient trap occupancy model for time evolution of NBTI during DC multi-cycle and AC stress, in *IEEE International Reliability Physics Symposium Proceedings* (2015)
36. A. Kerber, E. Cartier, Reliability challenges for CMOS Technology qualifications with hafnium oxide/titanium nitride gate stacks. IEEE Trans. Device Mater. Reliab. **9**, 147 (2009)

37. J. Mitard, X. Garros, L.P. Nguyen, C. Leroux, G. Ghibaudo, F. Martin, G. Reimbold, Large-scale time characterization and analysis of PBTI In HfO_2/metal gate stacks, in *IEEE International Reliability Physics Symposium Proceedings* (2006), p. 174
38. D. Heh, C.D. Young, G. Bersuker, Experimental evidence of the fast and slow charge trapping/detrapping processes in high-k dielectrics subjected to PBTI stress. IEEE Electr. Dev. Lett. **29**, 180 (2008)
39. S. Pae, M. Agostinelli, M. Brazier, R. Chau, G. Dewey, T. Ghani, M. Hattendorf, J. Hicks, J. Kavalieros, K. Kuhn, M. Kuhn, J. Maiz, M. Metz, K. Mistry, C. Prasad, S. Ramey, A. Roskowski, J. Sandford, C. Thomas, J. Thomas, C. Wiegand, J. Wiedemer, BTI reliability of 45 nm high-K+ metal-gate process technology, in *IEEE International Reliability Physics Symposium Proceedings* (2008), p. 352
40. J. Yang, M. Masuduzzaman, K. Joshi, S. Mukhopadhyay, J. Kang, S. Mahapatra, M.A. Alam, Intrinsic correlation between PBTI and TDDB degradations in nMOS HK/MG dielectrics, in *IEEE International Reliability Physics Symposium Proceedings* (2012), p. 5D.4.1
41. E. Cartier, A. Kerber, Stress-induced leakage current and defect generation in nFETs with HfO_2/TiN gate stacks during positive-bias temperature stress, in *IEEE International Reliability Physics Symposium Proceedings* (2009), p. 486
42. S. Zafar, A. Kerber, R. Muralidhar, Physics based PBTI model for accelerated estimation of 10 year lifetime. In *Symposium on VLSI Technology: Digest of Technical Papers* (2014). doi: 10.1109/VLSIT.2014.6894388
43. W. Liu, G. La Rosa, G. Tian, S. Boffoli, F. Guarin, W.L. Lai, V. Narayanan, H. Kothari, M. Jin, S. Uppal, W. McMahon, Process dependence of AC/DC PBTI in HKMG n-MOSFETs, in *IEEE International Reliability Physics Symposium Proceedings* (2014), p. XT.6.1
44. D.P. Ioannou, E. Cartier, Y. Wang, S. Mittl, PBTI response to interfacial layer thickness variation in Hf-based HKMG nFETs, in *IEEE International Reliability Physics Symposium Proceedings* (2010), p. 1044
45. E. Cartier, A. Kerber, T. Ando, M.M. Frank, K. Choi, S. Krishnan, B. Linder, K. Zhao, F. Monsieur, J. Stathis, V. Narayanan, Fundamental aspects of HfO_2-based high-k metal gate stack reliability and implications on tinv-scaling, in *IEEE International Electron Devices Meeting Technical Digest* (2011), p. 18.4.1
46. M.J. Bevan, R. Curtis, T. Guarini, W. Liu, S.C.H. Hung, H. Graoui, Ultrathin SiO_2 interface layer growth, in *International Conference on Advanced Thermal Processing of Semiconductors (RTP),* (2010), p. 154
47. E.P. Gusev, C. Cabral Jr, M. Copel, C. D'Emic, M. Gribelyuk, Ultrathin HfO_2 films grown on silicon by atomic layer deposition for advanced gate dielectrics applications. Microelectron. Eng. **69**, 145 (2003)
48. P.A. Kraus, K.Z. Ahmed, C.S. Olsen, F. Nouri, Physical models for predicting plasma nitrided Si-O-N gate dielectric properties from physical metrology. IEEE Electr. Dev. Lett. **24**, 559 (2003)
49. C. Olsen, Two-step post nitridation annealing for lower EOT plasma nitrided gate dielectrics. WO2004081984 A2, (2004)
50. G. Kapila, N. Goyal, V.D. Maheta, C. Olsen, K. Ahmed, S. Mahapatra, A comprehensive study of flicker noise in plasma nitride SiON p-MOSFETs: process dependence of pre-existing and NBTI stress generated trap distribution profiles. In *IEEE International Electron Devices Meeting Technical Digest* (2008). doi:10.1109/IEDM.2008.4796625
51. S. Mahapatra, S. De, K. Joshi, S. Mukhopadhyay, R.K. Pandey, K.V.R.M. Murali, Understanding process impact of hole traps and NBTI in HKMG p-MOSFETs using measurements and atomistic simulations. IEEE Electr. Dev. Lett. **34**, 963 (2013)
52. S. Takagi, N. Yasuda, A. Toriumi, Experimental evidence of inelastic tunneling in stress-induced leakage current. IEEE Trans. Electr. Dev. **46**, 335 (1999)
53. S. Takagi, N. Yasuda, A. Toriumi, A new I–V model for stress-induced leakage current including inelastic tunneling. IEEE Trans. Electr. Dev. **46**, 348 (1999)

54. M.A. Alam, SILC as a measure of trap generation and predictor of TBD in ultrathin oxides. IEEE Trans. Electr. Dev. **49**, 226 (2002)
55. S. Mahapatra, P. Bharath Kumar, M.A. Alam, Investigation and modeling of interface and bulk trap generation during negative bias temperature instability of p-MOSFETs". IEEE Trans. Electr. Dev. **51**, 1371 (2004)
56. M. Masuduzzaman, A.E. Islam, M.A. Alam, A multi-probe correlated bulk defect characterization scheme for ultra-thin high-κ dielectric, in *IEEE International Reliability Physics Symposium Proceedings* (2010), p. 1069

Chapter 4
Physical Mechanism of BTI Degradation—Modeling of Process and Material Dependence

Souvik Mahapatra, Kaustubh Joshi, Subhadeep Mukhopadhyay, Ankush Chaudhary and Nilesh Goel

Abstract This chapter proposes compact models to predict NBTI and PBTI degradation in p- and n-channel MOSFETs, respectively. Models are based on the physics of BTI degradation discussed in Chap. 3. Compact NBTI model has been verified against measurements in different SiON and HKMG gate insulator process-based planar Si devices, planar HKMG devices having SiGe channel without and with Si cap, and Si channel based HKMG FinFETs. The model can predict degradation in such wide variety of devices with a maximum of five, but in most cases just three, device dependent adjustable parameters. PBTI compact model has been verified to predict degradation in different HKMG gate insulator process-based planar Si devices with six device dependent parameters. Models for both NBTI and PBTI have three underlying subcomponents related to interface and bulk trap generation and charge trapping, overall degradation being the uncorrelated sum of these subcomponents. All subcomponents are independently verified using different measurement techniques, such as CP, DCIV, SILC, and flicker noise, discussed in Chap. 2.

4.1 Introduction

It has been discussed in Chap. 1 that Negative Bias Temperature Instability (NBTI) and Positive Bias Temperature Instability (PBTI) are well-known reliability issues respectively for p- and n-channel Metal Oxide Semiconductor Field Effect Transistors (MOSFETs) . NBTI has become an important issue as the industry migrated from conventional Silicon Dioxide (SiO_2) to Silicon Oxynitride (SiON)

S. Mahapatra (✉) · K. Joshi · S. Mukhopadhyay · A. Chaudhary · N. Goel
Department of Electrical Engineering, Indian Institute of Technology Bombay, Mumbai 400076, Maharashtra, India
e-mail: souvik@ee.iitb.ac.in; mahapatra.souvik@gmail.com

S. Mahapatra (ed.), *Fundamentals of Bias Temperature Instability in MOS Transistors*, Springer Series in Advanced Microelectronics 139,
DOI 10.1007/978-81-322-2508-9_4

gate insulator technology, and continues to remain as a concern for state-of-the-art High K Metal Gate (HKMG) planar and FinFET technologies. PBTI was never a serious concern for SiO_2 and SiON MOSFETs and only became important with the advent of HKMG technology, although PBTI is now perceived to be less of a concern for FinFET devices for sub 22 nm technology nodes. Refer to Chap. 1 for a brief overview of NBTI and PBTI degradation across different technology nodes.

It is now well known that BTI degradation is strongly impacted by the gate insulator process, as discussed in Chaps. 1 and 3 and also in this chapter. Therefore, process optimization is an important exercise to develop BTI resistant technologies. Note that BTI measurements are usually performed at accelerated stress conditions for short time, and it is important to develop suitable models to calculate the degradation at end-of-life (EOL) under normal operating condition. This is the primary objective of a reliability model from the perspective of technology qualification. It is important to note that a good model for any reliability phenomenon must have the following characteristics:

(a) It should adhere to the physics of the degradation mechanism.
(b) It should be able to predict degradation for differently processed devices.
(c) It should contain least possible fitting parameters so as to keep the model compact and simple to implement.
(d) It should be able to predict EOL degradation at use conditions to determine reliable operating lifetime.

Various models have been proposed over the years to explain the mechanism of NBTI and PBTI degradation respectively in p- and n-channel MOSFETs, and are briefly reviewed in Chap. 3. Note that there exist strong experimental evidences of interface trap generation (ΔN_{IT}) during NBTI stress, obtained from different independent measurement methods such as Charge Pumping (CP) [1–5], Gated Diode (DCIV) [5–13] and Low Voltage Stress Induced Leakage Current (LV-SILC) [7, 14], refer to Chaps. 2 and 3 for further details. However in spite of such irrefutable evidences, some authors have proposed hole trapping in pre-existing traps (ΔN_{HT}) as the sole NBTI physical mechanism [15–17], which is definitely not true. However, note that although trap generation is the dominant physical mechanism and is primarily responsible for EOL degradation for well-optimized devices and circuits [18], the contribution due to hole trapping should also be considered especially to model short-time degradation measured using ultra-fast methods [4, 10, 11, 12]. It is now well accepted that threshold voltage shift (ΔV_T) during NBTI stress is due to both ΔN_{IT} and ΔN_{HT}; although some authors have suggested strong coupling between the two [19], most evidences suggest they are mutually uncorrelated [1, 2, 4, 5, 10, 11, 12, 20, 21, 22], which will be discussed in detail and verified later in this chapter. In addition to ΔN_{IT} and ΔN_{HT}, contribution from TDDB like bulk trap generation (ΔN_{OT}) can also impact ΔV_T especially in thick gate insulators and high stress voltage ($V_{G\text{-}STR}$) [5, 23], and is also discussed later in this chapter. Different physical models for NBTI have been recently reviewed; refer to [10] for details.

Figure 4.1 illustrates different uncorrelated components that contribute to NBTI degradation in HKMG p-MOSFETs having SiO_2 or SiON interlayer (IL) and Hafnium Oxide (HfO_2) High-K layer [10], refer to Chap. 3 for details. Note that NBTI results in charging of the IL, primarily due to ΔN_{IT} at the channel and IL interface and also due to ΔN_{HT} in IL bulk . As discussed in detail in Chap. 2, CP and DCIV techniques can be used to independently estimate ΔN_{IT}. Flicker noise can be used to estimate pre-existing defects that give rise to ΔN_{HT}. Although ΔN_{OT} can be appreciable for thicker SiO_2 and SiON devices especially at higher $V_{G\text{-}STR}$ and can be independently estimated using conventional SILC as discussed in Chap. 2, it has relatively lower contribution in HKMG devices as shown later in this chapter [5, 10]. It is important to note that SILC is usually not observed for NBTI stress in HKMG p-MOSFETs due to band alignment issues [24], refer to Chap. 2, Fig. 2.33 for details.

As discussed in Chap. 1, a crucial difference between NBTI and PBTI is presence of transconductance degradation (Δg_m) for the former and its absence for the later, especially for well-optimized HKMG devices [25]. Historically, PBTI degradation has been assumed to be dominated by electron trapping (ΔN_{ET}) in High-K layer for early non-optimized gate insulator processes [26, 27]. Although electron trapping reduces with HKMG process improvement [25] and reduction in High-K layer thickness [28], some authors still propose ΔN_{ET} to be the sole contributor to ΔV_T [29], ignoring strong evidence of trap generation measured using independent methods such as SILC and DCIV [30–34]. In the past, PBTI has been correlated to trap generation measured using conventional SILC technique [30, 31]. However, it has been recently shown that stress time (t_{STR}), temperature (T) and $V_{G\text{-}STR}$ dependence of PBTI ΔV_T closely resembles the dependence of trap generation measured using DCIV and not SILC [13, 34], thereby suggesting the former to be primarily responsible for PBTI degradation in scaled and optimized HKMG stacks. As also illustrated in Fig. 4.1, it has been shown that PBTI degradation is due to uncorrelated contribution from trap generation at the interface between IL and

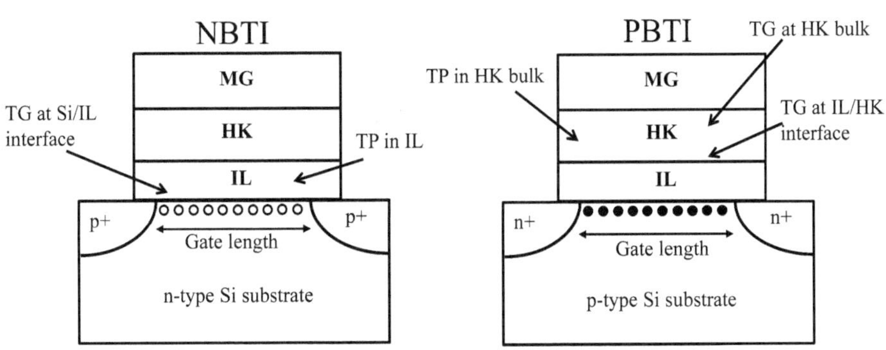

Fig. 4.1 Schematic of p- and n-channel HKMG MOSFETs showing different trap generation (TG) and trapping (TP) processes under NBTI and PBTI stress

High-K layers ($\Delta N_{IT\text{-}HK}$) and in High-K bulk ($\Delta N_{OT\text{-}HK}$), together with ΔN_{ET} in pre-existing bulk High-K defects [32–34], refer to Chap. 3 for further details.

In this chapter, compact models are developed for NBTI and PBTI degradation and validated respectively in differently processed SiON and HKMG p-MOSFETs and HKMG n-MOSFETs. The models are based on physical mechanism of BTI as discussed above and therefore have uncorrelated trap generation and trapping subcomponents, which are independently verified by different measurement methods. The impact of gate insulator processes on model parameters and underlying subcomponents are discussed in detail, which helps explain the material and process dependence of BTI degradation. The methodology developed in this chapter can be used to predict process impact of EOL degradation, which in-turn can be used to develop suitable gate insulator processes for BTI resistant technologies.

4.2 Compact Model and Device Description

4.2.1 NBTI Model

Table 4.1 lists the compact model to predict NBTI degradation at long t_{STR} values of ~seconds and higher. The model is based on results from NBTI studies that are summarized in Fig. 4.1. According to the model, at any given stress time $t = t_{STR}$, overall NBTI threshold voltage shift $\Delta V_T\ (t)$ is the uncorrelated sum of the following independent subcomponents [10, 32]:

(a) Interface trap generation, $\Delta V_{IT}(t) = q \cdot \Delta N_{IT}(t)/C_{OX}$ at device channel and IL interface.
(b) Hole trapping, $\Delta V_{HT}(t) = q \cdot \Delta N_{HT}(t)/C_{OX}$ in pre-existing, process related IL bulk traps.
c) Trap generation in IL bulk, $\Delta V_{OT}\ (t) = q \cdot \Delta N_{OT}(t)/C_{OX}$.

In the above expressions, q is the electronic charge and C_{OX} is oxide capacitance. Note that although the physical location of ΔN_{HT} and ΔN_{OT} are in the IL bulk, an effective interfacial density is assumed in the proposed model to calculate their respective contribution to overall ΔV_T. Also note that the model is equally applicable for both SiON and HKMG devices; the degradation will happen at the channel/SiON interface and SiON bulk for conventional SiON devices [10].

The interface trap generation component ΔN_{IT} is explained by the well-known Reaction Diffusion or RD model [10], refer to Chap. 5 for details. The model states that inversion layer holes tunnel into and subsequently break Si–H bonds[1] at the

[1]Note that RD model does not specify the exact type of bonds broken during NBTI stress. It assumes that there exists *H* passivated defects (*X–H*) at (or near) the Si channel and oxide interface that get broken during stress. The broken bonds (*X*−) show up as interface traps.

Table 4.1 NBTI compact model and fixed and device dependent model parameters

Interface trap generation	Hole trapping	Bulk trap generation
$\Delta N_{IT} = AE_{OX}e^{\Gamma_{IT}E_{OX}}e^{-\frac{E_{AIT}}{kT}}t^{1/6}$	$\Delta N_{HT} = BE_{OX}e^{\Gamma_{HT}E_{OX}}e^{-\frac{E_{AHT}}{kT}}$	$\Delta N_{OT} = C\left(1 - e^{-(t/w)\beta_{OT}}\right)$
$E_{AIT} = \frac{2}{3}(E_{AKF} - E_{AKR}) + \frac{E_{ADH2}}{6}$		$w = \eta(V_G - V_{T0} - \Delta V_T)^{-\frac{\Gamma_{OT}}{\beta_{OT}}}e^{\frac{E_{AOT}}{kT\beta_{OT}}}$
$E_{OX} = (V_{G-STR} - V_{T0} - \Delta V_T)/EOT$		
$\Delta V_T = \frac{q}{C_{OX}} * (\Delta N_{IT} + \Delta N_{HT} + \Delta N_{OT})$		
A, B, C, $\Gamma_{IT}(=\Gamma_{HT})$ and E_{AKF} are variable across devices		
Fixed parameters		
E_{AKR} = 0.2 eV	$E_{AHT} = 0.052\,eV$	$E_{AOT} = 0.16\,eV; \beta_{OT} = 0.33$
E_{ADH2} = 0.2 eV		$\Gamma_{OT} = 10; \eta = 5 \times 10^{12}$

Fixed parameters are constant across different type of SiON and HKMG gate insulator stacks, and different channel type and orientation of devices used in this study

MOSFET channel and IL interface to generate N_{IT}. The broken Si–H bonds release hydrogen (H) atoms that eventually diffuse away as molecular hydrogen (H_2) species; the diffusion of molecular H_2 determines long-time kinetics of N_{IT} generation. To achieve compactness, only the long-time solution of ΔN_{IT} is modeled as per RD model, which predicts power-law time dependence with time exponent (n) of 1/6. This assumption is consistent with measurement data shown in Chap. 3 and is also validated later in this chapter. ΔN_{IT} depends on the oxide field (E_{OX}) as $E_{OX} * \exp \cdot (\Gamma_{IT} \cdot E_{OX})$ with field acceleration factor Γ_{IT}, the linear and exponential terms representing inversion hole density and its tunneling, respectively. Arrhenius T activation is used with energy E_{AIT}, and A is a process dependent pre-factor. Although it is suffice to model the T activation with E_{AIT} as model parameter, the overall T activation is further divided using T activation corresponding to Si–H bond breaking (E_{AKF}) and passivation (E_{AKR}) and molecular H_2 diffusion (E_{ADH2}), to keep it closer to the actual physical mechanism; refer to Chaps. 5 and 6 for further details.

Note that the oxide field E_{OX} is calculated as ($V_{G\text{-}STR} - V_{T0} - \Delta V_T$)/EOT, where EOT is equivalent oxide thickness. Of particular interest is the appearance of ΔV_T in the expression, which is included to account for E_{OX} reduction effect at higher degradation. Although the insertion of this term makes the expression less compact, since ΔV_T needs to be calculated using iterative approach, this is a very important effect that governs the time exponent n at long t_{STR}, especially for situations where ΔV_T is appreciably high. Since ΔV_T has exponential E_{OX} dependence, reduction in E_{OX} at longer t_{STR} would cause a soft-saturation effect and as a consequence result in relatively lower n than expected. This aspect is also discussed in Chap. 1, and is used for both NBTI and PBTI compact model. It is also used in the composite modeling framework discussed in Chap. 6.

The hole-trapping component (ΔN_{HT}) in pre-existing traps can be calculated by the Two-Energy Well (2EW) model [11, 35], which predicts trapping saturation at long t_{STR}. Since the compact model is used for long t_{STR}, a fixed ΔV_{HT} is used with process dependent pre-factor B, field acceleration parameter Γ_{HT} and Arrhenius

T activation E_{AHT}. Note that hole trapping is presumed to be having weaker T activation compared to trap generation, which will be validated in this chapter. However, identical field acceleration is used for trap generation and trapping [10, 36].

The third component is generation of new bulk traps, ΔN_{OT}, which is negligible for HKMG devices unless very high $V_{G\text{-}STR}$ is used. However, it is appreciable for thicker SiO_2 and SiON devices and when stressed at high $V_{G\text{-}STR}$. Time evolution of ΔN_{OT} has a stretched exponential form based on theory for bulk trap generation [37], with Arrhenius temperature activation E_{AOT} and power-law voltage acceleration factor Γ_{OT} [5]. Note that unlike ΔN_{IT} that is dependent on oxide field, ΔN_{OT} is dependent on gate bias, the latter having much stronger bias dependence than the former [5, 23]. This component has a process dependent pre-factor C.

The pre-factors A, B, and C, field acceleration factor Γ_{IT} and temperature activation for interface trap generation component E_{AIT}, through variation in E_{AKF}, are process dependent and are used as fitting parameters to predict degradation in different SiON and HKMG p-MOSFETs. Identical values are used for Γ_{IT} and Γ_{HT}, so Γ_{HT} is not an independent parameter. All other parameters are constant and are shown in Table 4.1. Although the model in general has five free parameters, only four are needed for HKMG devices since ΔN_{OT} is negligible. Moreover, it will be shown that E_{AKF} remains almost constant unless IL process is drastically changed. Therefore, most HKMG devices can be successfully modeled with only three process dependent free parameters. Finally, CP and DCIV techniques have been used to independently estimate trap generation in SiON and HKMG devices respectively, while flicker noise is used to estimate pre-stress process related defects [10, 13, 33]. The compact model subcomponents are verified against these independent estimates to ascertain their correctness.

4.2.2 PBTI Model

A compact model to predict PBTI degradation is also developed along the same lines of NBTI model. This model is based on physical mechanism of PBTI summarized in Fig. 4.1. According to the model, overall PBTI degradation $\Delta V_T(t)$ at a time $t = t_{STR}$ is the sum of the following uncorrelated components [32–34]:

(a) Interface trap generation, $\Delta V_{IT\text{-}HK}(t) = q \cdot \Delta N_{IT\text{-}HK}(t)/C_{OX}$, at the interface between IL and High-K layers.
(b) Electron trapping, $\Delta V_{ET}(t) = q \cdot \Delta N_{ET}(t)/C_{OX}$ in pre-existing process related High-K bulk traps.
(c) Bulk trap generation in High-K layer, $\Delta V_{OT\text{-}HK}(t) = q \cdot \Delta N_{OT\text{-}HK}(t)/C_{OX}$.

In the above expressions, q is the electronic charge and C_{OX} is oxide capacitance. Note that PBTI stress in optimized HKMG devices does not result in charging of the IL layer, which is evident from negligible Δg_m [25]. It is important to remark that although the physical location of defects are at the IL/High-K interface and High-K bulk, for modeling purpose, an effective density is assumed at the channel/IL interface to compute their contribution to overall ΔV_T.

As shown in Chap. 3, trap generation measured using DCIV for NBTI and PBTI stress respectively in p- and n-channel HKMG MOSFETs demonstrates remarkably similar stress and recovery kinetics [13]. Therefore, interface generation component ($\Delta N_{IT\text{-}HK}$) under PBTI stress can also be modeled using the RD framework. Since IL degradation is negligible for PBTI stress, it is presumed that H passivated defects at the IL/High-K interface are broken by tunneling of inversion layer electrons , and released H atoms eventually diffuse as molecular H_2 species resulting in power-law time evolution of $\Delta N_{IT\text{-}HK}$ at long t_{STR} with exponent n of 1/6. The electron trapping component ΔN_{ET} is assumed to saturate at long t_{STR}, which is similar to the ΔN_{HT} component for NBTI stress; this assumption will be validated later in this chapter. Both interface trap generation and electron trapping are assumed to be field dependent phenomena and similar expressions have been used for NBTI and PBTI stress. However unlike NBTI, the field acceleration factor for interface trap generation ($\Gamma_{IT\text{-}HK}$) is not equal to that for electron trapping (Γ_{ET}). Both components have Arrhenius T activation, and once again, the T activation of electron trapping (E_{AET}) is assumed to be weaker than T activation for trap generation ($E_{AIT\text{-}HK}$), which will also be validated later in this chapter. Finally, the time evolution of bulk trap generation component ($\Delta N_{OT\text{-}HK}$) uses a stretched exponential form, has Arrhenius T activation ($E_{AOT\text{-}HK}$) and power-law $V_{G\text{-}STR}$ dependence ($\Gamma_{OT\text{-}HK}$) [31]. Although the bulk trap generation component has been found to be non-negligible for PBTI stress, however, due to stronger voltage acceleration factor its contribution becomes minimal at use condition.

The pre-factors for interface trap generation (A), electron trapping (B) and bulk trap generation (C), as well as $\Gamma_{IT\text{-}HK}$, Γ_{ET} and $E_{AIT\text{-}HK}$ are used as fitting parameters across different HKMG processes. All other parameters are kept constant for different processes and are listed in Table 4.2. Trap generation at IL/High-K

Table 4.2 PBTI compact model and fixed and device dependent model parameters

Interface trap generation	Hole trapping	Bulk trap generation
$\Delta N_{IT-HK} = AE_{OX}e^{\Gamma_{IT-HK}E_{OX}}e^{-\frac{E_{AIT-HK}}{kT}}t^{1/6}$	$\Delta N_{ET} = BE_{OX}e^{\Gamma_{ET}E_{OX}}e^{-\frac{E_{A-HT}}{kT}}$	$\Delta N_{OT-HK} = C\left(1 - e^{-(t/w)\beta_{OT-HK}}\right)$
$E_{OX} = (V_{G-STR} - V_{T0} - \Delta V_T)/EOT$		$w = \eta(V_G - V_{T0} - \Delta V_T)^{-\frac{\Gamma_{OT-HK}}{\beta_{OT-HK}}}e^{\frac{E_{AOT-HK}}{kT\beta_{OT-HK}}}$
$\Delta V_T = \frac{q}{C_{OX}} * (\Delta N_{IT-HK} + \Delta N_{ET} + \Delta N_{OT-HK})$		
A, B, C, Γ_{IT-HK}, Γ_{ET} and E_{AIT-HK} are variable across devices		
Fixed parameters		
	$E_{A-ET} = 0.052\,eV$	$E_{AOT-HK} = 0.32\,eV$; $\beta_{OT} = 0.35$
		$\Gamma_{OT} = 10$; $\eta = 5 \times 10^{12}$

Fixed parameters are constant across different type of HKMG devices used in this study interface is independently estimated using DCIV method [13], SILC is used to estimate trap generation in High-K bulk [30]. Flicker noise is used to estimate pre-existing bulk High-K trap density for different gate insulator processes [33]. The compact model components are verified against these independently estimated quantities to ascertain their correctness.

4.2.3 Process Dependence

Both NBTI and PBTI degradation are known to be extremely sensitive to the variations in gate insulator processes. The compact models are validated against large variety of SiON and HKMG MOSFETs, and the process impact on the underlying trap generation and trapping subcomponents is accessed. SiON devices were fabricated using Decoupled Plasma Nitridation (DPN) technique with peak Nitrogen (N) concentration away from Si/SiON interface [38]. All devices were subjected to proper two-step Post Nitridation Anneal (PNA) [39]. Devices with varying base oxide thickness and Nitrogen dosage were fabricated and are listed in Table 4.3; refer to Chap. 1 for additional details. NBTI degradation in SiON p-MOSFETs was measured using the On-The-Fly (OTF) technique with different time-zero (t_0) delay; measured data are suitably corrected for mobility degradation, refer to Chap. 2 for details. As mentioned before, PBTI degradation is not important for SiON devices and hence not studied.

Planar p- and n-channel MOSFETs with Silicon (Si) channel and having different HKMG gate insulator processes have been used to study and validate NBTI and PBTI compact models; device details are listed in Table 4.4. A gate first integration has been used; HKMG gate insulator stacks were fabricated using a novel cluster tool, which includes IL growth and Atomic Layer Deposition (ALD) technique based High-K deposition processes in a controlled vacuum environment to achieve Equivalent Oxide Thickness (EOT) down to 6 Å [33], see Chap. 3, Fig. 3.2 for additional details. IL was grown using a low temperature Rapid Thermal Process (RTP) and was followed by ALD HfO_2 deposition without any air break between the two processes. Devices with IL thickness having 5 and 3 Å were fabricated and are referred to as devices D1 and D2, respectively. To study the

Table 4.3 Details of SiON p-MOSFETs for NBTI measurements and analysis

Device no.	Type	Base oxide thickness (Å)	PNO N dose ($\times 10^{15}$ cm^{-2})	T_{XPS} (Å)	EOT (Å)	% N
W1	PNO	15	2.8	18.5	14.0	23
W4	PNO	20	6.8	24.4	14.6	43
W8	PNO	25	3.1	28.1	23.5	17
W9	PNO	25	5.5	28.5	21.4	29
W10	PNO	25	6.8	28.9	19.9	36

Refer to Chap. 1, Table 1.1 for additional information

Table 4.4 Details of HKMG p- and n-MOSFETs for NBTI and PBTI measurements and analysis

Device no.	IL thickness (Å)	High-K thickness (Å)	Remarks
D1	5	23	Low-T RTP
D2	3	23	Low-T RTP
D3	2.5	23	Low-T RTP, PHKN after high-K deposition
D4	1.5	23	Low-T RTP in N Ambient

Refer to Chap. 3, Fig. 3.2 for additional information

impact of Nitrogen, post High-K nitridation (PHKN) was performed on the 3 Å device by using DPN followed by PNA, these devices are referred to as D3. Extreme scaling of the IL (less than 2 Å) was achieved by performing RTP in Nitrogen environment followed by ALD High-K deposition; resulting devices labeled as D4. Ultra-Fast Measure-Stress-Measure (MSM) technique was used to measure BTI degradation in these devices, refer to Chaps. 1 and 2 for additional details.

NBTI compact model is also used to study planar HKMG p-MOSFETs having Silicon Germanium (SiGe) channel, with [40] and without [41] Si cap. The device details are shown in Table 4.5. Si capped devices have different Si cap thickness, SiGe composition and SiGe quantum well thickness, while non Si capped devices have different SiGe composition, refer to [40, 41] for further details. Finally, the NBTI compact model is also verified on FinFETs having Replacement Metal Gate

Table 4.5 Details of SiGe channel based HKMG p-MOSFETs for NBTI measurements and analysis

Device no.	Ge %	SiGe thickness (nm)	Si cap thickness (nm)	T_{INV} (nm)
SG1	0	0	0	1.55
SG2	45	7	1.3	1.4
SG3	55	3	1.3	1.4
SG4	55	5	0.65	1.28
SG5	55	5	1	1.32
SG6	55	5	1.3	1.4
SG7	55	5	2	1.59
SG8	55	7	1.3	1.4
SG9	0	0	0	1.23
SG10	20	7.5	0	1.29
SG11	25	7.5	0	1.33
SG12	35	7.5	0	1.51

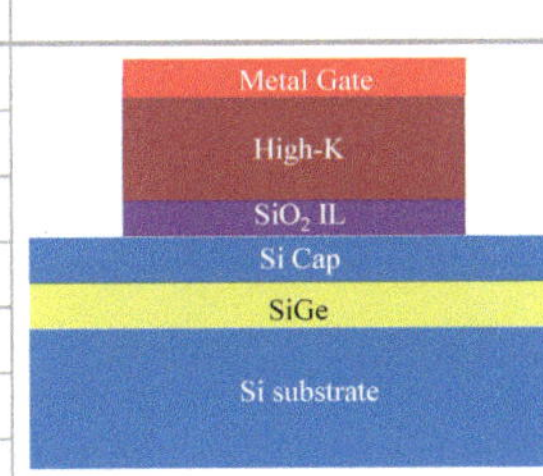

SG1 to SG8 data from [40], SG9 to SG12 data from [41]. The devices consist of the following layers: Si/SiGe/Si cap (optional)/IL /High-K/Metal Gate as shown

(RMG) based HKMG process [42]. One Spot Drop Down (OSDD) technique with sense bias ($V_{G\text{-}SNS}$) close to pre-stress $V_T(V_{T0})$ has been used (unless stated otherwise) to measure NBTI degradation in these devices, refer to Chap. 2 for details.

4.3 NBTI Process Dependence in SiON p-MOSFETs

Figure 4.2a through c show time evolution of threshold voltage shift ΔV_T at a particular stress voltage $V_{G\text{-}STR}$ and temperature T in SiON devices having different Nitrogen concentration (N%) and high EOT, measured using the OTF method having t_0 delay of 1 μs and 1 ms. NBTI degradation increases with increased N%; refer to Chap. 1 for additional details. Overall model prediction and the underlying subcomponents are also shown, see Table 4.1 for model and Table 4.3 for device details. Measured ΔV_T using different t_0 delay shows power-law time dependence, with lower degradation magnitude and higher time exponent n at higher t_0 delay. Time evolution of ΔV_T for different t_0 delay can be modeled by only varying the hole trapping concentration ΔV_{HT} and using identical trap generation subcomponents ΔV_{IT} and ΔV_{OT} as shown. This is consistent with the fact that delayed OTF measurements would capture somewhat smaller fast hole trapping subcomponent at the beginning of stress, since hole trapping is a fast process and gets triggered as soon as stress is applied, and therefore results in lower overall ΔV_T, see Chap. 2 for OTF measurement details. On the other hand, ΔV_{IT} and ΔV_{OT} subcomponents evolve at a much slower rate and are not impacted by t_0 delay, unless t_0 becomes unnaturally large. Moreover, since ΔV_{HT} saturates at longer t_{STR}, lower ΔV_{HT} contribution captured for larger t_0 delay would result in higher time exponent n and vice versa, which verifies the assumption that ΔV_{HT} saturates at long stress time.

Figure 4.2d through f show the time evolution of voltage shift corresponding to trap generation directly measured using CP method in different N% devices. Data are plotted after delay and band gap correction ; refer to Chap. 2 for CP measurement details. Time evolution of generated interface traps ΔV_{IT} and bulk traps ΔV_{OT} subcomponents as predicted by the compact model is also shown. Note that overall modeled trap generation subcomponent ($\Delta V_{IT} + \Delta V_{OT}$) shows an excellent agreement with directly measured contribution from trap generation for different SiON devices with varying Nitrogen concentration. Therefore, correctness of the trap generation subcomponents of the compact model is independently verified.

The compact model has also been validated against other SiON devices listed in Table 4.3 having different N% and EOT. Figure 4.3 shows the time evolution of ΔV_T for NBTI stress at different $V_{G\text{-}STR}$ and T, measured using OTF method having different t_0 delay. Overall model prediction is also shown, and it is important to note that different t_0 delay data can be predicted by varying only the ΔV_{HT} subcomponent, which again verifies that the ΔV_{HT} subcomponent is fast and saturates at longer t_{STR}. Therefore, the compact model of Table 4.1 can successfully predict

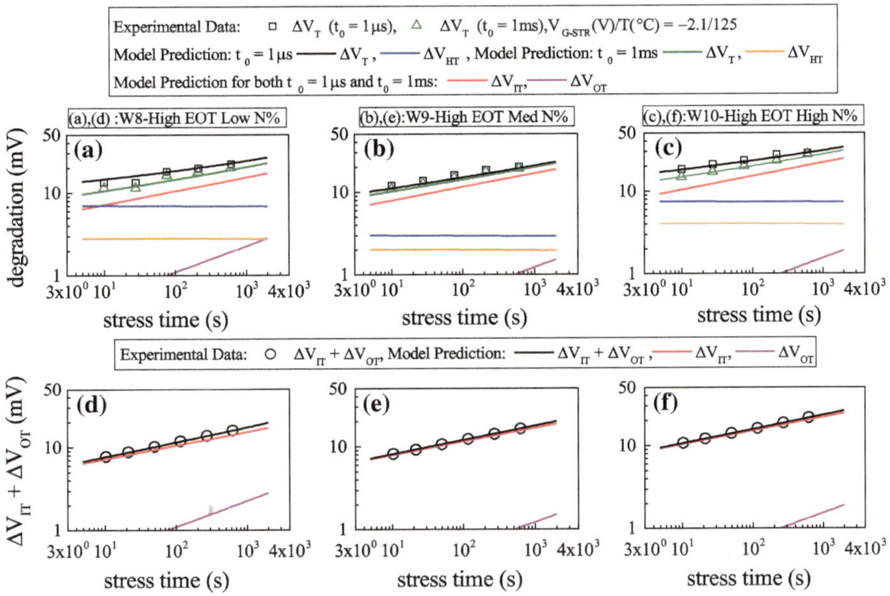

Fig. 4.2 **a–c** Time evolution of measured ΔV_T using OTF method with different time-zero delay (*symbols*), for NBTI stress in p-MOSFETs having different SiON gate insulator processes. Overall model prediction for ΔV_T and underlying ΔV_{IT}, ΔV_{HT}, and ΔV_{OT} subcomponents are also shown (*lines*). **d–f** Time evolution of measured trap generation using CP method after delay and band gap corrections (*symbols*) and model calculation for ΔV_{IT} + ΔV_{OT} (*lines*) for different SiON devices

ΔV_T time evolution measured in different SiON devices of Table 4.3 for various stress conditions with five process dependent variable parameters , which are listed in Table 4.6. All other parameters are kept fixed across devices and are listed in Table 4.1. The use of very few adjustable model parameters across wide variety of SiON devices and different stress and measurement conditions shows the unique predictive ability of the proposed model.

Figure 4.4 shows ΔV_T measured at a fixed stress time as a function of (a) oxide field[2] E_{OX} and (b) temperature T for NBTI stress in SiON devices having different N%. The compact model calculation is also shown. Note that increased N% leads to higher ΔV_T magnitude, but lowering of power-law field acceleration factor (Γ_E) and Arrhenius T activation energy (E_A), and the model can predict measured data reasonable well for different SiON processes using the fixed and variable

[2]For simplicity and ease of comparison with data shown elsewhere, the measured and calculated E_{OX} dependence is plotted in a log–log plot and power-law E_{OX} acceleration factor Γ_E is extracted. Note that the E_{OX} dependence of the compact model has a different form as it adheres to the underlying physical mechanism. The same is also done for NBTI and PBTI measured and calculated data in HKMG devices shown later in this chapter.

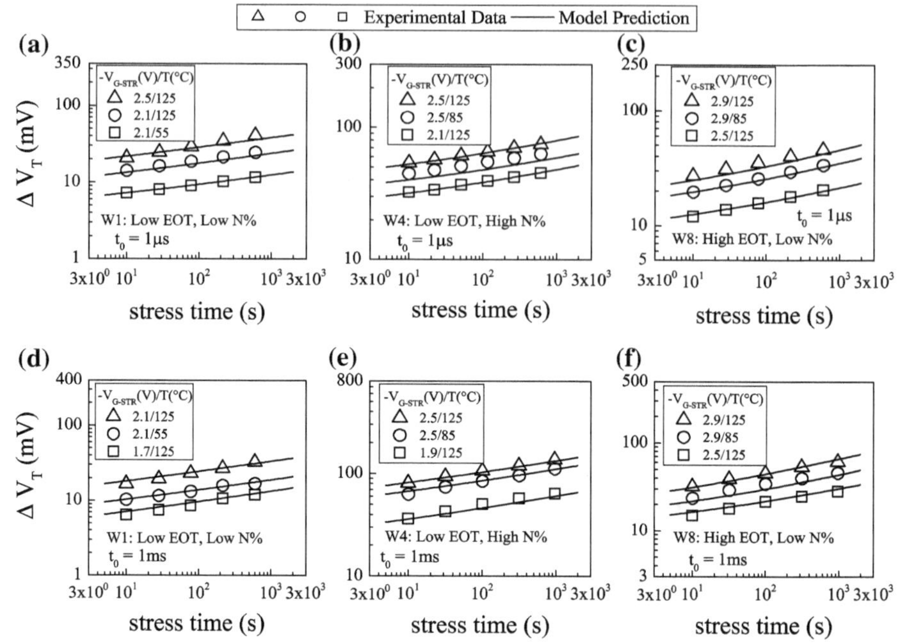

Fig. 4.3 Time evolution of measured ΔV_T using OTF method with different time-zero delay (*symbols*) and corresponding compact model prediction (*lines*), for NBTI stress at different $V_{G\text{-}STR}$ and T in p-MOSFETs having different SiON gate insulator processes. *Top panels* for t_0 = 1 μs, *bottom panels* for t_0 = 1 ms data

Table 4.6 SiON process dependent NBTI compact model parameters

	A	B	C	$\Gamma_{IT} = \Gamma_{HT}$	E_{AKF}
Units	$\frac{1}{\sqrt{cm}\, s^{1/6}}$	$\frac{1}{\sqrt{cm}}$	cm^{-2}	$\frac{cm}{MV}$	eV
Thin EOT low N%	7E9	1E8	1E11	0.25	0.185
Thin EOT high N%	3.5E10	2.5E9	1E11	0.2	0.15
Thick EOT	7E9	4.2E9	4.5E12	0.25	0.16

parameters listed respectively in Tables 4.1 and 4.6. Note that devices having higher N% show larger hole trapping component [4, 43], and the T activation of trapping is presumed to be small compared to the T activation of trap generation. Therefore, T activation of overall ΔV_T reduces with increased N% as shown. More work is needed to explain the reduction of Γ_E in devices having larger N%.

Figure 4.5 shows the correlation of measured and modeled (a) power-law time exponent n, (b) Arrhenius T activation energy E_A and (c) power-law E_{OX} acceleration factor Γ_E for NBTI stress in different SiON devices. Note that a power-law E_{OX} dependence is used for ease of comparison, although the compact model uses a

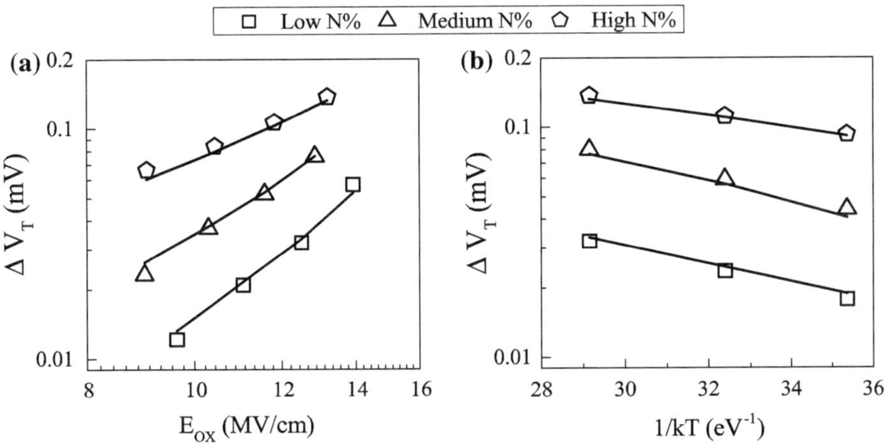

Fig. 4.4 UF-OTF measured ΔV_T at fixed stress time for NBTI stress in different SiON p-MOSFETs versus **a** stress E_{OX} and **b** stress T (*symbols*). E_{OX} dependence is plotted in a log–log scale, T dependence in a semi-log scale. *Lines* are model prediction

Fig. 4.5 Correlation of measured and compact model predicted **a** power-law time exponent n, **b** Arrhenius T activation energy E_A, and **c** power-law field acceleration factor Γ_E for NBTI stress in different SiON p-MOSFETs. Comparison has been done for different measurement delay. *Lines* represent 1:1 correlation

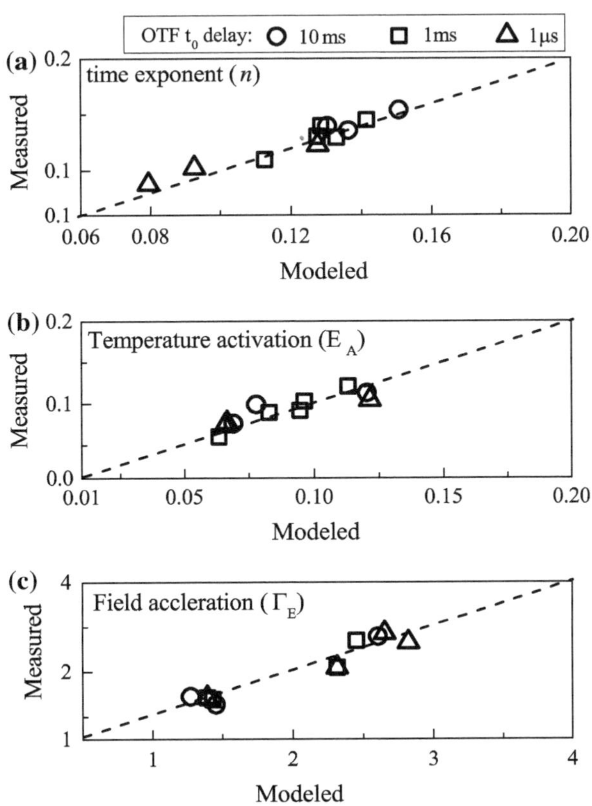

more physically based relation. Time exponent n is calculated using liner regression of measured ΔV_T time evolution in t_{STR} range of 10 s to 1 Ks, while Γ_E and E_A are calculated at fixed t_{STR}.

Note that excellent correlation is observed between measured data and model prediction for different measurement delay and for different SiON devices. Note that the relative contribution of trap generation and trapping subcomponents depends on SiON process and measurement delay. Since trap generation keeps on evolving at longer t_{STR} while trapping saturates, and since the former shows larger T activation than the latter, their relative dominance in particular decides the time exponent n and T activation E_A of overall NBTI degradation. Accurate prediction of measured parameters therefore suggests correctness of the underlying subcomponents of the model, which is further verified below using direct CP and flicker noise measurements.

Figure 4.6a shows the correlation of voltage shift contribution from trap generation measured using the CP technique to calculated trap generation subcomponents ($\Delta V_{IT} + \Delta V_{OT}$) calculated using the compact model for NBTI stress in SiON devices with different N%. Measured and modeled ΔV_{IT} time evolution has been obtained at different stress E_{OX} and T, although the data plotted in Fig. 4.6a are shown for a fixed t_{STR}. As discussed in Chap. 2, voltage shift due to trap generation from CP measurements is calculated after suitable delay and band gap correction. Excellent agreement between theory and measurements verify the accuracy of trap generation subcomponents across different devices. Furthermore, Fig. 4.6b plots the correlation of pre-existing trap density measured using the flicker noise technique to fractional hole trapping contribution ($\Delta V_{HT}/\Delta V_T$) calculated by the model for different SiON devices. Measured pre-existing trap density increases for SiON devices having higher N% and results in higher trapping magnitude during stress [43], and validates the hole-trapping subcomponent of NBTI compact model.

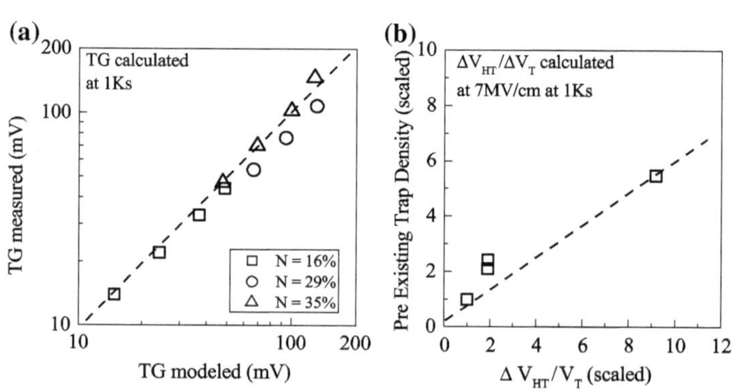

Fig. 4.6 Correlation of **a** trap generation (TG) measured using CP method after delay and band gap corrections and model prediction, and **b** pre-existing traps measured using flicker noise and calculated fractional hole trapping contribution, for NBTI stress in different SiON p-MOSFETs. *Lines* represent **a** 1:1 correlation and **b** guide to eye

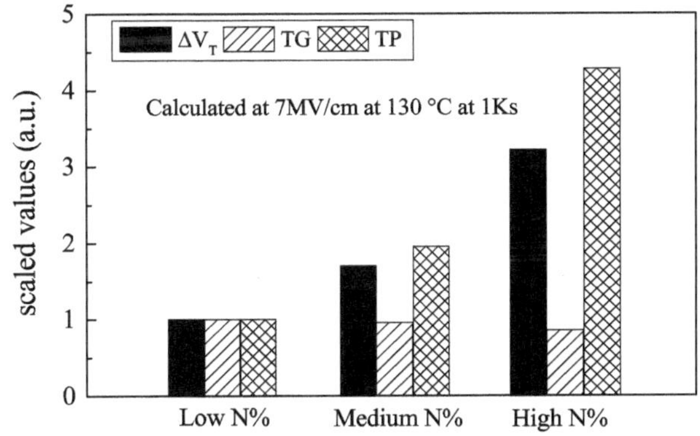

Fig. 4.7 Model calculated overall ΔV_T and underlying trap generation and trapping subcomponents of NBTI degradation in SiON devices having different N content in the gate insulator stack

It is important to assess the impact of N% on trap generation and trapping subcomponents of overall NBTI degradation. Figure 4.7 shows overall ΔV_T and the underlying ΔV_{HT} and ΔV_{IT} + ΔV_{OT} subcomponents for SiON devices having different N%. The compact model with parameters calibrated by prediction of stress data at different $V_{G\text{-}STR}$ and T is used to compute overall degradation and its subcomponents for identical stress condition across different N% devices. All data are normalized to the device having lowest N%. Note that ΔV_T increases with increase in N%, which is now a well-known result, and the increase is primarily due to increase in ΔV_{HT} subcomponent as shown [21, 22]. Interestingly, N% has negligible impact on trap generation as shown and verifies the uncoupled nature of trap generation and trapping subcomponents of overall degradation. Note that as plasma nitridation with proper PNA is used in devices having thicker gate oxides, the N density at the Si/SiON interface is unlikely to change significantly with variation in N dose, refer to Table 4.3. Therefore ΔV_{IT} does not change, while ΔV_{HT}, being a bulk phenomenon, gets affected, which is also independently corroborated by flicker noise measurements [43]. Moreover, the relative contribution of bulk trap generation ΔV_{OT} is lower compared to ΔV_{IT} for the stress conditions used in this study; hence overall trap generation remains unaffected by changes in bulk N concentration as shown.

4.4 NBTI Process Dependence in HKMG p-MOSFETs

4.4.1 Impact of HKMG Process

Figure 4.8 plots time evolution of UF-MSM measured threshold voltage shift ΔV_T for NBTI stress in differently processed planar HKMG Si channel devices listed in Table 4.4. Compared to non-nitrided D1 and D2 devices, nitrided D3 and D4

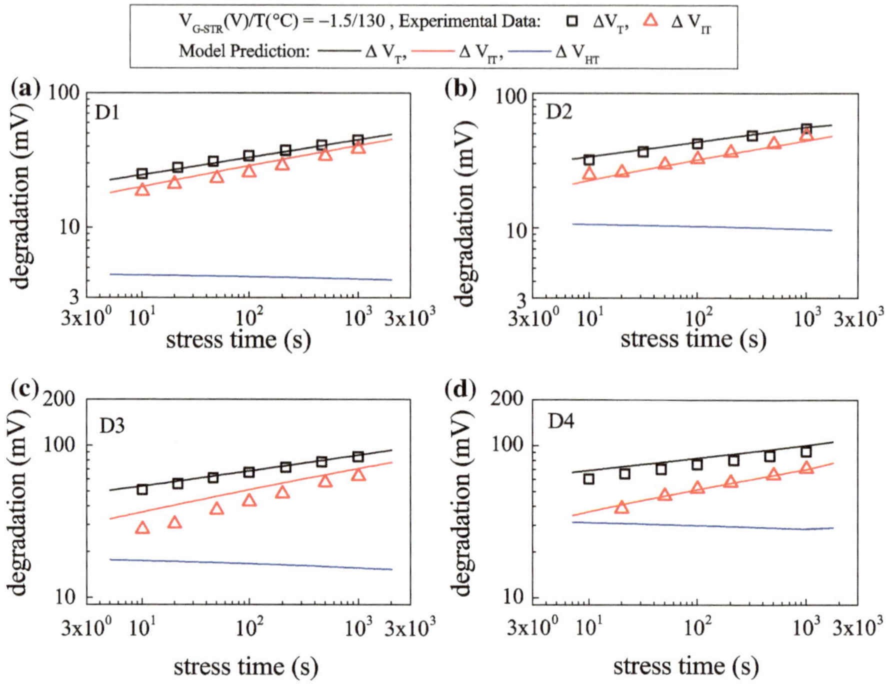

Fig. 4.8 Time evolution of UF-MSM measured ΔV_T and DCIV measured ΔV_{IT} after delay and band gap corrections (*symbols*) for NBTI stress in p-MOSFETs with different HKMG gate insulator stacks, refer to Table 4.4 for device details. Compact model prediction of overall ΔV_T and underlying trap generation and trapping subcomponents are shown (*lines*)

devices show higher NBTI degradation, which is consistent with SiON results; refer to Chap. 1 for further details. Time evolution of interface trap generation ΔN_{IT} is independently estimated by using DCIV technique, refer to Chap. 2 for details; the voltage shift ΔV_{IT} corresponding to trap generation extracted from DCIV measurements after delay and band gap correction is shown. ΔV_T modeled by three-component NBTI compact model proposed in Table 4.1 is shown along with underlying trap generation and trapping subcomponents. As mentioned before, bulk trap generation component for HKMG stacks is negligible and hence is not shown. Note that the overall model can predict measured ΔV_T time evolution for various devices while modeled ΔV_{IT} subcomponent agrees well with independently measured trap generation data. Also note that relative contribution of the saturated trapping subcomponent ΔV_{HT} increases in HKMG stacks D3 and D4 having additional N in the gate insulator stack, which is consistent with SiON data discussed earlier in this chapter.

Figure 4.9 shows time evolution of ΔV_T measured in different HKMG devices listed in Table 4.4 for NBTI stress using different $V_{G\text{-}STR}$ and temperature T. Calculated ΔV_T time evolution using the compact model of Table 4.1 is also shown. Note that the model can successfully predict measured data in such large

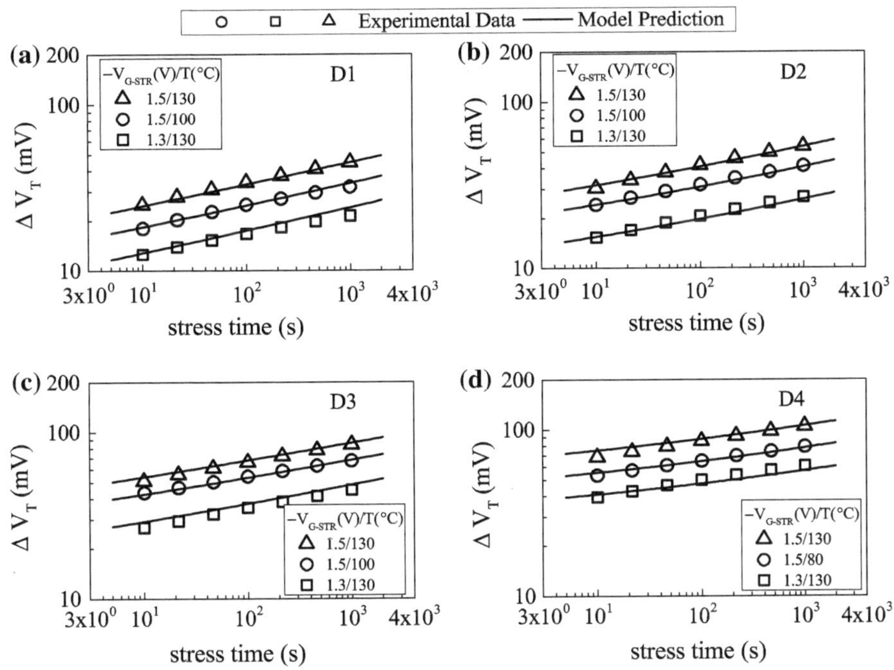

Fig. 4.9 Time evolution of UF-MSM measured ΔV_T (*symbols*) and corresponding compact model calculation (*lines*) for NBTI stress at different $V_{G\text{-STR}}$ and T in p-MOSFETs having different HKMG gate insulator stacks

variety of HKMG devices under different stress conditions with four adjustable parameters ; pre-factor A, temperature activation E_{AKF} and field acceleration Γ_{IT} for interface trap generation , and hole trapping pre-factor B, and are listed in Table 4.7. Unlike SiON, bulk trap generation pre-factor C is negligible in HKMG devices. Other parameters are kept constant across different devices and are listed in Table 4.1; identical fixed parameter values have been used for SiON and HKMG devices. Out of four adjustable parameters , Γ_{IT} and E_{AKF} are of interest as they depend on the IL type and quality, e.g., *N* density near the interface between Si channel and IL. However, similar Γ_{IT} and E_{AKF} values are used for a particular category of HKMG stacks but having different IL and HK thicknesses. Therefore

Table 4.7 HKMG process dependent NBTI compact model parameters

	A	B	C	$\Gamma_{IT} = \Gamma_{HT}$	E_{AKF}
Units	$\frac{1}{\text{V cm s}^{1/6}}$	$\frac{1}{\text{V cm}}$	cm^{-2}	$\frac{\text{cm}}{\text{MV}}$	eV
HKMG D1	9E10	6E9	0	0.245	0.215
HKMG D2	8E10	1.5E10	0	0.21	0.215
HKMG D3	2.8E11	5E10	0	0.16	0.21
HKMG D4	4E10	1E11	0	0.17	0.13

unless the process technology is drastically changed, small process variations can be predicted by only two adjustable parameters *A* and *B*. Once again, the use of few adjustable parameters shows the uniqueness of the model and signifies its importance in predicting degradation across wide variety of HKMG processes.

Figures 4.10 and 4.11 respectively show the oxide field E_{OX} and temperature T dependence of UF-MSM measured ΔV_T as well as DCIV measured ΔV_{IT} obtained after delay and band gap corrections, refer to Chap. 2 for details. The model calculated E_{OX} and T dependence of overall ΔV_T and underlying ΔV_{IT} and ΔV_{HT} subcomponents are also shown. As mentioned earlier, the contribution from bulk trap generation ΔV_{OT} is negligible and hence is not shown. Note that model calculated E_{OX} and T dependence of both ΔV_T and ΔV_{IT} is in excellent agreement with experiments for different HKMG processes, and is obtained using the fixed and process dependent parameters shown in Tables 4.1 and 4.7, respectively. Note that the relative contribution of hole trapping component ΔV_{HT} increases with increase in N content in the gate insulator stack. Also note that similar to SiON devices, Γ_{IT} equals Γ_{HT} and both reduce for stacks having larger N density in the IL, thereby reducing the field acceleration of overall ΔV_T, see Fig. 4.10. On the other hand,

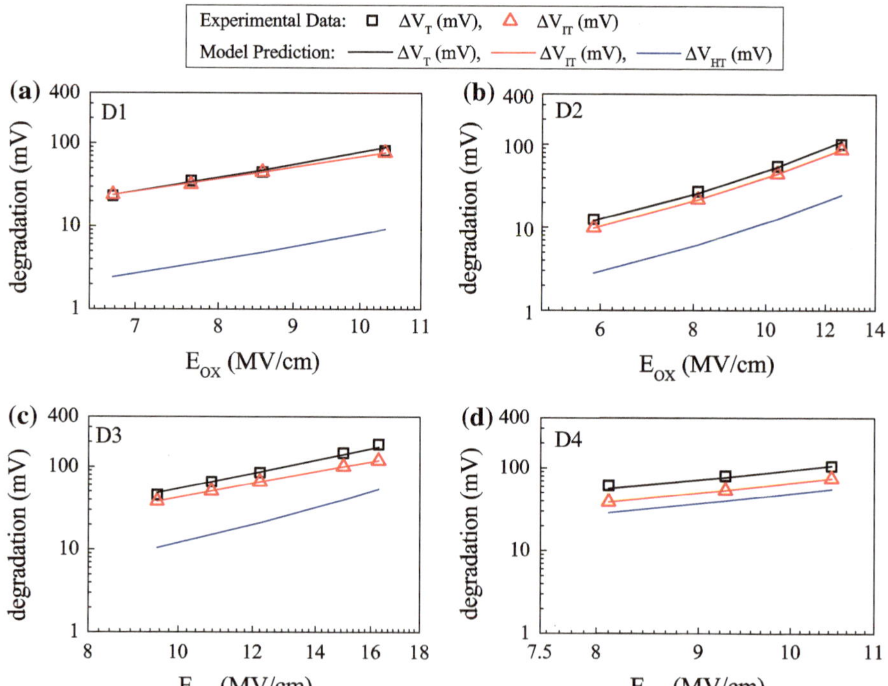

Fig. 4.10 UF-MSM measured ΔV_T and DCIV measured ΔV_{IT} after delay and band gap corrections (*symbols*) at fixed t_{STR} versus stress E_{OX} for NBTI stress in p-MOSFETs having different HKMG gate insulator stacks. Compact model prediction of ΔV_T and underlying trap generation and trapping subcomponents are shown (*lines*). E_{OX} dependence is plotted in a log–log scale

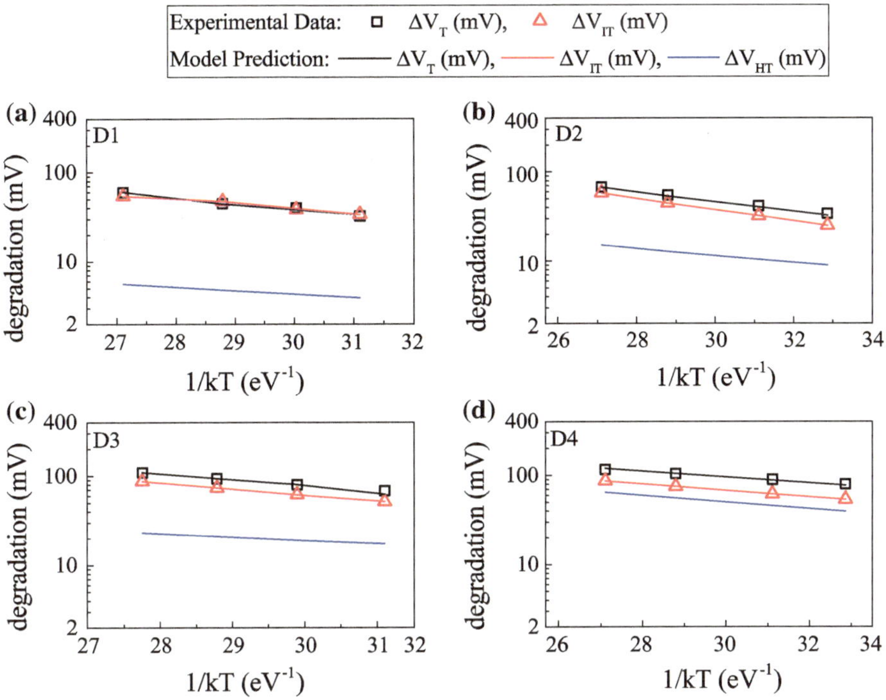

Fig. 4.11 UF-MSM measured ΔV_T and DCIV measured ΔV_{IT} after delay and band gap corrections (*symbols*) at fixed t_{STR} versus stress T for NBTI stress in p-MOSFETs having different HKMG gate insulator stacks. Compact model prediction of ΔV_T and underlying trap generation and trapping subcomponents are shown (*lines*)

ΔV_{HT} has weaker T activation compared to ΔV_{IT} as shown in Fig. 4.11, and larger relative ΔV_{HT} contribution results in reduced E_A of overall ΔV_T for devices having higher N content, and once again verifies the uncorrelated nature of trap generation and trapping processes.

Figure 4.12 shows the correlation of measured and modeled (a) power-law time exponent n, (b) Arrhenius T activation energy E_A and (c) power-law E_{OX} acceleration factor Γ_E for different HKMG devices listed in Table 4.4. Note that a power-law E_{OX} dependence is used for ease of comparison, although the compact model uses a more physically based relation. As before, the time exponent n is calculated using liner regression of measured ΔV_T time evolution in t_{STR} range of 10 s to 1 Ks, while Γ_E and E_A are calculated at fixed t_{STR}. Excellent correlation is obtained between measured data and model prediction for different measurement delay and across different HKMG gate insulator processes. Note that hole trapping saturates while interface trap generation keeps evolving with increase in t_{STR}, and the former has weaker T activation than the latter. Therefore as discussed before, prediction of time exponent n and T activation E_A implies precise determination of the underlying trap generation and trapping subcomponents for different HKMG

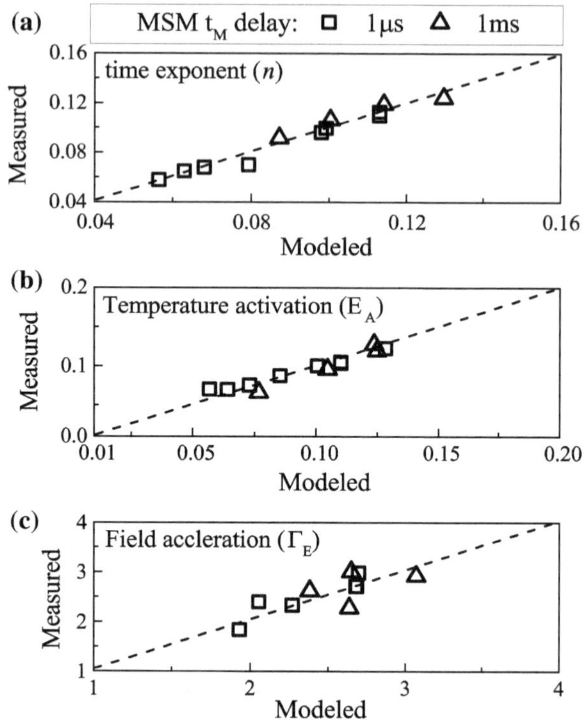

Fig. 4.12 Correlation of measured and compact model predicted **a** power-law time exponent n, **b** Arrhenius T activation energy E_A, and **c** power-law field acceleration factor Γ_E for NBTI stress in different HKMG p-MOSFETs. Comparison has been done for different measurement delay. *Lines* represent 1:1 correlation

processes and stress conditions. This is achieved with only four process dependent adjustable parameters as listed in Table 4.7. The individual model subcomponents are validated next using direct measurements.

Figure 4.13a shows the correlation of voltage shift corresponding to interface trap generation measured using DCIV to the model calculated ΔV_{IT} subcomponent for different HKMG devices mentioned in Table 4.4. Measured and modeled time evolution of ΔV_{IT} is obtained at different stress E_{OX} and T, although their correlation in Fig. 4.13a is shown at fixed t_{STR}. As mentioned before, measured DCIV data are corrected for delay and band gap , refer to Chap. 2 for details. Excellent correlation is observed between measurements and theory across different HKMG devices, which verifies the correctness of the calculated trap generation subcomponent. Figure 4.13b shows the correlation of measured pre-stress trap density using flicker noise technique to fractional hole trapping ($\Delta V_{HT}/\Delta V_T$) contribution calculated using the compact model for different HKMG devices. As mentioned before, increase in pre-stress trap density as measured using flicker noise for gate stacks having larger N content in the gate stack would result in higher hole trapping magnitude during NBTI stress. This is indicated by relatively larger fractional ΔV_{HT} concentration shown in Fig. 4.13b, and validates the hole-trapping subcomponent of NBTI compact model.

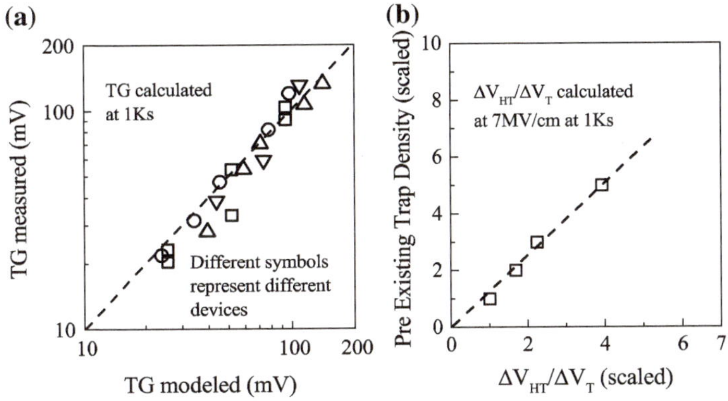

Fig. 4.13 Correlation of **a** trap generation (TG) measured using DCIV method after delay and band gap corrections and model prediction, and **b** pre-existing traps measured using flicker noise and calculated fractional hole trapping contribution, for NBTI stress in different HKMG p-MOSFETs. *Lines* represent **a** 1:1 correlation and **b** guide to eye

The EOT scaling of HKMG gate insulator stacks is of interest and is primarily achieved by IL thickness scaling, since High-K thickness scaling is detrimental from the viewpoint of increased gate leakage [33, 44]. The impact of IL thickness scaling on NBTI has been shown in Chap. 1. Figure 4.14 shows measured ΔV_{T} and ΔV_{IT} at fixed t_{STR} as a function of IL thickness, calculated at (a) fixed oxide field E_{OX} and

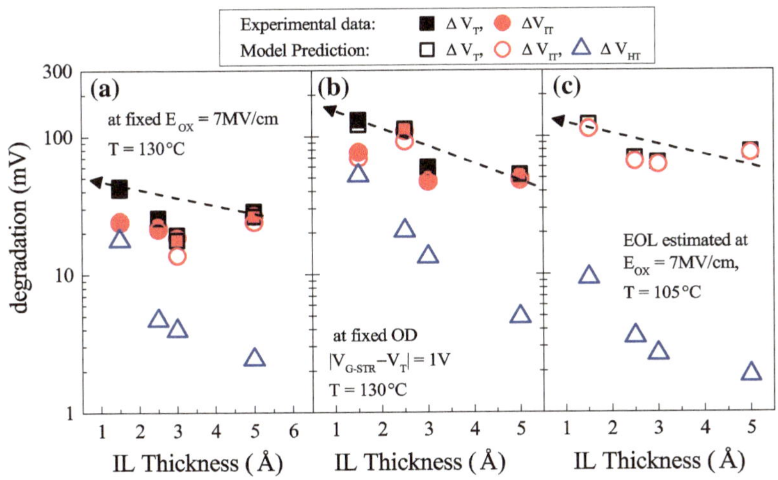

Fig. 4.14 Impact of IL thickness scaling on UF-MSM measured ΔV_{T} and DCIV measured ΔV_{IT} after delay and band gap corrections, for NBTI stress in HKMG p-MOSFETs at **a** constant E_{OX} and **b** constant gate overdrive . Model calculated overall ΔV_{T} and underlying ΔV_{IT} and ΔV_{HT} subcomponents are also shown. **c** IL thickness scaling impact on compact model calculated ΔV_{T} and underlying ΔV_{IT} and ΔV_{HT} subcomponents at 10 years. *Lines* are guide to eye

(b) fixed overdrive bias ($V_{G\text{-}STR} - V_{T0}$), V_{T0} being pre-stress V_T. As mentioned before, UF-MSM and DCIV measurements are, respectively, used for ΔV_T and ΔV_{IT}, the latter is obtained after delay and band gap corrections. Model calculated overall ΔV_T and underlying ΔV_{IT} and ΔV_{HT} subcomponents are also shown; excellent agreement has been observed between measured and modeled ΔV_T and ΔV_{IT} data. Note that ΔV_T at fixed E_{OX} increases as IL is scaled because of nitridation, refer to Table 4.4 for device details. However, increase in nitridation results in much larger relative increase in ΔV_{HT} when compared ΔV_{IT} as IL is scaled, refer to Fig. 4.14a. On the other hand, in addition to impact of *N*, increase in ΔV_T at fixed gate overdrive is also due to increased E_{OX} as IL is scaled, as shown in Fig. 4.14b.

Once again, a relatively larger increase in ΔV_{HT} is observed as nitridation is used for IL scaling . Therefore, similar to SiON data, N has larger impact on hole trapping than trap generation for HKMG devices as well. However unlike SiON devices, process variations causes variation in the Nitrogen content near the Si/IL interface for HKMG stacks, which impacts both ΔV_{IT} and ΔV_{HT} subcomponents. Figure 4.14c plots extrapolated ΔV_T and underlying ΔV_{IT} and ΔV_{HT} subcomponents at EOL value of 10 years for HKMG devices having different EOT, calculated using the NBTI compact model of Table 4.1 with calibrated parameters. Since ΔV_{HT} saturates quickly and only impacts short-time data, ΔV_T at EOL is completely dominated by ΔV_{IT} as shown. Therefore, the impact of hole trapping remains negligible at scaled EOT.

4.4.2 Impact of Channel Material

The compact model of Table 4.1 is also used to predict NBTI degradation in different SiGe channel based planar HKMG devices without and with Si cap, refer to Table 4.5 for device details [40, 41]. Time evolution of overall ΔV_T is modeled using ΔV_{IT} and ΔV_{HT} subcomponents, since ΔV_{OT} contribution has been found to be negligible in these devices [45]. Only three process dependent parameters have been used, i.e., interface trap generation and trapping pre-factors A and B as well as the field acceleration factor Γ_{IT} ($=\Gamma_{HT}$), refer to Tables 4.1 and 4.8. The parameter E_{AKF} related to *T* activation has been found to be constant across devices due to the use of very similar HKMG gate insulator stacks, and the pre-factor C for ΔV_{OT} is assumed to be negligible. Other fixed parameters are listed in Table 4.1, and identical values are used for Si and SiGe devices. As mentioned before, OSDD method has been used for ΔV_T time evolution measurements in these devices.

Figure 4.15 plots the time evolution of measured [40] and modeled ΔV_T in left panels, and underlying ΔV_{IT} and ΔV_{HT} subcomponents in right panels, for NBTI stress in Si capped SiGe devices having different Ge concentration; pure Si device is shown as reference. The process dependent adjustable parameters are listed in Table 4.8. SiGe devices show higher power-law field acceleration factor Γ_E compared to Si reference device [40]. NBTI degradation reduces with increased Ge% in the channel. ΔV_{IT} contribution to overall ΔV_T is much larger than ΔV_{HT}

Table 4.8 SiGe process dependent NBTI compact model parameters

Device no.	A $\left(\frac{1}{\text{V cm s}^{1/6}}\right)$	B $\left(\frac{1}{\text{V cm}}\right)$	$\Gamma_{IT} = \Gamma_{HT}$ $\left(\frac{\text{cm}}{\text{MV}}\right)$	E_{AKF} (eV)
SG1	2.15E11	2.9E10	0.15	0.214
SG2	1.2E10	5E8	0.27	
SG3	9E9	5E6		
SG4	1.15E9	5E6		
SG5	2.7E9	5E6		
SG6	6.3E9	5E6		
SG7	2.1E10	1E9		
SGS	4.9E9	5E6		
SG9	4.4E11	0	0.14	
SG10	1.25E11	0	0.19	
SG11	IE11	0		
SG12	4.E10	0		

Due to the absence of temperature dependent SiGe data [40, 41], identical E_{AKF} is used for Si and SiGe devices

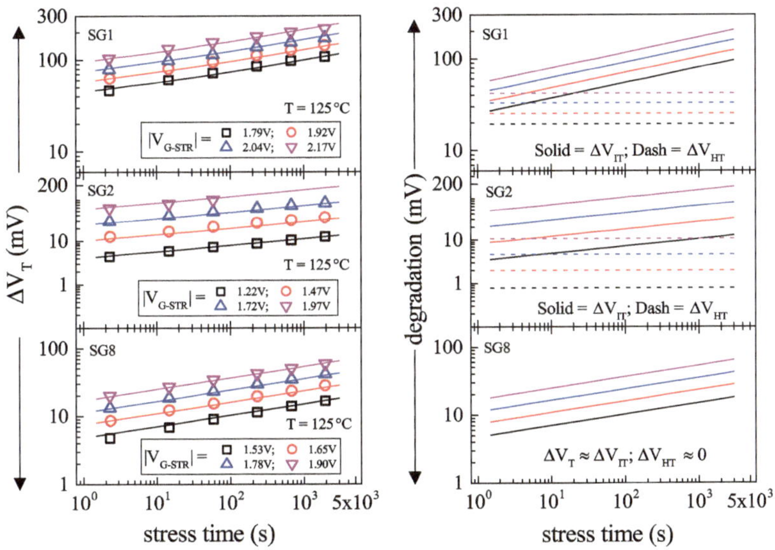

Fig. 4.15 Time evolution of ΔV_T from OSDD measurements and model prediction (*left panels*) and time evolution of calculated model subcomponents (*right panels*) for NBTI stress at different $V_{G\text{-}STR}$ in Si and SiGe based HKMG p-MOSFETs with different Ge concentration. SiGe devices have Si cap. Experimental data (*symbols*) from [40], *lines* are model calculation

contribution across all devices, and interestingly, the ΔV_{HT} contribution becomes negligible and hence ΔV_{IT} dominates overall ΔV_T as Ge concentration is increased beyond 50 %. Increased Ge concentration reduces the magnitude of ΔV_{IT} and hence that of ΔV_T, although the power-law time exponent for time evolution of ΔV_{IT} remains constant at $n \sim 1/6$ across different devices as shown. Therefore, power-law time exponent of overall ΔV_T becomes less than $n \sim 1/6$ for Si and low Ge% SiGe devices owing to non-negligible but saturated ΔV_{HT} contribution, although the $n \sim 1/6$ dependence is seen for the high Ge% SiGe devices due to complete ΔV_{IT} dominance. Note that an earlier analysis of these data suggested dominant ΔV_{HT} contribution [40], because the ΔV_{IT} subcomponent was obtained using CP measurement and is underestimated, since it was not corrected for measurement delay and band gap, refer to Chap. 2 for details [4].

Figure 4.16 shows time evolution of measured ΔV_T for SiGe devices without Si cap [41] and prediction by the compact model [45]; the Si device is also shown as reference. The device dependent adjustable parameters are listed in Table 4.8. It is important to note that higher Ge% devices exhibit noise and variability in measured data, which is due to process variability as discussed in [41]. The data plotted in Fig. 4.16 especially for higher Ge% devices are obtained by averaging different measurements. Once again, NBTI degradation reduces with increased Ge% in the channel. Moreover, the power-law field acceleration factor Γ_E is large for SiGe devices without Si cap as compared to pure Si device, which is similar to the Si capped data shown before. However, note that ΔV_T time evolution is completely dominated by ΔV_{IT} and the ΔV_{HT} subcomponent is negligible for these Si and SiGe

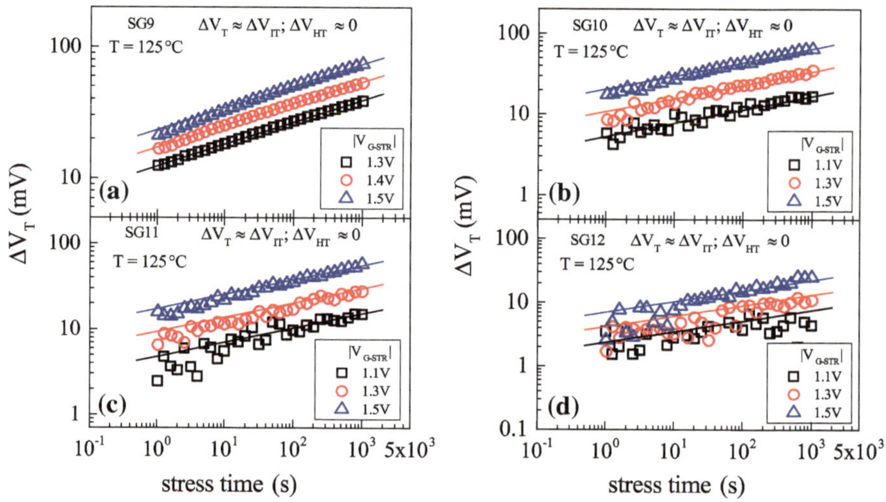

Fig. 4.16 Time evolution of ΔV_T from OSDD measurements and model prediction for NBTI stress at different $V_{G\text{-STR}}$ in Si and SiGe based HKMG p-MOSFETs with different Ge concentration. SiGe devices do not have Si cap . Experimental data (*symbols*) from [41], *lines* are model calculation

devices having different Ge concentration. Therefore, the ΔV_T time evolution exhibits power-law dependence with exponent $n \sim 1/6$, although the magnitude of ΔV_T reduces with increase in Ge concentration due to reduction in ΔV_{IT} magnitude, which is again similar to the Si capped results shown before.

Figure 4.17 shows time evolution of measured ΔV_T in Si capped SiGe devices having high Ge% but different Si cap thickness [40]. A reduction in NBTI degradation is observed with reduction in Si cap thickness. The compact model prediction is also shown [45]; adjustable parameters are listed in Table 4.8. Since these devices have high Ge concentration, ΔV_{IT} completely dominates time evolution of ΔV_T and ΔV_{HT} is negligible, hence the subcomponents are not plotted, for all but the thickest Si cap device. Since ΔV_{IT} dominates ΔV_T, a power-law time dependence with exponent $n \sim 1/6$ is observed for all but the thickest Si cap device. The time evolution of ΔV_{IT} and ΔV_{HT} subcomponents is shown for the thickest Si cap device. Although ΔV_{HT} is not negligible for thickest Si cap device, its contribution is still significantly lower than that of ΔV_{IT} as shown, and hence ΔV_{IT} still dominates the time evolution of overall ΔV_T especially at longer t_{STR}. However, no variation is observed in power-law field acceleration factor Γ_E for different Si cap thickness as shown in Table 4.8.

Figure 4.18 shows time evolution of measured ΔV_T in Si capped SiGe devices having high Ge% but different SiGe quantum well (QW) thickness [40]. All devices

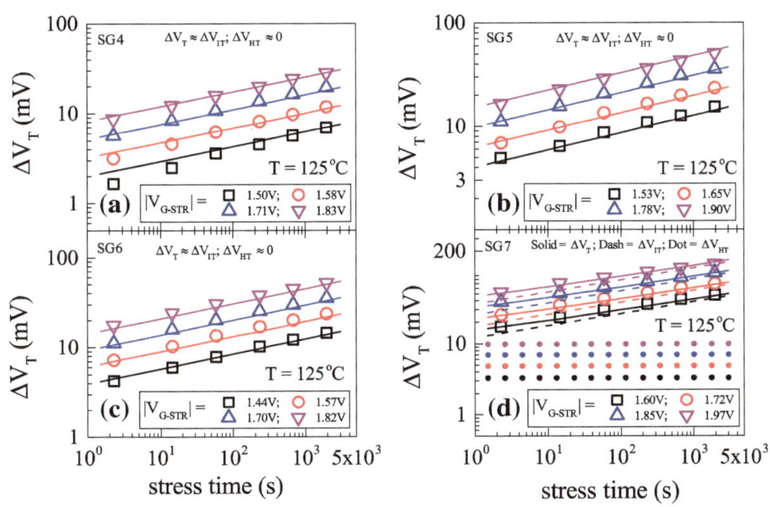

Fig. 4.17 Time evolution of ΔV_T from OSDD measurements and model prediction for NBTI stress at different $V_{G\text{-}STR}$ in Si capped SiGe based HKMG p-MOSFETs having different Si cap thickness. All devices, except the one with thickest Si cap, are modeled only using ΔV_{IT} subcomponent. The ΔV_{IT} and ΔV_{HT} subcomponents for the thickest Si cap device are shown. Experimental data (*symbols*) from [40], *lines* are model calculation

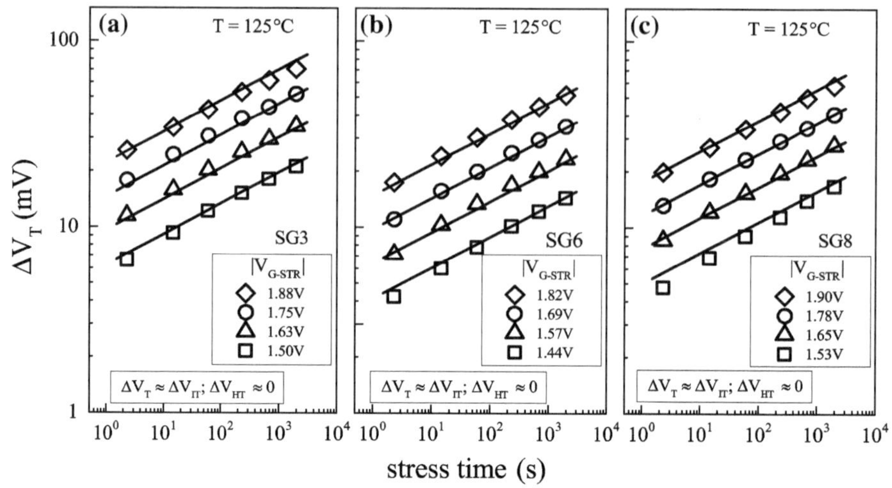

Fig. 4.18 Time evolution of ΔV_T from OSDD measurements and model prediction for NBTI stress at different $V_{G\text{-}STR}$ in Si capped SiGe based HKMG p-MOSFETs with different SiGe quantum well thickness. Experimental data (*symbols*) from [40], *lines* are model calculation

have identical moderately thick Si cap. Compact model prediction is shown; adjustable model parameters are listed in Table 4.8. Since these devices have high Ge concentration and not very thick Si cap, ΔV_{IT} completely dominates time evolution of ΔV_T and ΔV_{HT} is found to be negligible. Although the magnitude of ΔV_{IT} and hence that of ΔV_T reduces with increase in QW thickness, the power-law field acceleration Γ_E and time exponent n ($\sim 1/6$) remain unchanged as shown.

The relative contribution of ΔV_{IT} and ΔV_{HT} subcomponents to overall ΔV_T for different SiGe processes is of interest. Figure 4.19 plots ΔV_{IT} contribution normalized to overall ΔV_T, calculated at fixed E_{OX} for different Si and SiGe devices at (a–d) short t_{STR} and (e–h) t_{STR} of 10 years corresponding to end-of-life (EOL) degradation. The compact model shown in Table 4.1 is used for this calculation; model parameters are calibrated by prediction of measured data at different stress E_{OX}. A $\Delta V_{IT}/\Delta V_T$ value of 1 implies complete ΔV_{IT} domination, a value lower than 1 implies non-negligible ΔV_{HT} contribution.

Note that depending on gate insulator process, pure Si devices show [40] and do not show [41] ΔV_{HT} contribution. ΔV_T for SiGe devices without Si cap is completely dominated by ΔV_{IT} for different Ge concentration, and is also due to the particular gate insulator process. At shorter t_{STR}, SiGe devices with Si cap shows non-negligible ΔV_{HT} contribution only for lower Ge% or higher Si cap thickness; ΔV_{IT} completely dominates ΔV_T for all other process conditions leading to higher Ge surface concentration. However since ΔV_{HT} saturates at long time, EOL degradation is dominated by ΔV_{IT} for all devices as shown, which is similar to the HKMG process impact data shown earlier in this chapter.

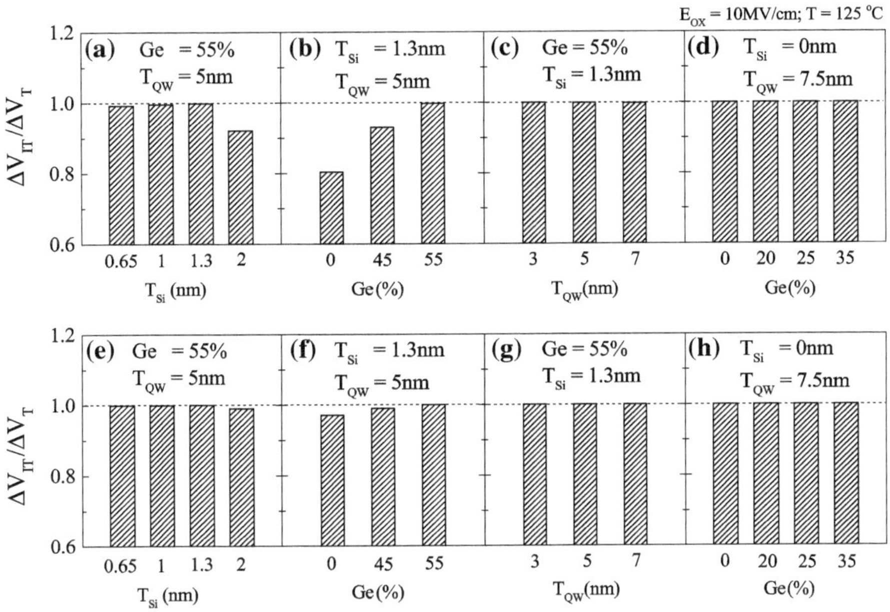

Fig. 4.19 Compact model calculated fractional contribution by generated interface traps ($\Delta V_{IT}/\Delta V_T$) at (*top panels*) short time and (*bottom panels*) end-of-life value of 10 years, for different Si and SiGe based HKMG p-MOSFETs listed in Table 4.5

4.4.3 Impact of Channel Orientation (FinFETs)

The compact model shown in Table 4.1 has also been used to predict NBTI degradation in Si channel based FinFETs with Replacement Metal Gate (RMG) HKMG gate insulator stacks. Devices with different channel length (L_{FIN}) and number of fins (N_{FIN}) are used. Measured time evolution of ΔV_T at different stress E_{OX} and T can be modeled using interface trap generation ΔV_{IT} and hole trapping ΔV_{HT} subcomponents. As mentioned before, the bulk trap generation subcomponent ΔV_{OT} is found to be negligible in these devices due to the use of HKMG gate insulators.

Figures 4.20 and 4.21 show time evolution of measured ΔV_T for L_{FIN} = 1 µm and L_{FIN} = 100 nm devices respectively, for NBTI stress at (a) different $V_{G\text{-}STR}$ but fixed T and (b) different T but fixed $V_{G\text{-}STR}$. Mobility corrected OSDD measurements are done at higher $V_{G\text{-}SNS}$ to reduce recovery. Both devices have identical N_{FIN} = 22. The compact model prediction of overall ΔV_T is also shown in panels (a) and (b), and the time evolution of underlying ΔV_{IT} and ΔV_{HT} subcomponents is shown for variation in (c) $V_{G\text{-}STR}$ and (d) stress T. The model can predict measured data reasonably well, and ΔV_T time evolution is dominated by ΔV_{IT} especially at long t_{STR}. Similar prediction of measured data is obtained for other combinations of $V_{G\text{-}STR}$ and T and in FinFETs having different N_{FIN}, not explicitly shown here for brevity.

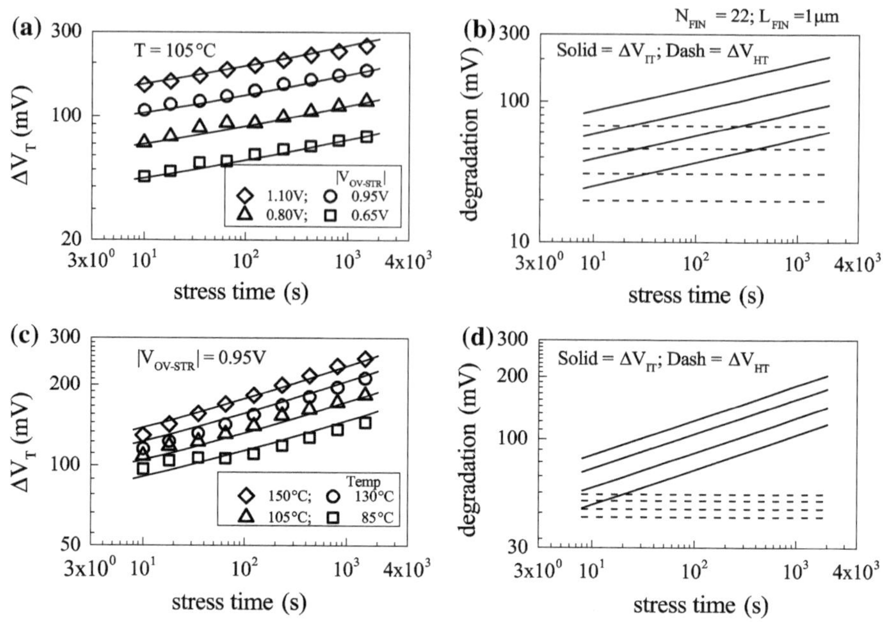

Fig. 4.20 Time evolution of ΔV_T from mobility corrected OSDD measurements at higher $|V_{G\text{-SNS}}|$ and model prediction (*left panels*) and time evolution of model subcomponents (*right panels*) for NBTI stress at different $V_{G\text{-STR}}$ and T in long channel HKMG p-FinFETs. *Symbols* measured data, *lines* model calculation

The L_{FIN} and N_{FIN} dependence of adjustable parameters are listed in Table 4.9. Measured ΔV_T time evolution can be predicted using only two adjustable parameters A and B corresponding to the pre-factors for ΔV_{IT} and ΔV_{HT} subcomponents, respectively. Other adjustable parameters Γ_{IT} ($=\Gamma_{HT}$) and E_{AKF} are kept fixed for different device dimensions due to the use of identical HKMG gate insulator process. Note that *A* and *B* reduce with reduction in L_{FIN}, however the relative ratio of trap generation and trapping remains constant across different L_{FIN} as expected. Moreover, negligible change is observed in parameters *A* and *B* with variation in N_{FIN} as shown.

Figure 4.22a shows time evolution of ΔV_T at fixed $V_{G\text{-STR}}$ and *T*, measured for NBTI stress in multiple small area FinFETs with L_{FIN} = 28 nm and N_{FIN} = 2. It is now well known that small-area devices show variable BTI degradation [46, 47], and results in distribution of ΔV_T magnitude and time exponent *n* as shown. The statistical aspects of NBTI degradation in these small-area devices are discussed in [42]. Figure 4.22b plots the time evolution of mean ΔV_T obtained by averaging measured data plotted in panel (a), and also shows overall model prediction and underlying ΔV_{IT} and ΔV_{HT} subcomponents. Only the pre-factors A and B are adjusted, while Γ_{IT} ($=\Gamma_{HT}$) and E_{AKF} are kept identical to larger area devices, refer to Table 4.9. Note that overall ΔV_T and hence the pre-factors A for ΔV_{IT} and *B* for ΔV_{HT} reduce as L_{FIN} is reduced and is consistent with larger area data. However, the ratio of *A* and *B*, i.e.,

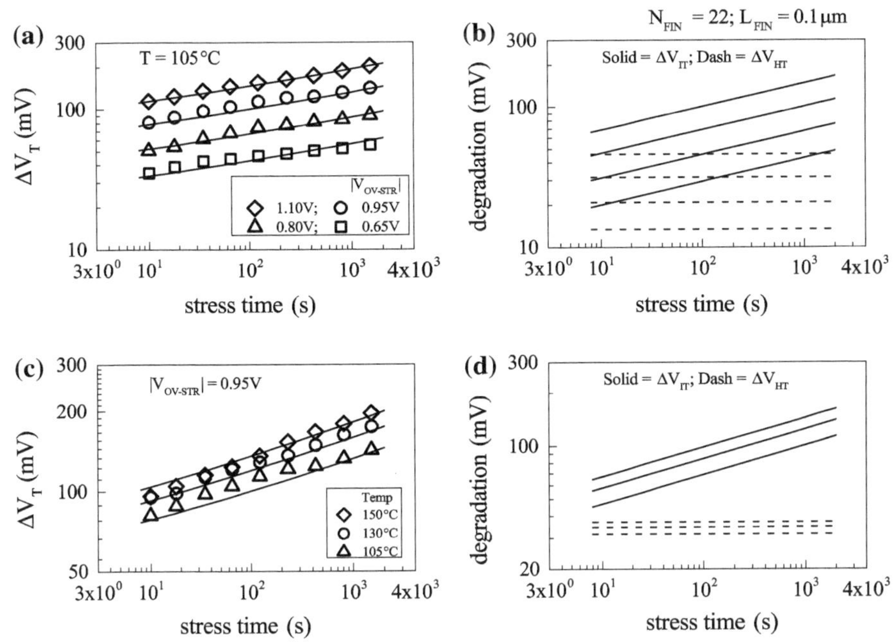

Fig. 4.21 Time evolution of ΔV_T from mobility corrected OSDD measurements at higher $|V_{G\text{-}SNS}|$ and model prediction (*left panels*) and time evolution of model subcomponents (*right panels*) for NBTI stress at different $V_{G\text{-}STR}$ and T in short channel HKMG p-FinFETs. *Symbols* measured data, *lines* model calculation

Table 4.9 NBTI compact model parameters for HKMG p-FinFETs, and compact model calculated fractional contribution by generated interface traps ($\Delta V_{IT}/\Delta V_T$) at short time and at 10 years

N_{FIN}	L_{FIN} (µm)	A $\left(\frac{1}{\text{V cm s}^{1/6}}\right)$	B $\left(\frac{1}{\text{V cm}}\right)$	$\Delta V_{IT}/\Delta V_T$ @ 1 Ks	$\Delta V_{IT}/\Delta V_T$ @ 10 years
2	0.028	3.5E11	5E10	0.79	0.97
22	0.1	4.8E11	7.5E10	0.78	0.97
	1	6E11	1.1E11	0.75	0.96
4	0.2	4.6E11	9E10	0.74	0.96
20		5.3E11	9E10	0.76	0.96
220		5.8E11	9E10	0.78	0.97

Note, Γ_{IT} (=Γ_{HT}) and E_{AKF} values are same as SG1 silicon device (Table 4.8)

the relative contribution of trap generation and trapping for small area very short channel devices remains similar to that observed for large area long and short channel devices. This is expected due to the use of identical HKMG process.

Table 4.9 also shows $\Delta V_{IT}/\Delta V_T$ fraction for FinFETs having different L_{FIN} and N_{FIN}, calculated using the calibrated compact model at short and long t_{STR} and at a particular E_{OX} and T. It is important to note that although ΔV_{IT} dominates, ΔV_{HT} is

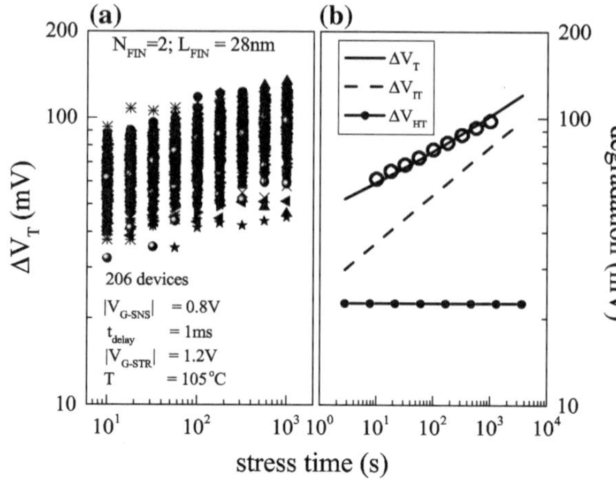

Fig. 4.22 Time evolution of ΔV_T using OSDD measurements in multiple small-area HKMG p-FinFETs, for NBTI stress at fixed $V_{G\text{-}STR}$ and T (*left panel*). Time evolution of measured mean ΔV_T, and compact model calculated ΔV_T and underlying ΔV_{IT} and ΔV_{HT} subcomponents (*right panel*). Identical $V_{G\text{-}SNS}$ is used as large area devices. *Symbols* measured data, *lines* model calculation

not negligible especially at shorter t_{STR}, and hence $\Delta V_{IT}/\Delta V_T < 1$. Note that identical $\Delta V_{IT}/\Delta V_T$ ratio is observed for all devices as expected. However, ΔV_{HT} saturates quickly and therefore the long-time ΔV_T is completely dominated by ΔV_{IT} as shown. Therefore, similar to planar Si and SiGe devices, NBTI degradation in FinFETs at EOL is also dominated by generation of interface traps; hole trapping in pre-existing defects have relatively minor contribution.

4.5 Process Impact on NBTI Parameters

In this section, impact of different gate insulator processes on NBTI parameters at long stress time; i.e., power-law time exponent (n), Arrhenius temperature activation (E_A) and power-law oxide field acceleration (Γ_E), is explored. Moreover, various assumptions mentioned earlier to achieve compactness of the proposed NBTI model are also verified.

4.5.1 Time Exponent

Figure 4.23a plots the time evolution of UF-MSM measured ΔV_T for different measurement delay in Si planar HKMG p-MOSFETs subjected to identical NBTI stress. As discussed in Chap. 1, Fig. 1.21, BTI degradation starts to recover as soon

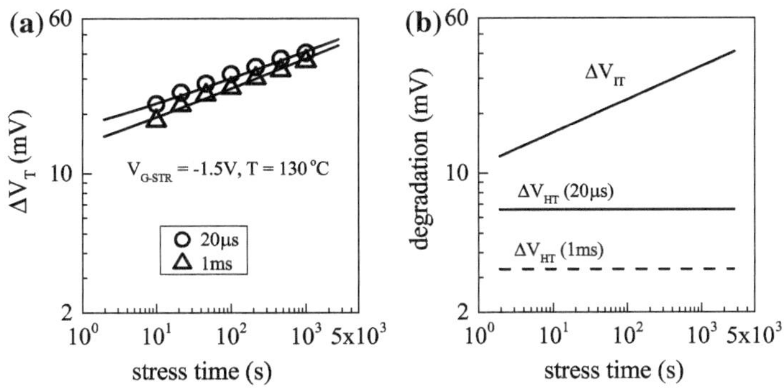

Fig. 4.23 Time evolution of (a) MSM measured ΔV_T for different measurement delay for NBTI stress in HKMG p-MOSFETs and model prediction, (b) underlying ΔV_{IT} and ΔV_{HT} subcomponents. *Symbols* measured data, *lines* model calculation

as the stress is interrupted for measurement, and larger measurement delay exhibits lower ΔV_T magnitude and higher power-law time exponent n, refer to Chap. 2 for additional details. The compact model calculation is also shown, and ΔV_T time evolution for different delay can be predicted by identical trap generation but different trapping subcomponents, plotted in Fig. 4.23b. As shown later in Chap. 6, the initial part of NBTI recovery is dominated by hole detrapping, so MSM measurement with larger delay would capture less trapped holes and vice versa. Therefore, MSM measured ΔV_T time evolution at different delay can be predicted using same ΔV_{IT} but different ΔV_{HT} subcomponents. Note that ΔV_{HT} saturates at longer t_{STR} and ΔV_{IT} shows power-law time dependence with $n \sim 1/6$. Therefore, larger ΔV_{HT} at smaller delay results in lower n of overall ΔV_T and vice versa as shown.

Note that although measurement delay has identical impact on MSM and OTF measurements, they are caused by different mechanisms as discussed in Chap. 2. It can be seen from Fig. 4.2 that larger t_0 delay for OTF measurements also results in lower ΔV_T magnitude and higher time exponent n. Since OTF technique measures degradation without interruption of stress, it does not suffer from recovery issues similar to the MSM method. However, the magnitude of degradation captured by OTF method at the initiation of stress depends on t_0 delay. Since early degradation is primarily caused by hole trapping , refer to Chap. 6, ΔV_T time evolution at different t_0 delay can be modeled using same ΔV_{IT} but different ΔV_{HT} subcomponents as shown in Fig. 4.2. Once again, as ΔV_{HT} saturates at longer t_{STR}, a relatively larger ΔV_{HT} contribution captured using OTF method for smaller t_0 delay, when added to ΔV_{IT} time evolution having $n \sim 1/6$, results in lower n of overall ΔV_T as shown.

Figure 4.24 plots power-law time exponent n calculated from ΔV_T time evolution data by linear regression in t_{STR} range of 10 s to 1 Ks, versus stress oxide field E_{OX}, for (a) SiON and (b) HKMG devices having different N content in gate stack. As mentioned before, SiON data are measured using OTF and HKMG data using UF-MSM method. Note that both SiON and HKMG devices having higher N content show lower time exponent n. As mentioned earlier in this chapter and also

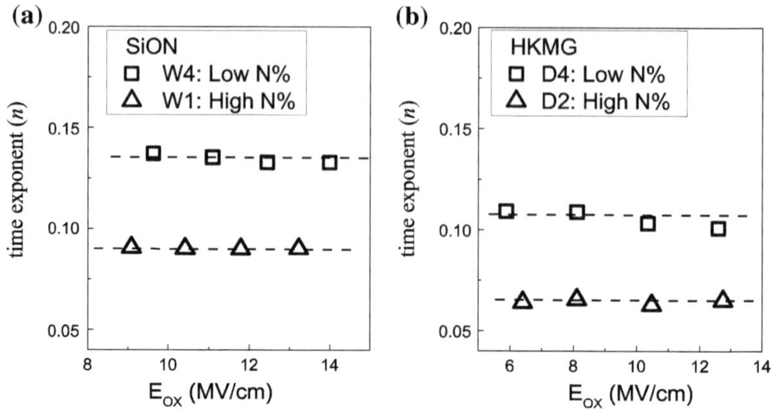

Fig. 4.24 Measured power-law time exponent n at long t_{STR} for NBTI stress using different E_{OX} in **a** SiON and **b** HKMG p-MOSFETs having different N content in the gate insulator stack. *Symbols* measured data, *lines* guide to eye

in Chap. 2, pre-stress flicker noise results in larger trap density in nitrided gate stacks, and results in larger hole trapping subcomponent ΔV_{HT} during NBTI stress. Therefore, a relatively larger saturated ΔV_{HT} contribution, when added to time evolution of ΔV_{IT} having $n \sim 1/6$ power-law time dependence, lowers the time exponent n of overall ΔV_T for nitrided devices as shown. This once again verifies that hole trapping and trap generation are two mutually independent subcomponents and the former saturates at longer stress time.

The correlation of N concentration in the gate stack and NBTI time exponent is further verified for SiON and HKMG devices. Figure 4.25a shows time exponent obtained using linear regression of OTF measured ΔV_T time evolution in t_{STR} range of 10 s to 1 Ks, as a function of N concentration in different plasma nitrided SiON devices [38]. N concentration is measured using X-ray Photoelectron Spectroscopy

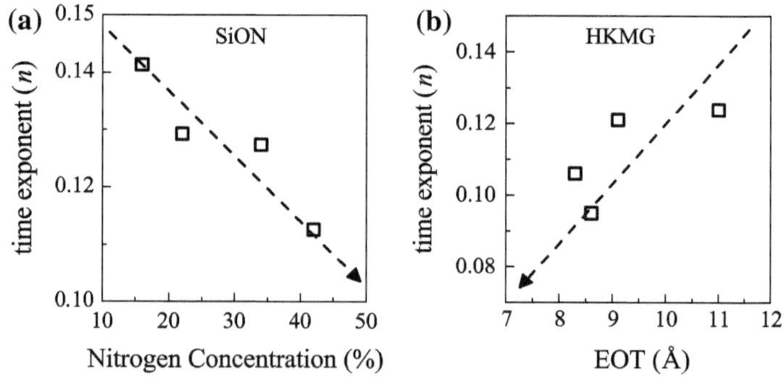

Fig. 4.25 Measured power-law time exponent n at long t_{STR} for NBTI stress in **a** SiON p-MOSFETs having different N content in the gate insulator stack, and **b** HKMG p-MOSFETs having different IL thickness. *Symbols* measured data, *lines* guide to eye

(XPS) technique [48]. Note that measured n reduces with increase in N concentration. Figure 4.25b shows time exponent obtained using linear regression of MSM measured ΔV_T time evolution in t_{STR} range of 10 s to 1 Ks, as a function of EOT of the gate stack for HKMG devices. EOT is estimated using Capacitance-Voltage (CV) method after necessary corrections [49]. As discussed earlier, refer to Table 4.4, EOT scaling is achieved either by nitridation after HK deposition or by using nitrided IL, therefore, EOT scaling also leads to reduction in time exponent n as shown. It has been shown using flicker noise method in SiON [43] and HKMG [33] p-MOSFETs that gate stacks with higher N concentration show higher pre-existing traps, and would result in higher hole trapping subcomponent ΔV_{HT} during NBTI stress. As discussed before, a relatively larger ΔV_{HT} contribution that saturates at longer t_{STR}, when added to ΔV_{IT} that continues to evolve with power-law time exponent $n \sim 1/6$, reduces time exponent n of overall ΔV_T. Once again, this verifies the uncoupled nature of trap generation and trapping subcomponents of NBTI degradation.

The impact of hole trapping on long-time exponent is further validated using a larger set of data; refer to [50] for additional details. Figure 4.26 plots the correlation of measured time exponent n and calculated fractional hole trapping contribution ($\Delta V_{HT}/\Delta V_T$) for various SiON and HKMG devices subjected to NBTI stress. Time exponent is calculated using linear regression of measured ΔV_T time evolution in t_{STR} range of 10 s to 1 Ks. The calibrated compact model shown earlier using Table 4.1 is used to calculate ΔV_{HT} and ΔV_T at fixed E_{OX} and T, and a large t_{STR} is used to capture saturated ΔV_{HT} magnitude. SiON and HKMG-A devices are listed earlier in Tables 4.3 and 4.4, respectively. HKMG-B and HKMG-C devices respectively have Si and SiGe channels, and both have Aluminum capped HKMG stacks and are listed in [50]. HKMG-D devices use HfO_2 and Hafnium Silicate (HfSiO) HKMG stacks, refer to [51] for device details. UF-OTF method is used for SiON and HKMG-D, UF-MSM for HKMG-A, and OSDD for HKMG-B and HKMG-C devices to obtain ΔV_T time evolution. Moreover, published data from

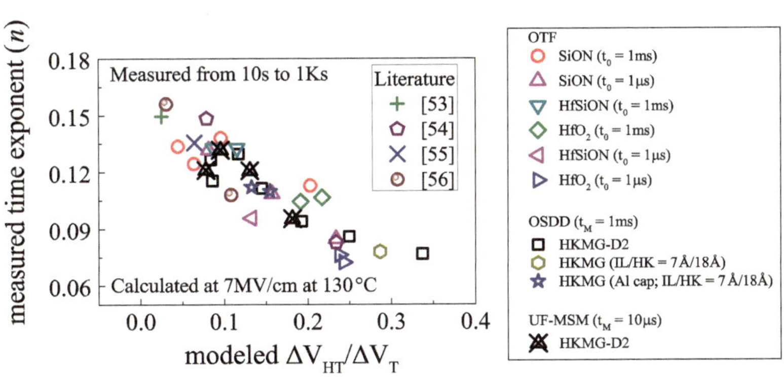

Fig. 4.26 Correlation of measured power-law time exponent n at long t_{STR} to compact model calculated fractional hole trapping contribution ($\Delta V_{HT}/\Delta V_T$) for NBTI stress in different SiON and HKMG devices

other reports [4, 52, 53, 54, 55] are also analyzed and plotted. An exceptional universality is observed for different devices and measurements, which unequivocally establishes that increase in fractional hole trapping contribution reduces the long-time power-law time exponent n of overall ΔV_T, which is due to saturation of ΔV_{HT} at long stress time. This also verifies uncorrelated nature of NBTI subcomponents.

Finally, note that the interface trap generation subcomponent shows power-law time dependence with time exponent $n \sim 1/6$, refer to Chap. 3 for additional details. As established above, the hole-trapping subcomponent saturates very early, and is strongly impacted by the quality of gate insulator stack. Moreover, owing to large voltage acceleration, the bulk trap generation subcomponent is negligible unless very large $V_{G\text{-}STR}$ is used for stress. Therefore, for well-processed devices and circuits, only the interface trap generation component would dominate overall NBTI degradation at moderate level of stress bias and longer stress time. As a validation, Fig. 4.27a shows $V_{G\text{-}STR}$ dependence of power-law time exponent extracted from long-time ΔV_T time evolution measurements in individual transistors [56–59], and Fig. 4.27b shows $V_{G\text{-}STR}$ dependence of power-law time exponent extracted from time evolution of Ring Oscillator frequency degradation [60, 61] and SRAM V_{MIN} shift [62, 63] measurements in circuits. It is important to remark that the universal $n \sim 1/6$ time exponent is observed across such large variety of devices and circuits and also across different industries [18]. This confirms that long-time NBTI degradation is dominated by interface trap generation and shows power-law time dependence with a time exponent of $n \sim 1/6$, which is readily predicted by H/H_2 Reaction-Diffusion (RD) model [5, 10]; refer to Chap. 5 for further details.

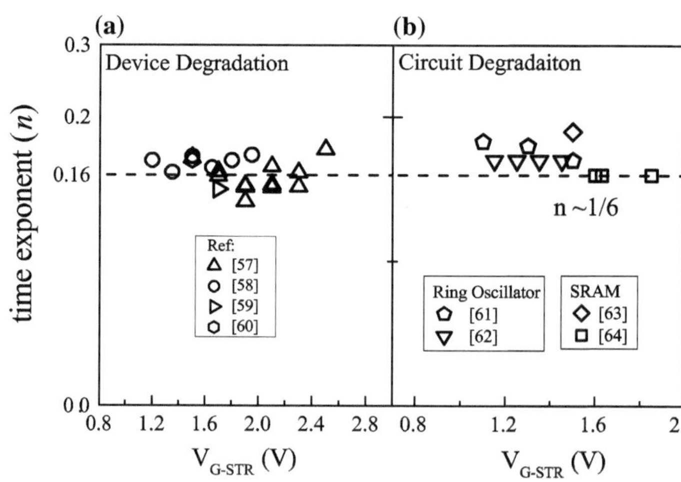

Fig. 4.27 Measured power-law time exponent n at long t_{STR} for NBTI stress in **a** p-MOSFETs across various technologies and **b** different type of circuits. *Symbols* measured data, *lines* guide to eye

4.5.2 Temperature Activation

Figure 4.28 shows measured ΔV_T at fixed NBTI stress time t_{STR} as a function of stress temperature T, for SiON and HKMG devices. SiON devices are measured using OTF method having different t_0 delay, Fig. 4.28a, and UF-OTF method is also used to measure SiON devices having different N content in gate insulator stack, Fig. 4.28b. UF-MSM method is used to study HKMG devices having different *N* content in gate insulator stack, Fig. 4.28c. Note that higher ΔV_T magnitude and lower *T* activation energy E_A are observed for faster OTF measurements and for SiON and HKMG devices having higher *N* content. As mentioned earlier, these situations favor relatively larger hole trapping subcomponent ΔV_{HT}, which is presumed to have lower *T* activation compared to trap generation subcomponent ΔV_{IT}, refer to Table 4.1 for details. Therefore, larger fractional ΔV_{HT} contribution reduces *T* activation of overall ΔV_T for faster measurements and for devices having higher N content as shown, and once again, verifies the uncorrelated nature of trap generation and trapping.

The correlation of temperature activation energy and N concentration is further validated for SiON and HKMG devices. Figure 4.29 shows temperature activation energy E_A for (a) SiON devices having different N content and (b) HKMG devices with EOT scaling achieved by increased N concentration in the gate stack. Once again, *T* activation is obtained from ΔV_T measured at fixed t_{STR}, using UF-OTF method for SiON and UF-MSM for HKMG devices. As shown earlier in Fig. 4.25, increased N% for SiON and EOT scaling for HKMG gate stacks result in higher fractional ΔV_{HT} contribution. Since hole trapping has lower *T* activation compared to trap generation, relatively larger ΔV_{HT} reduces *T* activation of overall ΔV_T for

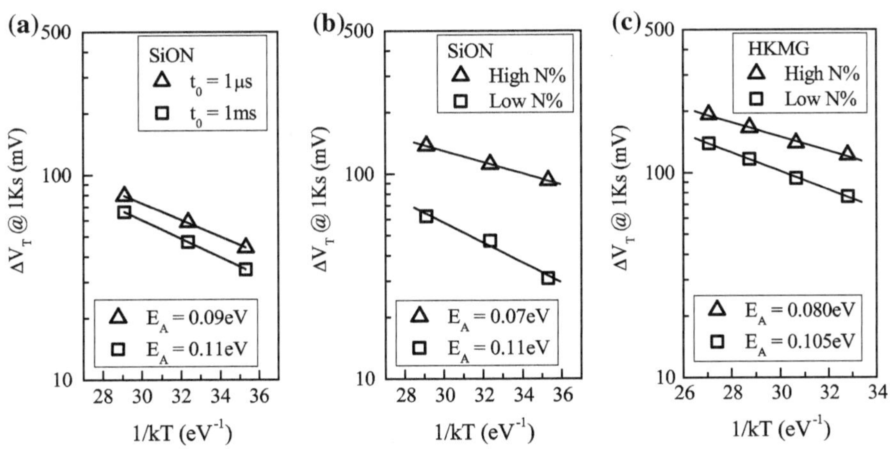

Fig. 4.28 Measured ΔV_T at fixed stress time for NBTI stress at different T, in **a** SiON p-MOSFETs for different OTF time-zero delay and in **b** SiON, and **c** HKMG p-MOSFETs having different *N* content in the gate insulator stack. *Symbols* measured data, *lines* model calculation

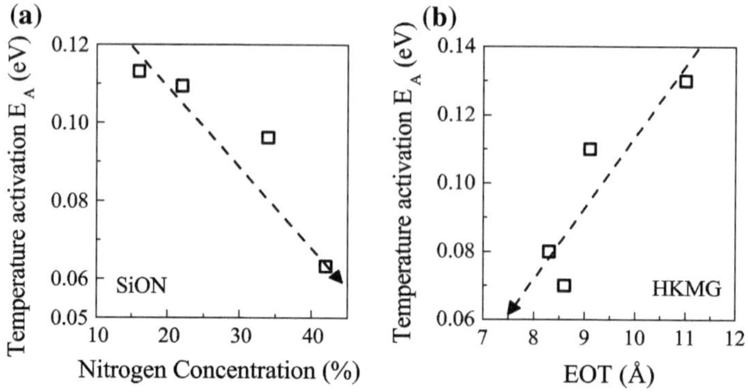

Fig. 4.29 Measured Arrhenius T activation energy E_A at fixed stress time for NBTI stress in **a** SiON p-MOSFETs having different *N* content in gate insulator stack, and **b** HKMG p-MOSFETs having different IL thickness. *Symbols* measured data, *lines* guide to eye

higher N% SiON and lower EOT HKMG devices. Once again, this validates the model assumption of lower *T* activation for the hole trapping subcomponent.

As discussed before, hole trapping saturates at longer t_{STR} and results in lower power-law time exponent *n* for time evolution of overall ΔV_T in situations involving relatively larger ΔV_{HT} contribution, such as faster NBTI measurement and for gate stacks having higher *N* content. Moreover as shown above, hole trapping also shows lower *T* activation and reduces E_A of overall ΔV_T for these situations. The similarity is further illustrated in Fig. 4.30, showing the correlation of time exponent *n* and *T* activation E_A for SiON devices with different *N* content and HKMG devices having different EOT achieved by increased *N* content in the gate stack. Time exponent *n* is obtained using linear regression of ΔV_T time evolution data in

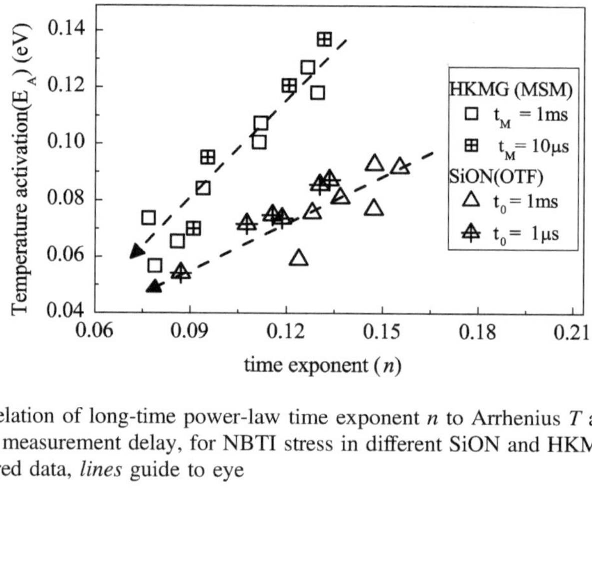

Fig. 4.30 Correlation of long-time power-law time exponent *n* to Arrhenius *T* activation energy E_A for different measurement delay, for NBTI stress in different SiON and HKMG p-MOSFETs. *Symbols* measured data, *lines* guide to eye

t_{STR} range of 10 s to 1 Ks, and T activation is obtained at fixed t_{STR}. OTF method with different t_0 delay is used for SiON, while MSM method with different measurement delay is used for HKMG devices. As expected, n and E_A of overall ΔV_T are nicely correlated for both SiON and HKMG devices and for different measurement delay. However, the correlation line has been found to be different between SiON and HKMG devices, due to higher T activation energy E_{AKF} associated with the ΔV_{IT} subcomponent for HKMG devices as shown earlier in the chapter, refer to Tables 4.6 and 4.7.

According to the RD model discussed in detail in Chap. 5, ΔV_{IT} time evolution at longer stress time is governed by molecular H_2 diffusion and shows power-law time dependence with exponent $n \sim 1/6$ [10] and T activation $E_{ADH2} \sim 0.6$ eV [64]. It can be observed from Table 4.1 that for situations where E_{AKF} is similar to E_{AKR}, i.e., Si–H bond dissociation and passivation are similarly activated; the temperature activation of ΔV_{IT} at a fixed t_{STR} would be given by $E_{AIT} = E_{ADH2}/6$. On the other hand, temperature activation of the time to reach a particular ΔV_{IT} magnitude at different T would be equal to E_{ADH2}. Therefore, time evolution of ΔV_{IT} obtained at different stress T and plotted in a log–log plot can be scaled along time or X-axis to merge with each other, and the scale factor (t_{SCALE}) can be determined at every T; the temperature activation of $1/t_{SCALE}$ would be equal to E_{ADH2}. For situations when ΔV_{HT} is negligible, such as low N% devices, relatively slower measurement, long t_{STR}, and ΔV_{OT} is also negligible such as not too large $V_{G\text{-}STR}$, ΔV_T would be dominated by ΔV_{IT}, and a similar exercise involving ΔV_T time evolution data at different stress T would also provide E_{ADH2}.

Figure 4.31 shows T activation of $1/t_{SCALE}$ calculated from ΔV_T time evolution measured in pure SiO_2 as well as different plasma nitrided SiON devices with low

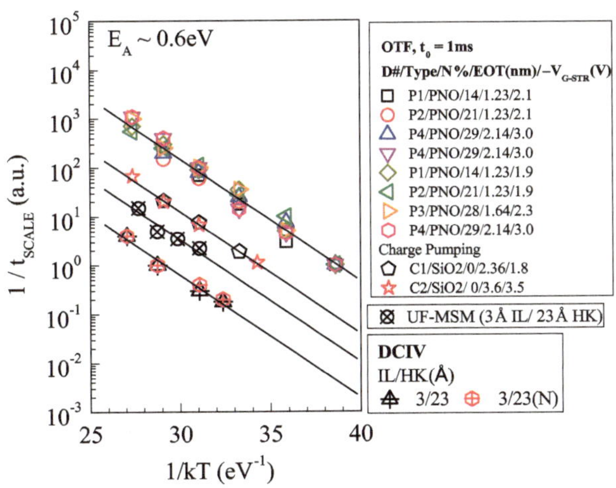

Fig. 4.31 T activation of time to reach a particular degradation for SiO_2, SiON, and HKMG p-MOSFETs under NBTI stress, measured using OTF I_{DLIN}, CP, and DCIV methods (*symbols*). CP and DCIV data are plotted after delay and band gap corrections. *Lines* are fit to data

N content in the gate insulator stack. Measurements are performed using $t_0 = 1$ ms OTF method and moderate $V_{G\text{-}STR}$ is used for stress, ensuring negligible contribution from ΔV_{HT} and ΔV_{OT}, respectively; hence ΔV_T is dominated by ΔV_{IT}. Similar data are also shown for HKMG devices with low N content, measured using MSM method for stress at moderate $V_{G\text{-}STR}$. Moreover, direct ΔV_{IT} time evolution measurements are done using CP in SiO_2 devices and using DCIV in HKMG devices, and the corresponding T activation values are also shown. It is important to note that a remarkable consistency in T activation of $1/t_{SCALE}$ is observed for such diverse set of devices and measurement conditions, and obtained E_A is consistent with that reported for molecular H_2 diffusion [64], which is also consistent with observed power-law time evolution with exponent $n \sim 1/6$ at long stress time. This verifies the time and T dependence of ΔV_{IT} subcomponent of the compact model. Refer to Chap. 6 for further discussion in this topic.

4.5.3 Field Acceleration

The oxide field dependence of interface trap generation and hole trapping comes from inversion layer holes tunneling to Si–H bonds and pre-existing traps respectively [10], and therefore is related to the product of inversion hole density ($\sim E_{OX}$) and its tunneling probability ($\sim \exp \cdot (\Gamma \cdot E_{OX})$) as shown in Table 4.1 [57]. Identical field acceleration factor is assumed for trap generation (Γ_{IT}) and trapping (Γ_{HT}), as they occur close to the channel and gate insulator interface [4, 10]. However for simplicity and without loss of generality, power-law E_{OX} dependence with acceleration factor Γ_E is often used, as shown elsewhere in the book. Bulk trap generation depends on stress gate voltage and has power-law dependence [5], however, this component is often negligible as discussed earlier in this chapter. Hence in most situations, interface trap generation and hole-trapping processes determine field acceleration for overall ΔV_T. It is important to study the impact of gate insulator process on E_{OX} dependence, as it determines the extrapolation of degradation from stress to use condition.

Figure 4.32 shows measured ΔV_T for NBTI stress at fixed t_{STR} as a function of stress oxide field E_{OX} for SiON and HKMG devices. SiON devices are measured using OTF method having different t_0 delay, Fig. 4.32a, and UF-OTF method is used to measure for SiON devices with different N content in gate insulator stack, Fig. 4.32b. UF-MSM method is used for HKMG devices having different N content in gate insulator stack, Fig. 4.32c. As discussed before, magnitude of ΔV_T is smaller for higher t_0 delay as lower hole trapping magnitude is captured. However, both measurements show identical field acceleration factor Γ_E as shown, which is due to similar E_{OX} dependence for ΔV_{IT} and ΔV_{HT} subcomponents. This is contrary to the T dependence shown earlier in Fig. 4.28a, where higher E_A is observed for larger t_0 delay when lower ΔV_{HT} is captured, due to relatively lower T activation E_{AHT} of the hole trapping component. However, both SiON and HKMG gate stacks having larger N content show larger ΔV_T magnitude but lower Γ_E. As discussed earlier in

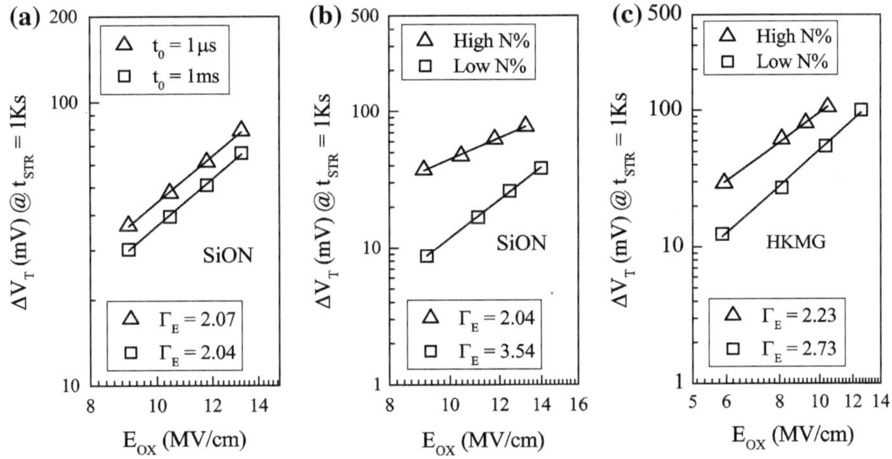

Fig. 4.32 Measured ΔV_T at fixed stress time for NBTI stress at different E_{OX}, in **a** SiON p-MOSFETs for different OTF time-zero delay and in **b** SiON and **c** HKMG p-MOSFETs having different N content in the gate insulator stack. Field dependence is plotted in a log–log scale. *Symbols* measured data, *lines* model calculation

the chapter, higher ΔV_T is primarily due to increased ΔV_{HT} contribution for SiON devices. For HKMG devices, increased *N* results in increase in both ΔV_{IT} and ΔV_{HT}, although ΔV_{HT} increases much more than ΔV_{IT}, and is also shown earlier in this chapter. More work is needed to understand the reduction in Γ_E for higher *N* content in the gate insulator stack for both SiON and HKMG devices as shown, which is beyond the scope of the present book.

Figure 4.33 shows power-law field acceleration factor Γ_E for (a) SiON devices with different *N* concentration and (b) HKMG devices having different EOT resulting from different *N* concentration in the gate stack. Once again, E_{OX}

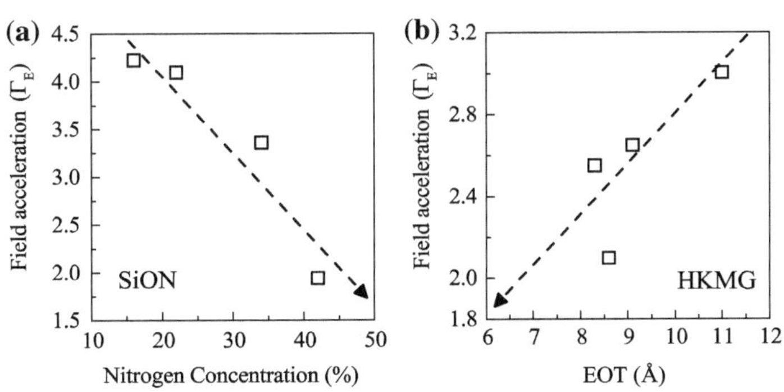

Fig. 4.33 Measured power-law field acceleration Γ_E at fixed stress time for NBTI stress in **a** SiON p-MOSFETs having different N content in gate insulator stack, and **b** HKMG p-MOSFETs having different IL thickness. *Symbols* measured data, *lines* guide to eye

dependent ΔV_T measurements are performed at fixed t_{STR} to determine Γ_E, using UF-OTF method for SiON and UF-MSM for HKMG devices. Note that Γ_E reduces as the N concentration is increased for SiON devices and also with EOT scaling in HKMG devices achieved using higher N content in the gate stack. It is interesting to remark that although the time and T dependence of ΔV_{IT} and ΔV_{HT} subcomponents are different from each other, they show very similar E_{OX} dependence, and hence Γ_E can be used to monitor N concentration in the gate stack and is shown below.

As discussed before, increased N concentration preferentially increases the hole trapping process compared to interface trap generation, and reduces parameters n, E_A and Γ_E for NBTI stress in both SiON and HKMG devices. Reduction in time exponent n is attributed to saturation of ΔV_{HT} subcomponent at longer t_{STR}, while reduction in T activation E_A is due to lower E_{AHT} associated with ΔV_{HT}. As mentioned before, the physical mechanism responsible for Γ_E reduction is not yet understood. However, Γ_E can be correlated to other parameters, and Fig. 4.34 plots (a) time exponent n and (b) temperature activation E_A as a function of power-law field acceleration Γ_E for SiON and HKMG devices. Note that reduction in n or E_A nicely correlates to reduction in Γ_E for both SiON and HKMG devices containing larger N density in the gate insulator stack. As mentioned earlier in this chapter, increased N concentration also results in larger ΔV_T during NBTI stress. Therefore, Γ_E can be used as an overall monitor of gate insulator quality, with higher Γ_E signifying better quality and vice versa.

Figure 4.35 plots the oxide field dependence of ΔV_T for NBTI stress in different production quality SiON and HKMG devices [18], extracted from published reports across multiple technologies [22, 24, 25, 65, 66]. Although physics-based analysis would suggest the correct field dependent expression as shown in Table 4.1 [57], a power-law E_{OX} dependence is used for ease of comparison across published reports. Power-law E_{OX} dependence having acceleration factor of $\Gamma_E \sim 4$ is universally observed for different SiON and gate first (GF) HKMG devices, and Γ_E becomes

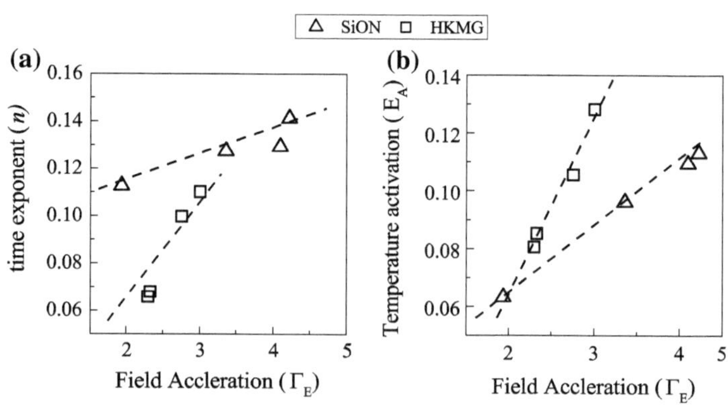

Fig. 4.34 Correlation of **a** long-time power-law time exponent n and **b** Arrhenius T activation energy E_A to power-law field acceleration Γ_E for NBTI stress in different SiON and HKMG p-MOSFETs. *Symbols* measured data, *lines* guide to eye

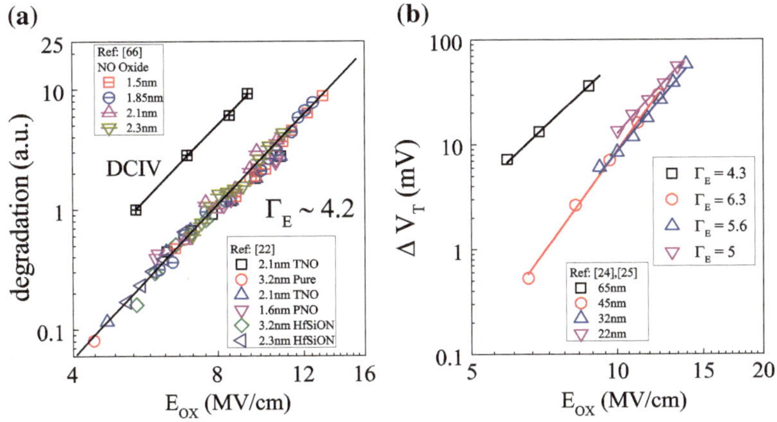

Fig. 4.35 Fixed time ΔV_T versus stress E_{OX} for **a** different SiON and gate first HKMG p-MOSFETs and for **b** SiON and different RMG HKMG p-MOSFETs. DCIV measured ΔN_{IT} data are also plotted after delay and band gap corrections for gate first HKMG p-MOSFET as a reference. *Symbols* measured data, *lines* fit to data

larger for RMG HKMG devices. As shown in Fig. 4.35b, increase in Γ_E is associated with reduction in ΔV_T magnitude for RMG HKMG compared to SiON devices. However, note that Γ_E reduces slightly with slight increase in ΔV_T magnitude as RMG HKMG technology is scaled. Moreover as a comparison, the E_{OX} dependence of ΔN_{IT} obtained using DCIV method in gate first HKMG devices after delay and band gap correction is also plotted in Fig. 4.35a [13]. A power-law dependence of $\Gamma_E \sim 4$ is also obtained as shown, which is consistent with ΔV_T measurement data, and verifies identical E_{OX} dependence for ΔV_{IT} and ΔV_{HT} subcomponents.

4.6 PBTI Process Dependence in HKMG n-MOSFETs

In this section, the physics-based PBTI compact model discussed in Table 4.2 is validated against different HKMG devices listed in Table 4.4. The model has trap generation subcomponents located at IL/High-K interface ($\Delta V_{IT\text{-}HK}$) as well as in High-K bulk ($\Delta V_{OT\text{-}HK}$), refer to Fig. 4.1, which are independently verified using DCIV and SILC measurements respectively, refer to Chap. 3 for additional details. As discussed in Chap. 2 and shown in Fig. 4.36, DCIV method probes traps aligned with the center of the Si band gap and SILC probes traps that are closer to High-K conduction band. Moreover as also discussed in Chap. 2, DCIV estimated traps are corrected for both measurement delay and band gap,[3] while SILC estimated traps

[3]The energy zone corresponding to trap generation at IL/High-K interface during PBTI stress is not known. Since kinetics of trap generation at channel/IL interface for NBTI and IL/High-K interface for PBTI is remarkably similar, see Chap. 3, identical band gap correction factor is assumed for

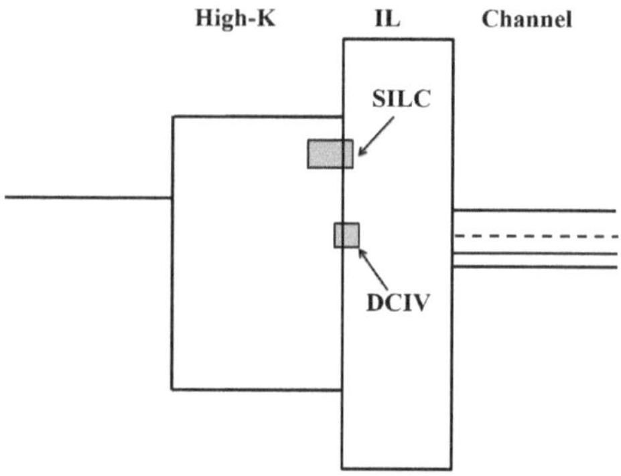

Fig. 4.36 Energy band diagram of a HKMG gate stack illustrating traps probed by DCIV and SILC methods

are corrected for measurement delay. Pre-stress High-K bulk trap density is independently estimated using flicker noise measurements, refer to Chap. 2, and these traps are responsible for electron trapping (ΔV_{ET}) during PBTI stress. Overall ΔV_T is measured using UF-MSM method and modeled as uncorrelated sum of $\Delta V_{IT\text{-}HK}$, ΔV_{ET} and $\Delta V_{OT\text{-}HK}$ subcomponents [32, 34], see Table 4.2 for model equations and list of fixed and process dependent parameters.

Figure 4.37 shows the time evolution of UF-MSM measured ΔV_T (left panels) as well as of DCIV measured $\Delta V_{IT\text{-}HK}$ and SILC measured $\Delta V_{OT\text{-}HK}$ (right panels) in different HKMG n-MOSFETs for PBTI stress at constant $V_{G\text{-}STR}$ and T. Contrary to NBTI data shown earlier, D3 and D4 devices having nitrided gate insulator stacks show lower PBTI degradation and lower trap generation compared to non-nitrided D1 and D2 devices. Long-time degradation has power-law time dependence, with similar time exponent n for ΔV_T and $\Delta V_{IT\text{-}HK}$, but large n for $\Delta V_{OT\text{-}HK}$. Model prediction for ΔV_T and underlying $\Delta V_{IT\text{-}HK}$, ΔV_{ET} and $\Delta V_{OT\text{-}HK}$ subcomponents are also shown. The model can successfully estimate measured time evolution of ΔV_T, $\Delta V_{IT\text{-}HK}$ and $\Delta V_{OT\text{-}HK}$, and uses a saturated value for ΔV_{ET}. Device dependent variable parameter values are listed in Table 4.10 and fixed parameters are listed in Table 4.2. Note that $\Delta V_{IT\text{-}HK}$ dominates overall ΔV_T for non-nitrided devices D1 and D2. Although ΔV_{ET} increases significantly and $\Delta V_{IT\text{-}HK}$ reduces slightly for nitrided devices D3 and D4, $\Delta V_{IT\text{-}HK}$ still dominates ΔV_T especially at longer t_{STR}. For all devices, $\Delta V_{OT\text{-}HK}$ contribution is much less as shown.

Figure 4.38 plots UF-MSM measured time evolution of ΔV_T at different $V_{G\text{-}STR}$ and T for PBTI stress different HKMG n-MOSFETs. Compact model calculations are also plotted; fixed and device dependent parameters are listed in Tables 4.2 and 4.10, respectively. Note that six device dependent parameters are used; pre-factors A, B and C corresponding to $\Delta V_{IT\text{-}HK}$, ΔV_{ET} and $\Delta V_{OT\text{-}HK}$ subcomponents, oxide field acceleration factors $\Gamma_{IT\text{-}HK}$ and Γ_{ET} associated with $\Delta V_{IT\text{-}HK}$ and ΔV_{ET}, and

(Footnote 3 continued)

NBTI and PBTI stress. Universal recovery expression is used for measurement delay correction. Refer to Chap. 2 for details.

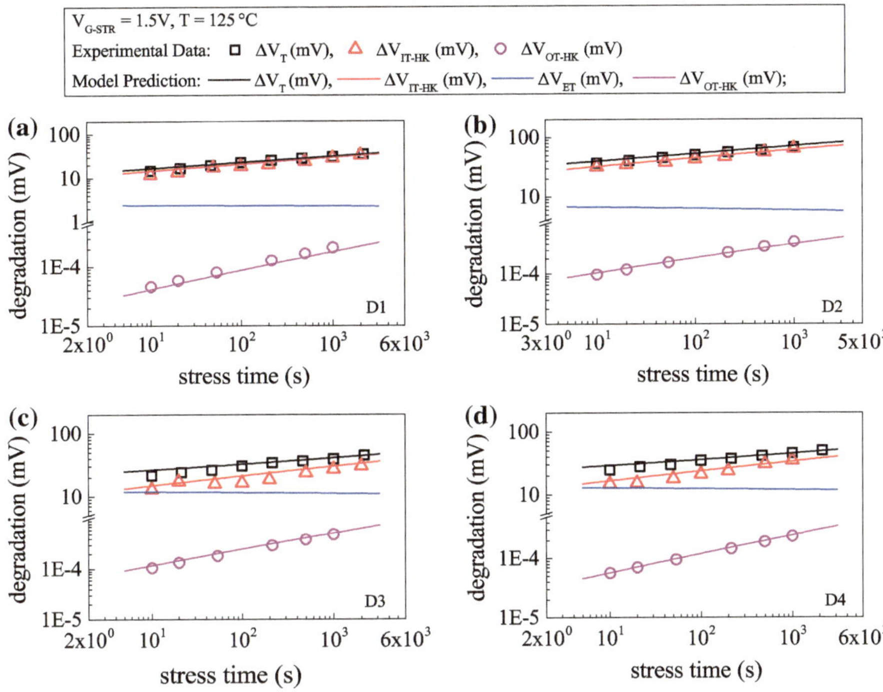

Fig. 4.37 Time evolution of UF-MSM measured ΔV_T, DCIV measured $\Delta V_{IT\text{-}HK}$ as well as SILC measured $\Delta V_{OT\text{-}HK}$ (*symbols*) for PBTI stress in different HKMG n-MOSFETs. Refer to Table 4.4 for device details. Compact model calculation (*lines*) of overall ΔV_T and underlying trapping and trap generation subcomponents are shown. DCIV data plotted after delay and band gap correction, SILC data after delay correction

Table 4.10 HKMG process dependent PBTI compact model parameters

Device no.	A $\left(\frac{1}{\text{V cm s}^{1/6}}\right)$	B $\left(\frac{1}{\text{V cm}}\right)$	C (cm^{-2})	Γ_{IT-HT} $\left(\frac{\text{cm}}{\text{MV}}\right)$	Γ_{ET} $\left(\frac{\text{cm}}{\text{MV}}\right)$	$E_{AIT\text{-}HK}$ (eV)
HKMG D1	8E10	3E09	4E13	0.23	0.21	0.24
HKMG D2	8.65E10	3E09	2.5E13	0.18	0.206	0.206
HKMG D3	1.1E10	6.5E09	2E13	0.205	0.19	0.19
HKMG D4	4.3E10	7E09	2.1E13	0.188	0.206	0.206

T activation $E_{AIT\text{-}HK}$ for $\Delta V_{IT\text{-}HK}$ [32, 34]. Unlike NBTI, the field acceleration parameters $\Gamma_{IT\text{-}HK}$ and Γ_{ET} associated with trap generation and trapping respectively at IL/High-K interface and High-K bulk are found to be somewhat different, which likely indicates different physical mechanisms for these processes and need further investigation. It is important to remark that trap generation reduces while trapping increases for nitrided devices, which gets reflected in device dependent pre-factors A, B, and C as evident from Table 4.10. The compact model can predict measured data quite well as shown.

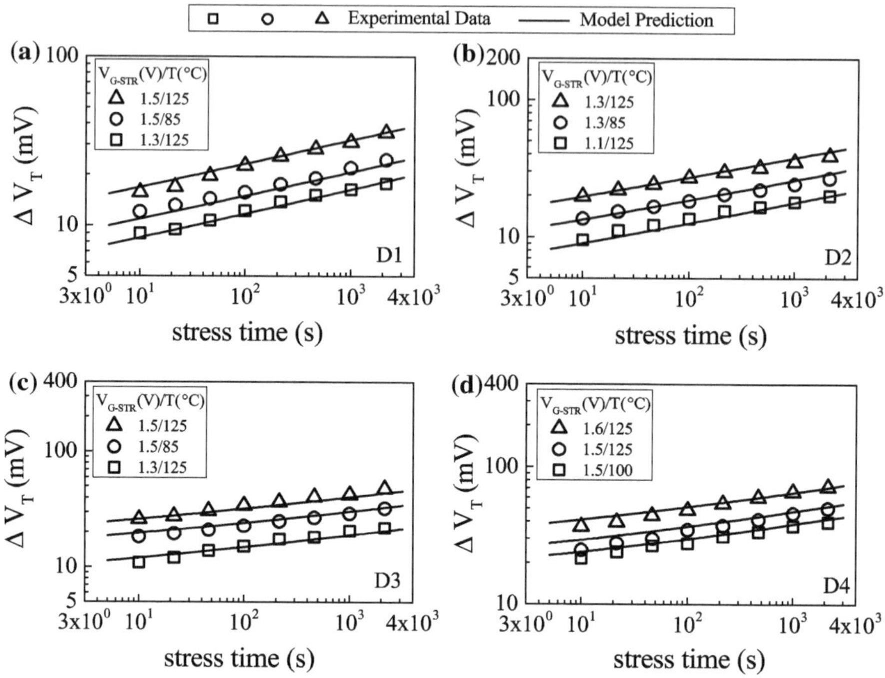

Fig. 4.38 Time evolution of UF-MSM measured ΔV_T (*symbols*) and compact model calculation (*lines*) for PBTI stress at different $V_{G\text{-}STR}$ and T in n-MOSFETs having different HKMG gate insulator stacks

Figures 4.39 and 4.40 respectively plots E_{OX} and T dependence of UF-MSM measured ΔV_T, DCIV measured $\Delta V_{IT\text{-}HK}$ and SILC measured $\Delta V_{OT\text{-}HK}$ at fixed t_{STR} for PBTI stress in different HKMG devices. As mentioned before, as-measured DCIV and SILC data are corrected for comparison with UF-MSM data.

Note that the magnitude, E_{OX} and T dependence of ΔV_T and $\Delta V_{IT\text{-}HK}$ are remarkably similar, and this similarity holds good for all devices, however, the corresponding values for $\Delta V_{OT\text{-}HK}$ are significantly different as shown. This reaffirms the dominance of IL/High-K interface trap generation on PBTI degradation. The model prediction of ΔV_T and its underlying $\Delta V_{IT\text{-}HK}$, ΔV_{ET} and $\Delta V_{OT\text{-}HK}$ subcomponents are shown, calculated by using fixed and device dependent parameters shown in Tables 4.2 and 4.10 respectively; calculated degradation is in good agreement with measurements. Note that $\Delta V_{IT\text{-}HK}$ is always larger than ΔV_{ET}, although the former reduces slightly while the latter increases for nitrided devices D3 and D4 as shown. Since ΔV_{ET} has lower T activation than $\Delta V_{IT\text{-}HK}$ as evident from listed model parameters, relative increase in contribution from ΔV_{ET} reduces the overall T activation of ΔV_T for nitrided devices, and verifies the uncoupled nature of trap generation and trapping processes.

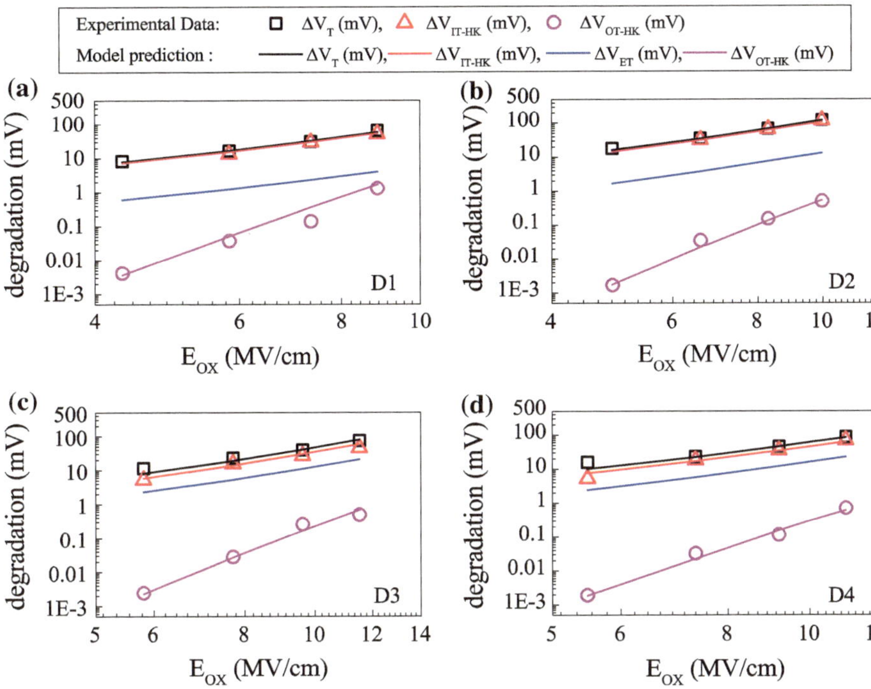

Fig. 4.39 UF-MSM measured ΔV_T, DCIV measured $\Delta V_{IT\text{-}HK}$ after delay and band gap corrections and SILC measured $\Delta V_{OT\text{-}HK}$ after delay correction (*symbols*) at fixed t_{STR} as a function of stress E_{OX} for PBTI stress in n-MOSFETs having different HKMG gate insulator stacks. Compact model prediction (*lines*) of ΔV_T and underlying trap generation and trapping subcomponents are shown. E_{OX} dependence is plotted in a log–log scale

Figure 4.41 correlates independently measured (a) $\Delta V_{IT\text{-}HK}$ from DCIV and (b) $\Delta V_{OT\text{-}HK}$ from SILC at fixed t_{STR} to the corresponding values calculated using the compact model of Table 4.2, for PBTI stress using different E_{OX} and T in different HKMG devices. Once again, as-measured DCIV and SILC data are suitably corrected. Figure 4.41c correlates pre-stress High-K trap density measured using flicker noise method to model calculated fractional electron trapping ($\Delta V_{ET}/\Delta V_T$) for different devices. Model calculation agrees well with measurement and hence verifies the accuracy of underlying subcomponents and overall PBTI degradation.

Note that pre-stress High-K trap density depends on gate insulator process and determines the ΔV_{ET} subcomponent of overall ΔV_T as shown in Fig. 4.41c. Since ΔV_{ET} saturates at long t_{STR} as described in Table 4.2 while $\Delta V_{IT\text{-}HK}$ and $\Delta V_{OT\text{-}HK}$ evolve in time with power-law dependence having exponent $n \sim 1/6$ and $n \sim 1/3$, respectively, larger relative contribution from ΔV_{ET} would reduce the time exponent n of overall ΔV_T. Figure 4.42a plots the correlation of measured time exponent n to calculated fractional electron trapping contribution ($\Delta V_{ET}/\Delta V_T$) for different HKMG devices listed in Table 4.4. Data published in different reports [31, 67, 68, 69] are also

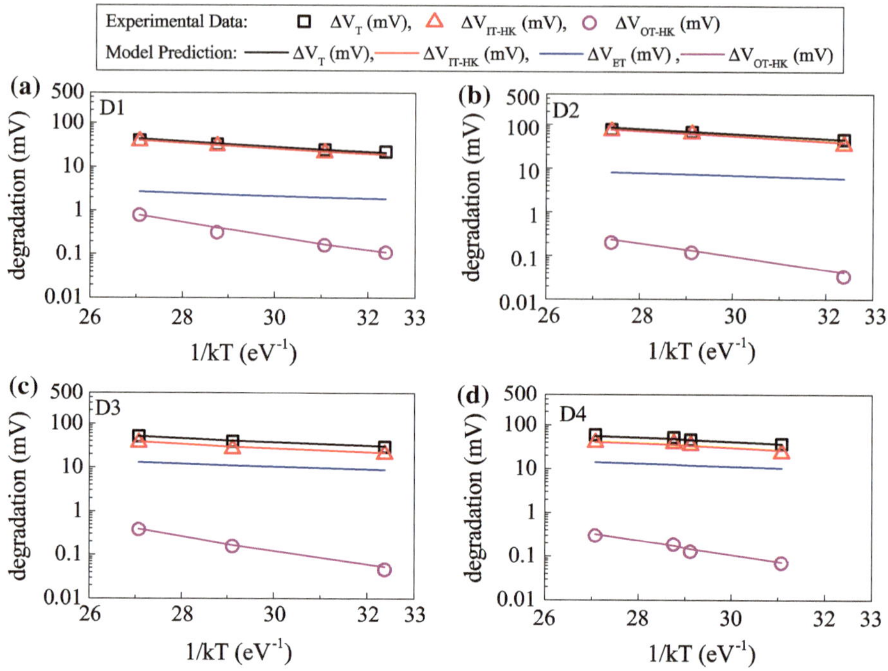

Fig. 4.40 UF-MSM measured ΔV_T, DCIV measured $\Delta V_{IT\text{-}HK}$ after delay and band gap corrections and SILC measured $\Delta V_{OT\text{-}HK}$ after delay correction (*symbols*) at fixed t_{STR} as a function of stress T for PBTI stress in n-MOSFETs having different HKMG gate insulator stacks. Compact model prediction (*lines*) of ΔV_T and underlying trap generation and trapping subcomponents are shown

analyzed and shown. Time exponent n is calculated by linear regression of measured data in t_{STR} range of 10 s to 1 Ks, and $\Delta V_{ET}/\Delta V_T$ is calculated at fixed E_{OX}, T and t_{STR} using the calibrated compact model. Note that measured n reduces with increased trapping fraction as expected, and a remarkable consistency is observed across different devices that verifies the assumption that electron trapping saturates at longer stress time. Moreover, since ΔV_{ET} has lower T activation than $\Delta V_{IT\text{-}HK}$ and $\Delta V_{OT\text{-}HK}$, larger fractional ΔV_{ET} contribution would reduce the T activation of overall ΔV_T, hence n and E_A would reduce in tandem for devices having higher as-processes HK defects. Figure 4.42b correlates measured n and E_A for different HKMG devices of Table 4.4 and verifies the above hypothesis.

The impact of EOT scaling of HKMG gate insulator stacks on PBTI degradation and its parameters is of interest, and is discussed in Chap. 1, Fig. 1.28. As mentioned earlier in this chapter, EOT scaling of HKMG devices listed in Table 4.4 is achieved either by IL thickness scaling (D2), or by post High-K nitridation (D3), or by nitrided IL (D4). Figure 4.43a plots measured ΔV_T, $\Delta V_{IT\text{-}HK}$ and $\Delta V_{OT\text{-}HK}$ obtained respectively by UF-MSM, delay and band gap corrected DCIV and delay corrected SILC measurements at a particular E_{OX}, T, and t_{STR} as a function of IL thickness for different HKMG devices. The ΔV_{ET} subcomponent is estimated from

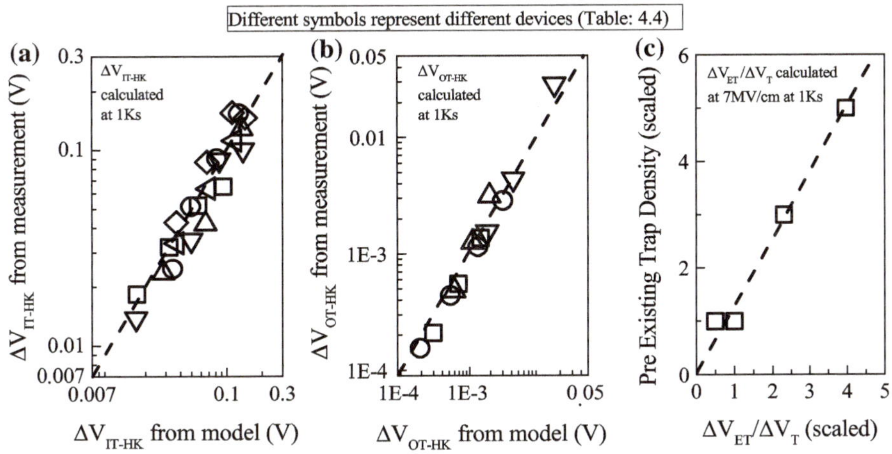

Fig. 4.41 Correlation of trap generation measured **a** using DCIV method after delay and band gap corrections and **b** using SILC method after delay correction to trap generation calculated using compact model. **c** Correlation of pre-existing traps measured using flicker noise method and calculated fractional electron trapping contribution, for PBTI stress in different HKMG n-MOSFETs. *Lines* represent **a**, **b** 1:1 correlation, **c** guide to eye

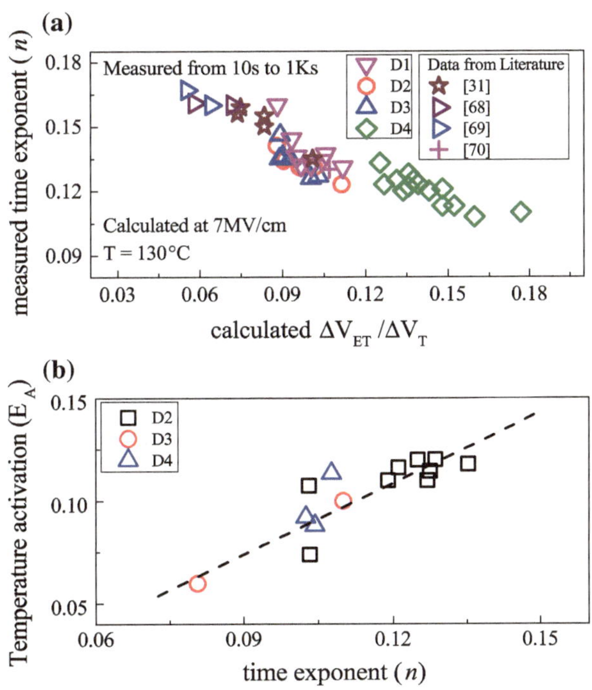

Fig. 4.42 **a** Correlation of measured power-law time exponent n at long t_{STR} to fractional electron trapping contribution ($\Delta V_{ET}/\Delta V_T$) calculated using the compact model for PBTI stress in different HKMG n-MOSFETs. **b** Correlation of measured n and T activation E_A also for PBTI stress. *Line* is guide to eye

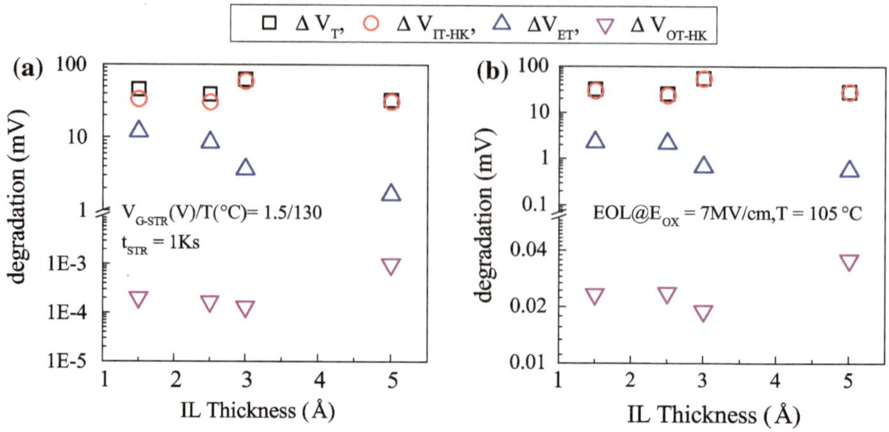

Fig. 4.43 Impact of IL thickness scaling on **a** UF-MSM measured ΔV_T, DCIV measured $\Delta V_{IT\text{-}HK}$ after delay and band gap correction and SILC measured $\Delta V_{OT\text{-}HK}$ after delay correction for PBTI stress in HKMG p-MOSFETs at fixed t_{STR}; estimated ΔV_{ET} subcomponent is also shown. **b** IL thickness scaling impact on compact model calculated ΔV_T, and underlying $\Delta V_{IT\text{-}HK}$, $\Delta V_{OT\text{-}HK}$, and ΔV_{ET} subcomponents at end-of-life (10 years)

the difference and is also shown. Figure 4.43b shows IL thickness dependence of ΔV_T and its underlying $\Delta V_{IT\text{-}HK}$, ΔV_{ET}, and $\Delta V_{OT\text{-}HK}$ subcomponents at the end-of-life of 10 years, calculated using the calibrated compact model at a fixed E_{OX} and T. As shown, ΔV_{ET} increases as IL is scaled especially for gate stacks D3 and D4 having higher N density, refer to [13] for additional details. However, since $\Delta V_{IT\text{-}HK}$ dominates ΔV_T and more so at end-of-life; a reduction in $\Delta V_{IT\text{-}HK}$ causes reduction in ΔV_T, in spite of increase in ΔV_{ET}, for nitrided devices having scaled IL. The $\Delta V_{OT\text{-}HK}$ subcomponent remains negligible as shown. Therefore, unlike NBTI, IL scaling achieved by nitridation reduces PBTI degradation, and has significant impact on technology scaling.

4.7 Summary

The physical mechanism responsible for NBTI and PBTI degradation respectively in p- and n-channel MOSFETs is explored in devices having different gate insulator processes. A compact model is developed to predict time evolution of measured ΔV_T at different stress bias and temperature, for NBTI stress in Si planar devices having differently processed SiON and HKMG gate insulators, SiGe planar devices without and with Si cap as well as in Si channel FinFETs, having HKMG gate insulators, and for PBTI stress in differently processed Si planar HKMG devices. The NBTI compact model consists of trap generation at channel/gate insulator interface and in gate insulator bulk, although the latter is found to be negligible for HKMG devices, as well as hole trapping in gate insulator bulk. Transconductance degradation is negligible that suggests creation of defects away from channel for PBTI stress. Therefore, the PBTI compact model for HKMG stacks consists of trap generation at the IL/High-K interface and High-K bulk, and electron trapping in High-K bulk, and IL remains undegraded. Overall, ΔV_T is modeled using uncorrelated sum of underlying trap generation and trapping subcomponents.

Trap generation during NBTI stress in SiON and HKMG p-MOSFETs is independently estimated using CP and DCIV techniques, respectively. For PBTI stress, trap generation at the IL/High-K interface and High-K bulk is estimated respectively using DCIV and SILC techniques. Time evolution of as-measured generated traps from CP and DCIV is corrected for measurement delay and band gap, while that from SILC is corrected for measurement delay, for proper comparison with measured time evolution of overall ΔV_T. Pre-stress trap density is estimated using the flicker noise technique and is responsible for hole and electron trapping respectively for NBTI and PBTI stress.

The compact model with consistent set of calibrated model parameters can predict ΔV_T time evolution for NBTI and PBTI stress in different devices and under different stress conditions, and calculated underlying model subcomponents agree well with independently measured contribution from trap generation and trapping. It is shown that trap generation at channel/gate insulator interface for NBTI and at the IL/High-K interface for PBTI demonstrate remarkably similar time dependent kinetics of power-law time dependence with time exponent $n \sim 1/6$. These generated traps dominate the time evolution of overall ΔV_T and explains similar kinetics observed between NBTI and PBTI stress. The trapping subcomponent saturates at longer time and has weaker T activation as compared to trap generation, and reduces the time exponent and T activation of overall ΔV_T, especially for processes leading to large pre-existing gate insulator defects. However, since trapping saturates at long stress time and bulk trap generation becomes negligible at low voltage use condition due to large voltage de-acceleration factor, the end-of-life BTI degradation is always determined by kinetics of interface trap generation for all process variations studied in this chapter.

Acknowledgments The authors would like to acknowledge Vrajesh Maheta for SiON measurements, Gautam Kapila and Bijesh Rajamohanan for flicker noise measurements, Dhanoop Varghese for charge pumping measurements, Applied Materials for providing devices used in this work, Jacopo Franco, Ben Kaczer, and Purushothaman Srinivasan for providing SiGe data.

References

1. Y. Mitani, M. Nagamine, H. Satake, Akira Toriumi, NBTI mechanism in ultra-thin gate dielectric-nitrogen-originated mechanism in SiON, in *IEEE International Electron Devices Meeting Technical Digest* (2002), p. 509
2. V. Huard, M. Denais, F. Perrier, N. Revil, C. Parthasarathy, A. Bravaix, E. Vincent, A thorough investigation of MOSFETs NBTI degradation. Microelectron. Reliab. **45**, 83 (2005)

3. D. Varghese, D. Saha, S. Mahapatra, K. Ahmed, F. Nouri, M.A. Alam, On the dispersive versus Arrhenius temperature activation of NBTI time evolution in plasma nitrided gate oxides: measurements, theory, and implications, in *IEEE International Electron Devices Meeting Technical Digest* (2005), p. 684
4. S. Mahapatra, K. Ahmed, D. Varghese, A.E. Islam, G. Gupta, L. Madhav, D. Saha, M.A. Alam, On the physical mechanism of NBTI in silicon oxynitride p-MOSFETs: can differences in insulator processing conditions resolve the interface trap generation versus hole trapping controversy? in *IEEE International Reliability Physics Symposium Proceedings* (2007), p. 1
5. S. Mahapatra, A. Islam, S. Deora, V. Maheta, K. Joshi, A. Jain, M. Alam, A critical re-evaluation of the usefulness of R-D framework in predicting NBTI stress and recovery, in *IEEE International Reliability Physics Symposium Proceedings* (2011), p. 6A.3.1
6. G. Chen, K.Y. Chuah, M.-F. Li, D.S.H. Chan, C.H. Ang, J.Z. Zheng, Y. Jin, D.L. Kwong, Dynamic NBTI of PMOS transistors and its impact on device lifetime, in *IEEE International Reliability Physics Symposium Proceedings* (2003), p. 196
7. J.H. Stathis, G. La Rosa, A. Chou, Broad energy distribution of NBTI-induced interface states in p-MOSFETs with ultra-thin nitrided oxide, in *IEEE International Reliability Physics Symposium Proceedings* (2004), p. 1
8. J.P. Campbell, P.M. Lenahan, A.T. Krishnan, S. Krishnan, NBTI: an atomic-scale defect perspective, in *IEEE International Reliability Physics Symposium Proceedings*, (2006), p. 442
9. A. Neugroschel, G. Bersuker, R. Choi, C. Cochrane, P. Lenahan, D. Heh, C. Young, C.Y. Kang, B.H. Lee, R. Jammy, An accurate lifetime analysis methodology incorporating governing NBTI mechanisms in high-k/SiO_2 gate stacks, in *IEEE International Electron Devices Meeting Technical Digest* (2006). doi:10.1109/IEDM.2006.346772
10. S. Mahapatra, N. Goel, S. Desai, S. Gupta, B. Jose, S. Mukhopadhyay, K. Joshi, A. Jain, A.E. Islam, M.A. Alam, A comparative study of different physics-based NBTI models. IEEE Trans. Electron Devices **60**, 901 (2013)
11. S. Desai, S. Mukhopadhyay, N. Goel, N. Nanaware, B. Jose, K. Joshi, S. Mahapatra, A comprehensive AC/DC NBTI model: stress, recovery, frequency, duty cycle and process dependence, in *IEEE International Reliability Physics Symposium Proceedings*, (2013), p. XT.2.1
12. N. Goel, S. Mukhopadhyay, N. Nanaware, S. De, R.K. Pandey, K.V.R.M. Murali, S. Mahapatra, A comprehensive DC/AC model for ultra-fast NBTI in deep EOT scaled HKMG p-MOSFETs, in *IEEE International Reliability Physics Symposium Proceedings* (2014), p. 6A.4.1
13. S. Mukhopadhyay, K. Joshi, V. Chaudhary, N. Goel, S. De, R.K. Pandey, K.V.R.M. Murali, S. Mahapatra, Trap generation in IL and HK layers during BTI/TDDB stress in scaled HKMG N and P MOSFETs, in *IEEE International Reliability Physics Symposium Proceedings* (2014), p. GD.3.1
14. A.T. Krishnan, C. Chancellor, S. Chakravarthi, P.E. Nicollian, V. Reddy, A. Varghese, R.B. Khamankar, S. Krishnan, Material dependence of hydrogen diffusion: implications for NBTI degradation, in *IEEE International Electron Devices Meeting Technical Digest* (2005), p. 688
15. D.S. Ang, S. Wang, Recovery of the NBTI-stressed ultrathin gate p-MOSFET: the role of deep-level hole traps. IEEE Electron Device Lett. **27**, 914 (2006)
16. T. Grasser, B. Kaczer, W. Goes, T. Aichinger, P. Hehenberger, M. Nelhiebel, Understanding negative bias temperature instability in the context of hole trapping. Microelectron. Eng. **86**, 1876 (2009)
17. H. Reisinger, T. Grasser, K. Ermisch, H. Nielen, W. Gustin, C. Schlunder, Understanding and modeling AC BTI, in *IEEE International Reliability Physics Symposium Proceedings* (2011), p. 6A.1.1
18. S. Mahapatra, V. Huard, A. Kerber, V. Reddy, S. Kalpat, A. Haggag, Universality of NBTI-From devices to circuits and products, in *IEEE International Reliability Physics Symposium Proceedings* (2014), p. 3B.1.1

19. T. Grasser, B. Kaczer, W. Goes, T. Aichinger, P. Hehenberger, M. Nelhiebel, A two stage model for negative bias temperature instability, in *IEEE International Reliability Physics Symposium Proceedings* (2009), p. 33
20. J.H. Lee, W.H. Wu, A.E. Islam, M.A. Alam, A.S. Oates, Separation method of hole trapping and interface trap generation and their roles in NBTI Reaction-Diffusion model, in *IEEE International Reliability Physics Symposium Proceedings* (2008), p. 745
21. S. Mahapatra, V.D. Maheta, A.E. Islam, M.A. Alam, Isolation of NBTI stress generated interface trap and hole-trapping components in PNO p-MOSFETs. IEEE Trans. Electron Devices **56**, 236 (2009)
22. V. Huard, Two independent components modeling for negative bias temperature instability, in *IEEE International Reliability Physics Symposium Proceedings* (2010), p. 33
23. S. Mahapatra, P. Bharath, P. Bharath Kumar, M.A. Alam, Investigation and modeling of interface and bulk trap generation during negative bias temperature instability of p-MOSFETs. IEEE Trans. Electron Devices **51**, 1371 (2004)
24. S. Ramey, A. Ashutosh, C. Auth, J. Clifford, M. Hattendorf, J. Hicks, R. James, A. Rahman, V. Sharma, A. St.Amour, C. Wiegand, Intrinsic transistor reliability improvements from 22 nm tri-gate technology, in *IEEE International Reliability Physics Symposium Proceedings* (2013), p. 4C.5.1
25. S. Pae, M. Agostinelli, M. Brazier, R. Chau, G. Dewey, T. Ghani, M. Hattendorf, J. Hicks, J. Kavalieros, K. Kuhn, M. Kuhn, J. Maiz, M. Metz, K. Mistry, C. Prasad, S. Ramey, A. Roskowski, J. Sandford, C. Thomas, J. Thomas, C. Wiegand, J. Wiedemer, BTI reliability of 45 nm high-K + metal-gate process technology, in *IEEE International Reliability Physics Symposium Proceedings* (2008), p. 352
26. J. Mitard, X. Garros, L.P. Nguyen, C. Leroux, G. Ghibaudo, F. Martin, G. Reimbold, Large-scale time characterization and analysis of PBTI In HfO_2/metal gate stacks, in *IEEE International Reliability Physics Symposium Proceedings* (2006), p. 174
27. D. Heh, C.D. Young, G. Bersuker, Experimental evidence of the fast and slow charge trapping/detrapping processes in high-k dielectrics subjected to PBTI stress. IEEE Electron Dev. Lett., **29**, 180 (2008)
28. A. Kerber, E. Cartier, Reliability Challenges for CMOS Technology Qualifications With Hafnium Oxide/Titanium Nitride Gate Stacks. IEEE Trans. Device Mater. Reliab. **9**, 147 (2009)
29. S. Zafar, A. Kerber, R. Muralidhar, Physics based PBTI model for accelerated estimation of 10 year lifetime, in *Symposium on VLSI Technology: Digest of Technical Papers* (2014). doi: 10.1109/VLSIT.2014.6894388
30. E. Cartier, A. Kerber, Stress-induced leakage current and defect generation in nFETs with HfO2/TiN gate stacks during positive-bias temperature stress, in *IEEE International Reliability Physics Symposium Proceedings* (2009), p. 486
31. J.Q. Yang, M. Masuduzzman, J.F. Kang, M.A. Alam, SILC-based reassignment of trapping and trap generation regimes of positive bias temperature instability, in *IEEE International Reliability Physics Symposium Proceedings* (2011), p. 3A.3.1
32. K. Joshi, S. Mukhopadhyay, N. Goel, S. Mahapatra, A consistent physical framework for N and P BTI in HKMG MOSFETs, in *IEEE International Reliability Physics Symposium Proceedings* (2012), p. 5A.3.1
33. K. Joshi, S. Hung, S. Mukhopadhyay, V. Chaudhary, N. Nanaware, B. Rajamohanan, T. Sato, M. Bevan, A. Wei, A. Noori, B. McDougall, C. Ni, C. Lazik, G. Saheli, P. Liu, D. Chu, L. Date, S. Datta, A. Brand, J. Swenberg, S. Mahapatra, HKMG process impact on N, P BTI: role of thermal IL scaling, IL/HK integration and post HK nitridation, in *IEEE International Reliability Physics Symposium Proceedings* (2013), p. 4C.2.1
34. S. Mukhopadhyay, K. Joshi, S. Mahapatra, A compact model for estimating end of life PBTI degradation. IEEE Trans. Electron Devices (under review) (2015)
35. T. Grasser, B. Kaczer, W. Goes, An energy-level perspective of bias temperature instability, in *IEEE International Reliability Physics Symposium Proceedings* (2008), p. 28

36. N. Goel, K. Joshi, S. Mukhopadhyay, N. Nanaware, S. Mahapatra, A comprehensive modeling framework for gate stack process dependence of DC and AC NBTI in SiON and HKMG p-MOSFETs. Microelectron. Reliab. **54**, 491 (2014)
37. D. Varghese, Ph.d. Dissertation, Multi-probe experimental and 'bottom-up' computational analysis of correlated defect generation in modern nanoscale transistors, Purdue University (2009)
38. V.D. Maheta, C. Olsen, K. Ahmed, S. Mahapatra, The impact of nitrogen engineering in silicon oxynitride gate dielectric on negative—bias temperature instability of p-MOSFETs: a study by ultrafast on-the-fly technique. IEEE Trans. Electron Devices **55**, 1630 (2008)
39. C. Olsen, Two-step post nitridation annealing for lower EOT plasma nitrided gate dielectrics, WO2004081984 A2 (2004)
40. J. Franco, B. Kaczer, J. Mitard, M. Toledano-Luque, P.J. Roussel, L. Witters, T. Grasser, G. Groeseneken, NBTI reliability of SiGe and ge channel pMOSFETs with SiO_2/HfO_2 dielectric stack. IEEE Trans. Device Mater. Reliab. **13**, 497 (2013)
41. P. Srinivasan, J. Fronheiser, K. Akarvardar, A. Kerber, L.F. Edge, R.G. Southwick, E. Cartier, H. Kothari, SiGe composition and thickness effects on NBTI in replacement metal gate/high-κ technologies, in *IEEE International Reliability Physics Symposium Proceedings*, (2014), p. 6A.3.1
42. A. Chaudhary, B. Kaczer, P.J. Roussel, T. Chiarella, N. Horiguchi, S. Mahapatra, Time dependent variability in RMG-HKMG FinFETs: impact of extraction scheme on stochastic NBTI, in *IEEE International Reliability Physics Symposium Proceedings*, (2015), p. 3B.4
43. G. Kapila, N. Goyal, V.D. Maheta, C. Olsen, K. Ahmed, S. Mahapatra, A comprehensive study of flicker noise in plasma nitride SiON p-MOSFETs: process dependence of pre-existing and NBTI stress generated trap distribution profiles, in *IEEE International Electron Devices Meeting Technical Digest* (2008). doi:10.1109/IEDM.2008.4796625
44. K. Choi, H. Jagannathan, C. Choi, L. Edge, T. Ando, M. Frank, P. Jamison, et al., Extremely scaled gate-first high-k/metal gate stack with EOT of 0.55 nm using novel interfacial layer scavenging techniques for 22 nm technology node and beyond, in *Symposium on VLSI Technology: Digest of Technical Papers* (2009), p. 138
45. A. Chaudhary, IIT Bombay, Private Communication
46. A. Kerber, T. Nigam, Challenges in the characterization and modeling of BTI induced variability in metal gate/high-k CMOS technologies, in *IEEE International Reliability Physics Symposium Proceedings*, (2013), p. 2D.4.1
47. C. Prasad, M. Agostinelli, J. Hicks, S. Ramey, C. Auth, K. Mistry, S. Natarajan, P. Packan, I. Post, S. Bodapati, M. Giles, S. Gupta, S. Mudanai, K. Kuhn, Bias temperature instability variation on SiON/Poly, HK/MG and trigate architectures, in *IEEE International Reliability Physics Symposium Proceedings*, (2014), p. 6A.5.1
48. XPS: X-Ray Photoelectron Spectroscopy, *Physical Electronics*. [Online]. Available: http://www.phi.com/surface-analysis-techniques/xps.html
49. S. Guha, V. Narayanan, High-k/metal gate science and technology. Annu. Rev. Mater. Res. **39**, 181 (2009)
50. K. Joshi, S. Mukhopadhyay, N. Goel, N. Nanware, S. Mahapatra, A detailed study of gate insulator process dependence of NBTI using a compact model. IEEE Trans. Electron Devices **61**, 408 (2014)
51. S. Deora, V.D. Maheta, G. Bersuker, C. Olsen, K.Z. Ahmed, R. Jammy, S. Mahapatra, A comparative NBTI study of HfO_2, HfSiOx and SiON p-MOSFETs using UF-OTF I_{DLIN} technique. IEEE Electron Device Lett. **30**, 152 (2009)
52. H. Reisinger, O. Blank, W. Heinrigs, A. Muhlhoff, W. Gustin, C. Schlunder, Analysis of NBTI degradation- and recovery-behavior based on ultra fast VT-measurements, in *IEEE International Reliability Physics Symposium Proceedings* (2006), p. 448
53. M. Jo, M. Chang, S. Kim, H.S. Jung, R. Choi, H. Hwang, Contribution of interface states and oxide traps to the negative bias temperature instability of high-k pMOSFETs. IEEE Electron Device Lett. **30**, 291 (2009)

54. T. Grasser, P. Wagner, H. Reisinger, T. Aichinger, G. Pobegen, M. Nelhiebel, B. Kaczer, Analytic modeling of the bias temperature instability using capture/emission time maps, in *IEEE International Electron Devices Meeting Technical Digest* (2011), p. 27.4
55. C. Shen, M.F. Li, C.E. Foo, T. Yang, D.M. Huang, A. Yap, G.S. Samudra, Y.C. Yeo, Characterization and physical origin of fast Vth transient in NBTI of pMOSFETs with SiON dielectric, in *IEEE International Electron Devices Meeting Technical Digest* (2006), p. 12.5
56. C.L. Chen, Y.M. Lin, C.J. Wang, and K. Wu, "A new finding on NBTI lifetime model and an investigation on NBTI degradation characteristic for 1.2 nm ultra thin oxide, in *IEEE International Reliability Physics Symposium Proceedings* (2005), p. 704
57. A. Islam, G. Gupta, S. Mahapatra, A. T. Krishnan, K. Ahmed, F. Nouri, A. Oates, M.A. Alam, Gate leakage vs. NBTI in plasma nitrided oxides: characterization, physical principles, and optimization, in *IEEE International Electron Devices Meeting Technical Digest* (2006). doi: 10.1109/IEDM.2006.346775
58. K. Hofmann, H. Reisinger, K. Ermisch, C. Schlünder, W. Gustin, T. Pompl, G. Georgakos, K. V. Arnim, J. Hatsch, T. Kodytek, T. Baumann, C. Pacha, Highly accurate product level aging monitoring in 40 nm CMOS, in *Symposium on VLSI Technology: Digest of Technical Papers* (2010), p. 27
59. A. Kerber, Characterization of BTI induced variability in scaled metal gate/highK CMOS technologies, in *Variability Workshop* (2013)
60. A.T. Krishnan, F. Cano, C. Chancellor, V. Reddy, Z. Qi, P. Jain, J. Masin, S. Zuhoski, S. Krishnan, J. Ondrusek, Product drift from NBTI: guardbanding, circuit and statistical effects, in *IEEE International Electron Devices Meeting Technical Digest* (2010), p. 4.3
61. D.P. Ioannou, S. Mittl, D. Brochu, Burn-in stress induced BTI degradation and post-burn-in high temperature anneal (Bake) effects in advanced HKMG and oxynitride based CMOS ring oscillators, in *IEEE International Reliability Physics Symposium Proceedings* (2012), p. 5C.2.1
62. A.T. Krishnan, V. Reddy, D. Aldrich, J. Raval, K. Christensen, J. Rosal, C. O. Brien, R. Khamankar, A. Marshall, W. Loh, R. Mckee, S. Krishnan, SRAM cell static noise margin and VMIN sensitivity to transistor degradation, in *IEEE International Electron Devices Meeting Technical Digest* (2006). doi:10.1109/IEDM.2006.346778
63. A. Haggag, G. Anderson, S. Parihar, D. Burnett, G. Abeln, J. Higman, M. Moosa, Understanding SRAM high-temperature-operating-life NBTI: statistics and permanent vs. recoverable damage, in *IEEE International Reliability Physics Symposium Proceedings* (2007), p. 452
64. M.L. Reed, J.D. Plummer, Chemistry of Si–SiO_2 interface trap annealing. J. Appl. Phys. **63**, 5776 (1988)
65. S. Tsujikawa, Y. Akamatsu, H. Umeda, J. Yugami, Two concerns about NBTI issue: gate dielectric scaling and increasing gate current, in *IEEE International Reliability Physics Symposium Proceedings* (2004), p. 28
66. S. Pae, T. Ghani, M. Hattendorf, J. Hicks, J. Jopling, J. Maiz, K. Mistry, J. O'Donnell, C. Prasad, J. Wiedemer, J. Xu, Characterization of SILC and its end-of-life reliability assessment on 45 nm high-K and metal-gate technology, in *IEEE International Reliability Physics Symposium Proceedings* (2009), p. 499
67. J. Yang, M. Masuduzzaman, K. Joshi, S. Mukhopadhyay, J. Kang, S. Mahapatra, M.A. Alam, Intrinsic correlation between PBTI and TDDB degradations in nMOS HK/MG dielectrics, in *IEEE International Reliability Physics Symposium Proceedings* (2012), p. 5D.4.1
68. K. Zhao, J.H. Stathis, A. Kerber, E. Cartier, PBTI relaxation dynamics after AC vs. DC stress in high-k/metal gate stacks, in *IEEE International Reliability Physics Symposium Proceedings* (2010), p. 50
69. W. Liu, G. La Rosa, G. Tian, S. Boffoli, F. Guarin, W.L. Lai, V. Narayanan, H. Kothari, M. Jin, S. Uppal, W. McMahon, Process dependence of AC/DC PBTI in HKMG n-MOSFETs, in *IEEE International Reliability Physics Symposium Proceedings* (2014), p. XT.6.1

Chapter 5
Reaction-Diffusion Model

Ahmad Ehteshamul Islam, Nilesh Goel, Souvik Mahapatra and Muhammad Ashraful Alam

Abstract This chapter discusses different aspects of the Reaction-Diffusion (RD) model used to interpret dissociation and reformation of the Si–H bonds (in other words, creation and reverse-anneal of dangling Si-bonds or interface traps) present at the silicon/oxide interfaces of a CMOS transistor. The theory presented in this chapter is later combined with other features of NBTI to interpret measurements in Chap. 6. The reaction part of the R-D model interprets the chemical reactions like Si–H bond dissociation and reformation taking place at the interface, while the diffusion part interprets the transport of Hydrogen species in the oxide and the gate medium. In the NBTI stress phase with a particular stress bias, the Si–H bond dissociation initiates generation of interface traps over time, which later reaches quasi-equilibrium with the diffusive components. After that the diffusion of Hydrogen species defines the time evolution of interface trap generation with a power law. The power-law time exponent is a unique signature of the diffusing species and it shows no variation with the change in stress conditions (bias, temperature and frequency). In the NBTI relaxation phase, the diffusion of Hydrogen species also defines the time evolution of interface trap repassivation. This single dependence of the time evolution, in both stress and relaxation phases, only on the diffusing species leads to frequency independence of interface trap generation—a distinct feature of NBTI measured over a wide range of transistors.

A.E. Islam (✉)
Materials and Manufacturing Directorate, Air Force Research Laboratory, Dayton, OH 45433, USA
e-mail: aeislam@ieee.org; aeislam@gmail.com

A.E. Islam
National Research Council, National Academy of Sciences, Washington DC 20001, USA

N. Goel · S. Mahapatra
Department of Electrical Engineering, Indian Institute of Technology Bombay, Mumbai 400076, India

M.A. Alam
Department of Electrical and Computer Engineering, Purdue University, West Lafayette, IN 47907, USA

S. Mahapatra (ed.), *Fundamentals of Bias Temperature Instability in MOS Transistors*, Springer Series in Advanced Microelectronics 139,
DOI 10.1007/978-81-322-2508-9_5

5.1 Introduction

The silicon/oxide interface dictates charge transport in modern CMOS transistors. This interface is generally passivated with Hydrogen, which converts the interface traps (or dangling Si- bonds) having a density of $\sim 10^{12}$ cm^{-2} into Si–H bonds [1–7]. The residual trap density after Hydrogen passivation is $\sim 10^{10}$ cm^{-2} or even below [8–10]. Application of negative gate bias $V_{\text{G-STR}}$ at an elevated temperature during NBTI stress, stretches the Si–H bonds and recreates the interface traps [11–15]. The dissociation of Si–H bond (*reaction*) generates Hydrogen species that move away from the interface (*diffusion*)—a phenomenon that can be easily captured using a Reaction-Diffusion (RD) framework. Starting from the pioneering work of Jeppson et al., successive refinements of RD model have interpreted a broad range of NBTI experiments [11–25]. Differences in these refinements depend on how one interprets: (a) the Si–H bond dissociation and (b) the type of Hydrogen species moving away from the interface. In the following sections, we start with the RD model proposed by Jeppson and Svensson [12], which has been reinterpreted by Alam and Mahapatra [14], and later explain the existing variants of this original proposal. In general, RD model seeks to capture the following key aspects of NBTI as observed in measurements over a wide range of transistors with variations in $V_{\text{G-STR}}$, temperature T, stress cycle, gate dielectric, and channel strain:

- RD model captures the time evolution of the interface trap component (ΔV_{IT}) of threshold voltage degradation (ΔV_{T}). As elaborated in Chaps. 4 and 6, one should decompose ΔV_{T} appropriately into its components before comparison with RD theory (Fig. 5.1).
- In the stress phase, the RD theory studies the origin of power-law change in ΔV_{IT} over time (i.e., $\Delta V_{\text{IT}} \sim t^n$). As discussed in Chap. 3, the time exponent (n) in stress phase is a unique feature of NBTI.

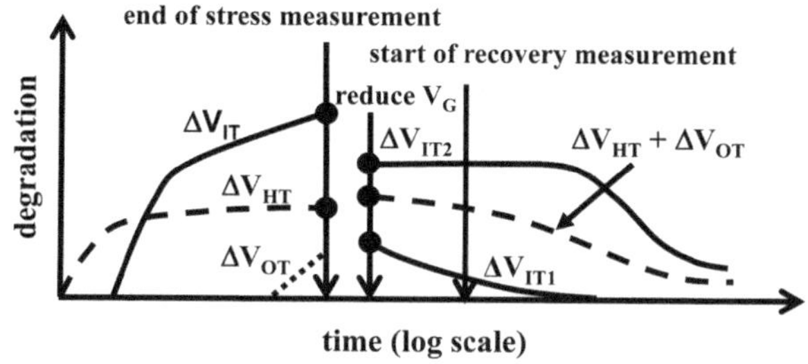

Fig. 5.1 Schematic illustration of different components of threshold voltage shift (ΔV_{T}), measured during NBTI stress and relaxation experiments. This chapter discusses the theory of interface trap component (ΔV_{IT} and ΔV_{IT2}), while Sect. 6.3 discusses the extension of the theory to include fast electron capture into the interface traps going below the Fermi level (ΔV_{IT1}), hole trapping into pre-existing (ΔV_{HT}) and generated (ΔV_{OT}) oxide defects. In the stress phase, ΔV_{IT} versus time follows power-law with time exponent (n); while in the relaxation phase, ΔV_{IT2} responds slowly compared to the other components

- In the relaxation phase, the RD theory studies the nature of interface trap passivation. As discussed in Sect. 6.3, NBTI experiments suggest this passivation (ΔV_{IT2} of Fig. 5.1) to be slower than the fast electron capture by the interface traps going below the Fermi level (ΔV_{IT1} of Fig. 5.1) and the hole trapping components (ΔV_{HT} and ΔV_{OT} in Fig. 5.1).

In the following sections, we will elaborate on the evolution of interface traps in the stress and relaxation phase of NBTI based on the RD theory.

5.2 RD Formalism for the NBTI Stress Phase

This first proposal of RD model by Jeppson and Svensson [12] for interface trap generation considers the following dissociation reaction:

$$Si_3 \equiv Si - H + Y \quad \leftrightarrow \quad Si_3 \equiv Si- + X \tag{5.1}$$

where a chemical species Y reacts with the Si−H bond at the interface and creates interface defect (the dangling Si- bond) and a neutral species X (atomic Hydrogen H) that diffuses away from the interface (Fig. 5.2a). The following set of differential equations captures the reaction and diffusion steps:

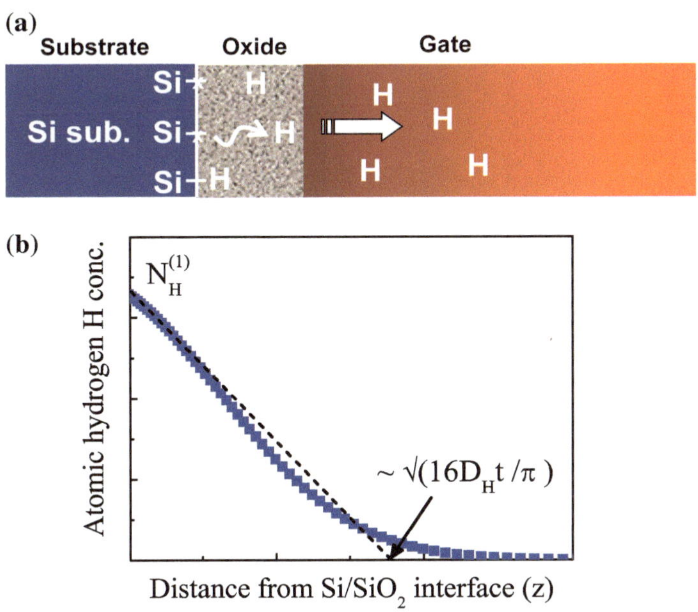

Fig. 5.2 **a** Schematic representation of Si−H bond dissociation, as per (5.1), and diffusion of species X (=H). **b** Diffusion front at time t reaches $z \sim \sqrt{D_{H}t}$. The diffusion profile (*symbols*; obtained from numerical simulation) can be approximated using the *dotted line* that facilitates calculation of total diffusing species within the medium, as performed in (5.6)

$$\frac{\partial N_{IT}}{\partial t} = k_F(N_0 - N_{IT}) - k_R N_{IT} N_x^{(1)} \tag{5.2}$$

$$\frac{\partial N_x}{\partial t} = D_x \frac{\partial^2 N_x}{\partial z^2} \tag{5.3}$$

where k_F is the Si–H bond dissociation rate (influence of Y as a catalyst is considered within this factor), k_R is the backward reaction or Si- annealing rate, N_0 is the initial ($t = 0$) concentration of Si–H bonds, N_{IT} is the interface trap density, N_X is the concentration of X along the z-axis (superscript (1) indicates near Si/SiO$_2$ interface quantities), and D_X is the diffusion co-efficient of X within the diffusing media (like oxide and gate). The fluxes of (5.2)–(5.3) are balanced by the following boundary condition:

$$\frac{\partial}{\partial t}\left[\frac{1}{2}\delta N_X^{(1)}\right] = \frac{\partial N_{IT}}{\partial t} - D_X \frac{\partial N_X^{(1)}}{\partial z} \tag{5.4}$$

where δ is the interfacial layer thickness approximately equal to the Si–H bond length $\delta \sim 1.5$ Å [12], and superscript (1) indicates near Si/SiO$_2$ interface quantities. Equation (5.4) suggests that N_X is increased by Si–H bond dissociation and is decreased by the outward diffusion flux.

At $t \gg 0$, when the rate of diffusion defines the overall kinetics, the surface reaction is in quasi-equilibrium, i.e., $\partial N_{IT}/\partial t \sim 0$ and hence (5.2) yields:

$$k_F N_0 \cong k_R N_X^{(1)} N_{IT} \tag{5.5}$$

The solution of (5.3) is an error function $\sim \mathrm{erfc}\left[z/\sqrt{(D_X t)}\right]$ that can be approximated using a triangular profile with significant values of N_X from $z = 0$ to $\sqrt{(16 D_X t/\pi)}$ (Fig. 5.2b) [16, 17, 26, 27]. The integral of X from $z = 0$ to ∞ equals N_{IT}, so that one obtains:

$$N_{IT} = \int_{z=0}^{z=\infty} N_x(z,t)\mathrm{d}z \cong \frac{1}{2} N_x^{(1)} \sqrt{16 D_x t/\pi} \tag{5.6}$$

Combining (5.5) and (5.6), we have:

$$N_{IT} \cong \left(\frac{k_F N_0}{k_R}\right)^{1/2} (4 D_x t/\pi)^{1/4} \tag{5.7}$$

which suggests interface trap generation to have a power-law time exponent, $n = 1/4$.[1] When the forward and reverse reactions, as well as the diffusion process, have Arrhenius temperature activation in the form of $\sim\exp[-E_A/kT]$, where E_A is the activation energy, (5.7) suggests:

$$E_{A,IT} = \frac{1}{2}\left(E_{A,kF} - E_{A,kR} - aE_{OX}\right) + \frac{1}{4}E_{A,DX}, \tag{5.8}$$

where, $E_{A,kF}$, $E_{A,kR}$, $E_{A,DX}$, and $E_{A,IT}$ are the activation energies for k_F, k_R, D_X, and interface trap, respectively; and aE_{OX} represents the field-induced barrier reduction for Si−H bond dissociation with a being the effective dipole moment. We will discuss this field-induced barrier reduction in more detail later in Sect. 5.6.

5.2.1 Diffusing/Drifting Species in RD Formalism

Jeppson's model did not explicitly identify the diffusing species or the diffusion mechanism. The Hydrogen released from the Si−H bond can diffuse or drift as a neutral atom H, or as a proton H^+, or as a molecule H_2, or in any of their combinations. As shown below, these species and their diffusion mechanism define critical parameters of the RD solution.

5.2.1.1 H and H_2 Diffusion with Direct H–H_2 Conversion

In a general sense, atomic H released from Si−H dissociation can diffuse away from the interface, dimerize with another H, and form H_2 that can also diffuse away from the interface (see Fig. 5.3 for a schematic representation). (An alternate method of second interface assisted H $\rightarrow$ H_2 conversion is discussed in the next section.) The following equations define the related kinetics [16, 17, 19, 20]:

$$\frac{dN_{IT}}{dt} = k_F(N_0 - N_{IT}) - k_R N_{IT} N_H^{(1)} \tag{5.9}$$

$$\begin{aligned} \frac{\delta}{2}\frac{dN_H^{(1)}}{dt} &= D_H \frac{dN_H^{(1)}}{dz} + \frac{dN_{IT}}{dt} - \delta k_H \left[N_H^{(1)}\right]^2 + \delta k_{H2} N_{H2}^{(1)} \\ \frac{\delta}{2}\frac{dN_{H2}^{(1)}}{dt} &= D_{H2} \frac{dN_{H2}^{(1)}}{dz} + \frac{\delta}{2} k_H \left[N_H^{(1)}\right]^2 - \frac{\delta}{2} k_{H2} N_{H2}^{(1)} \end{aligned} \tag{5.10}$$

[1] This solution is intriguing, as among many differential equations used to describe natural phenomena in this universe, RD is a unique one that gives a robust power-law time exponent for several decades in time, whose value depends only on nature of diffusion.

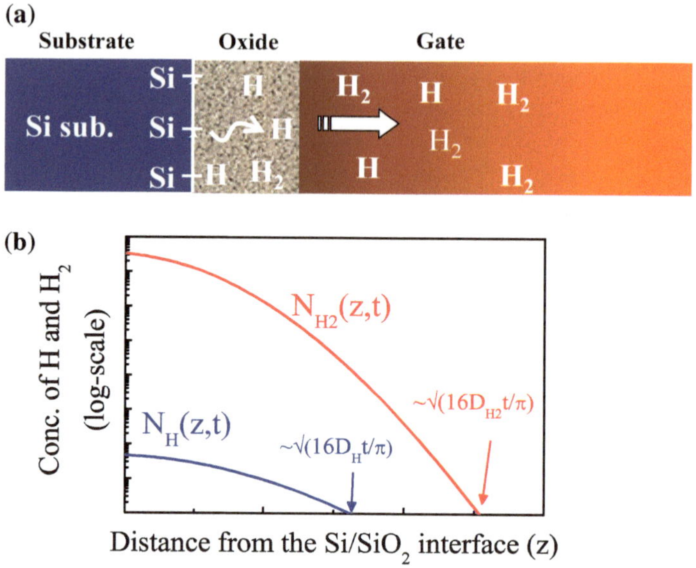

Fig. 5.3 **a** Schematic illustration of Si–H bond dissociation and H, H_2 diffusion during NBTI stress. **b** At long stress time, the concentration of H (N_H) is small compared to that for H_2 (N_{H2}) and hence diffusion of H_2 dominates the kinetics of interface trap generation

$$\begin{aligned}\frac{dN_H}{dt} &= D_H\frac{d^2N_H}{dz^2} - k_H(N_H)^2 + k_{H2}N_{H2},\\ \frac{dN_{H2}}{dt} &= D_{H2}\frac{d^2N_{H2}}{dz^2} + \frac{1}{2}k_H(N_H)^2 - \frac{1}{2}k_{H2}N_{H2},\end{aligned} \tag{5.11}$$

As before, (5.9) represents the passivation/de-passivation of Si–H bond, (5.10) the conservation of fluxes near the interface, and (5.11) the diffusion of H and H_2 within the oxide and gate. The $k_H(N_H)^2$ and $k_{H2}N_{H2}$ terms in (5.10) and (5.11) describe the H–H_2 conversion. Among the new symbols, k_H, k_{H2} represent generation and dissociation rates of H_2; D_H, D_{H2} represent the diffusion coefficients for H and H_2; and N_H, N_{H2} represent the concentration of H and H_2.

To obtain an analytical solution of (5.9, 5.10, 5.11), we assume $N_0 \gg N_{IT}$ [20–22], $dN_{IT}/dt \sim N_{IT}/t$ [14, 23], $dN_H^{(1)}/dt \sim 0$ and negligible diffusion for H. All of these assumptions are based on detailed numerical simulation [20, 22]. The assumptions simplify (5.9) and (5.10) to the following equations:

$$N_H^{(1)} = \frac{k_F N_0 - N_{IT}/t}{k_R N_{IT}} \tag{5.12}$$

$$\frac{N_{IT}}{t} = \delta k_H\left(N_H^{(1)}\right)^2 - \delta k_{H2}N_{H2}^{(1)} \tag{5.13}$$

Moreover, the numerical solution of H_2 profile (Fig. 5.3b) allows a triangular approximation and in a manner similar to (5.6), we can write:

$$N_{IT} \approx N_{H2}^{(1)}\sqrt{6D_{H2}t} \tag{5.14}$$

Note that the extent of diffusion profile $\sqrt{6D_{H2}t}$ used in (5.14) exactly captures the solution of diffusion equation [20] and is consistent with $\sqrt{(16D_{H2}t/\pi)}$ used to capture the diffusion profile [26, 27]. Eliminating $N_{H}^{(1)}$ and $N_{H2}^{(1)}$ from (5.12) to (5.14), we find:

$$\frac{N_{IT}}{t} - \frac{\delta k_{H}(k_F N_0 - N_{IT}/t)^2}{k_R^2 N_{IT}^2} + \frac{\delta k_{H2} N_{IT}}{\sqrt{6D_{H2}t}} = 0. \tag{5.15}$$

Figure 5.4 compares the numerical solution of (5.9)–(5.11) with the (implicit) analytical solution of (5.15) that agrees within 20 % over ~ 10 decades in stress time. Equation (5.15) reduces to—(i) the reaction-limited solution $N_{IT} = k_F N_0 t$, when the first and third terms become negligible at very short stress time, (ii) the H_2 diffusion limited solution $N_{IT} = (k_H/k_{H2})^{1/3}(k_F N_0/k_R)^{2/3}(6D_{H2}t)^{1/6}$, when N_{IT}/t becomes negligible at the long stress time. At an intermediate stress time, the diffusive term (third term) in (5.15) is negligible and $k_F N_0 \gg N_{IT}/t$, which yields the solution of $N_{IT} = (k_F N_0/k_R)^{2/3}(\delta k_H t)^{1/3}$—the signature of H–$H_2$ conversion [20]. Using the approximation of (5.15) at long stress time (when H_2 diffusion dominates the kinetics), one can estimate the following expression of activation energy for N_{IT}:

$$E_{A,IT} = \frac{2}{3}\left(E_{A,kF} - E_{A,kR} - aE_{ox}\right) + \frac{1}{6}E_{A,DH2} + \frac{1}{3}\left(E_{A,kH} - E_{A,kH2}\right) \tag{5.16}$$

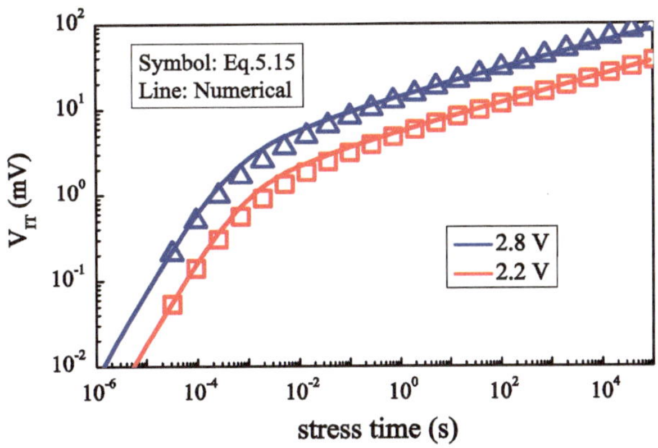

Fig. 5.4 Comparison of numerical solution of ΔN_{IT} (*lines*) from (5.9) to (5.11) with analytical solution using (5.15) (*symbols*). Simulation is performed for a transistor having effective oxide thickness: 2.2 nm, stressed at T = 125 °C and at the voltages mentioned in the legend

5.2.1.2 H and H_2 Diffusion with a Second-Interface Assisted Dimerization

As an alternative to the direct H to H_2 dimerization discussed in the last section, dimerization can also result from reactions involving both the first Si/SiO_2 interface and at a second interface (like the SiO_2/poly gate interface [21, 24]), as follows:

$$\begin{aligned} &Si_3{\equiv}Si-H+h^+ \quad \leftrightarrow \quad Si_3{\equiv}Si^+ + H \quad (\text{at Si/SiO}_2 \text{ interface or first interface}) \\ &Si_3{\equiv}Si-H+H \quad \leftrightarrow \quad Si_3{\equiv}Si- + H_2 \quad (\text{at second interface}) \end{aligned} \tag{5.17}$$

The creation of dangling Si- bonds at a second interface like SiO_2/poly has been confirmed using low-voltage stress induced leakage current (LV-SILC) experiments [23, 24]. The following rate equations describe the kinetics of the reactions presented in (5.17):

$$\frac{dN_{IT(1)}}{dt} = k_{F(1)}\left(N_{0(1)} - N_{IT(1)}\right) - k_{R(1)}N_{IT(1)}N_H^{(1)} \tag{5.18}$$

$$\frac{dN_{IT(2)}}{dt} = k_{F(2)}\left(N_{0(2)} - N_{IT(2)}\right)N_H^{(2)} - k_{R(2)}N_{IT(2)}N_{H2}^{(2)} \tag{5.19}$$

where variables with (1) and (2) represent the quantities at the first and second interfaces, respectively. A numerical solution of (5.18)–(5.19) with appropriate boundary conditions suggests 1/6 time exponent at long stress time, similar to the single interface H–H_2RD model with direct H to H_2 dimerization (Fig. 5.5), with some differences near the reaction ($n \sim 1$) to diffusion ($n \sim 1/6$) transitions [21, 23]. For transistors with SiO_2/high-K gate dielectric, the second interface for the above two-interface reactions will be at the SiO_2/high-K interface, as illustrated

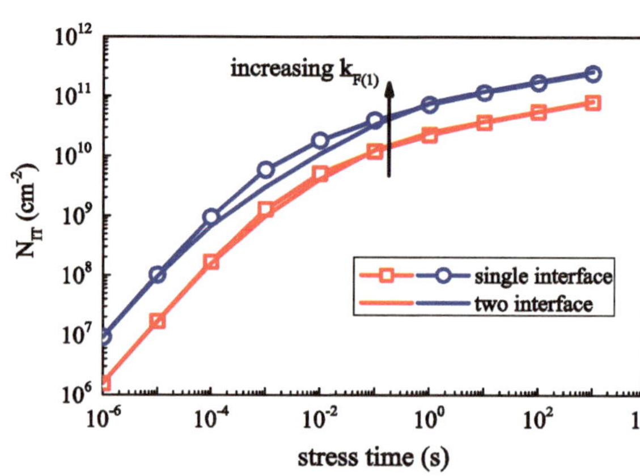

Fig. 5.5 Comparision of ΔN_{IT} time evolution of single-interface and two-interface RD models for two different k_F values corresponding to first interface reaction

in Fig. 6.4. The RD analysis for this SiO_2/high-K interface, however, remains the same as that for the SiO_2/poly interface presented above.

5.2.1.3 H_2-Only Diffusion

As an approximation to the generalized H–H_2 based RD models discussed above, we assume H to H_2 dimerization only at the silicon/oxide interface using the following chemical reaction:

$$2\mathrm{Si_3{\equiv}Si-H+2h^+} \quad \leftrightarrow \quad 2\mathrm{Si_3{\equiv}Si^+ + H_2}. \tag{5.20}$$

The kinetics of (5.20) can be captured using the following rate equation, where we consider $\left[\mathrm{N_H^{(1)}}\right]^2 \sim \left[\mathrm{N_{H2}^{(1)}}\right]$ for the bimolecular reaction:

$$\frac{\partial N_{\mathrm{IT}}}{\partial t} = k_{\mathrm{F}}(N_0 - N_{\mathrm{IT}}) - k_{\mathrm{R}} N_{\mathrm{IT}} \sqrt{N_{\mathrm{H2}}^{(0)}} \tag{5.21}$$

The same procedure used in (5.2)–(5.7) yields,

$$N_{\mathrm{IT}} \sim \left(\frac{k_{\mathrm{F}} N_0}{k_{\mathrm{R}}}\right)^{2/3} (4D_{\mathrm{H2}} t/\pi)^{1/6}. \tag{5.22}$$

The RD framework with H_2 diffusion, therefore, predicts a power-law time exponent, $n \sim 1/6$ and the following activation energy [steps are same as that in (5.8)] for the interface trap generation:

$$E_{\mathrm{A,IT}} = \frac{2}{3}\left(E_{\mathrm{A,kF}} - E_{\mathrm{A,kR}} - aE_{\mathrm{ox}}\right) + \frac{1}{6} E_{\mathrm{A,DH_2}}. \tag{5.23}$$

5.2.1.4 H-Only Diffusion

Analysis for RD framework with H-only diffusion is same as Jeppson's RD analysis in (5.1)–(5.8) with the replacement of X with H and Y with h^+.

5.2.1.5 Drift of Proton (H^+)

Protons may be produced from the dissociation of Si–H bonds as follows:

$$\begin{aligned} &\mathrm{Si_3{\equiv}Si-H+h^+} \quad \leftrightarrow \quad \mathrm{Si_3{\equiv}Si^+ + H} \\ &\mathrm{H+h^+} \quad \leftrightarrow \quad \mathrm{H^+} \end{aligned} \tag{5.24}$$

Just as in (5.5), consideration of dynamic equilibrium between dissociation and annealing for $t \gg 0$ yields:

$$k_F N_0 \cong k_R N_{H^+}^{(0)} N_{IT}. \tag{5.25}$$

Under NBTI stress condition, protons will drift faster than diffusion, so that we can estimate the amount of N_{IT} at $t \gg 0$ (Fig. 5.6) using [16]:

$$N_{IT} = \int_{z=0}^{z=\infty} N_{H^+}(z,t)\mathrm{d}z \sim N_{H^+}^{(0)} \mu_{H^+} E_{ox} t \tag{5.26}$$

where μ_{H+} is the mobility of H^+ and E_{OX} is the applied electric field under NBTI stress. Combining (5.25) and (5.26), we obtain:

$$N_{IT} \sim \left(\frac{k_F N_0}{k_R} \mu_{H^+} E_{ox} \right)^{1/2} t^{1/2}, \tag{5.27}$$

that suggests $n = ½$ for proton diffusion.

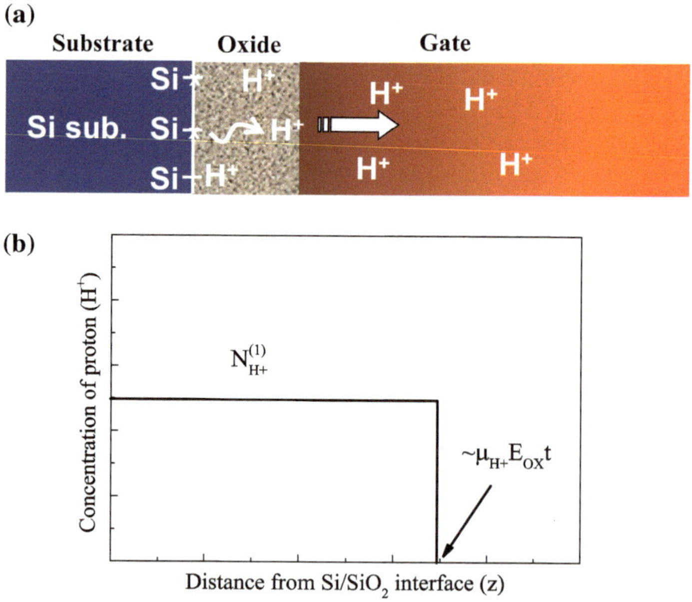

Fig. 5.6 **a** Schematic illustration of Si–H bond dissociation and drift (that dominates over diffusion) of H^+. **b** Drift front at time t reaches $\mu_{H+} E_{OX} t$ and can be approximated using a *rectangle* [16], as used in (5.26)

5.2.2 *Features of RD Formalism for NBTI Stress Phase*

Figure 5.7 shows a generalized schematic for the time evolution of ΔV_{IT} for the NBTI stress phase in a log-log scale. For short stress time, $\Delta V_{IT} \sim N_{IT} = k_F N_0 t$ for any diffusing species that explains the unity time exponent in that region. For a higher stress time, the diffusion of Hydrogen species and H to H_2 dimerization reduces the time exponent from unity. The time exponent in this transient phase depends on the type of RD model under consideration (e.g., equals 1/3 for the RD model with H and H_2diffusion and direct H to H_2 dimerization [18]). At long stress time, time exponent settles down to n and its magnitude is a unique signature of the diffusing species under consideration. For atomic H, molecular H_2 and proton, n is 1/4, 1/6, and 1/2, respectively, irrespective of the mechanism of H to H_2 conversion, stress voltage, temperature, gate dielectric, and channel strain. However, when a transistor is stressed at a fixed $V_{G\text{-}STR}$ for a long time, a typical case for NBTI experiments, there is a reduction of time exponent from the n values mentioned above [20]. A long-duration NBTI stress at a fixed $V_{G\text{-}STR}$ increases the threshold voltage of a transistor and reduces the oxide electric field from its value at the beginning of NBTI stress. This reduced electric field breaks the Si–H bond at a lower rate and hence causes reduction in the time exponent. This particular aspect of soft-saturation at a long stress time is very important to predict NBTI lifetime, as a reduction in time exponent can significantly increase the lifetime of a transistor at a certain operating condition, thus relax the reliability criteria for transistor designers. Figure 5.8 shows variation in the time exponent as a function of stress time for various stress bias for the transistor analyzed in Fig. 5.4 using H–H_2RD model. The time exponent reduces from the robust $n \sim 1/6$ value at a rate of $\delta n \sim 0.005$/decades with the reduction being higher for higher $V_{G\text{-}STR}$.

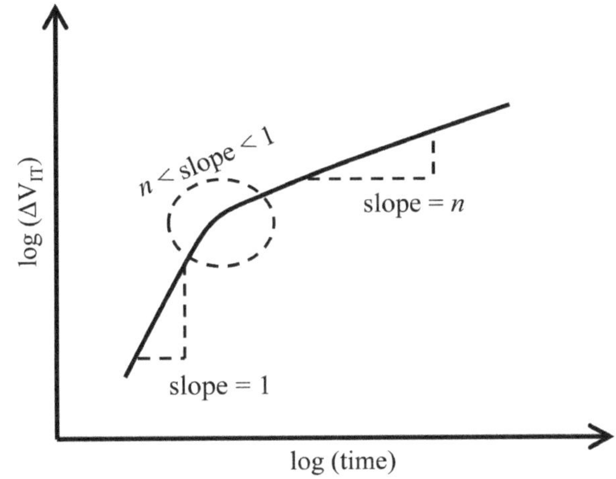

Fig. 5.7 A generalized schematic for temporal evolution of ΔV_{IT} in a log–log scale. The slope of the degradation changes from 1 to n over time. The magnitude of n and also the slope in the transition region (*highlighted using dashed circle*) depend on the diffusing species

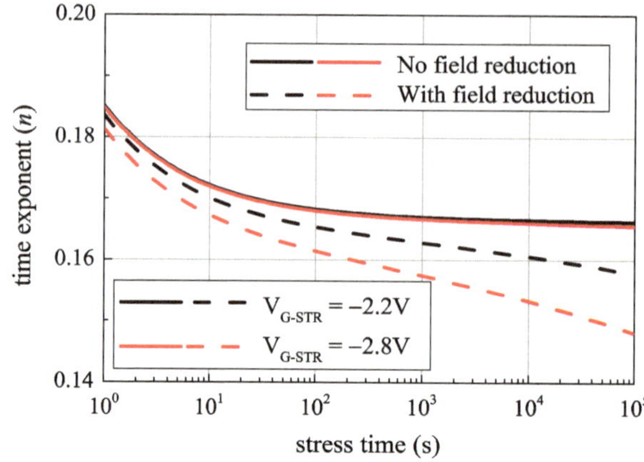

Fig. 5.8 Decrease in long-term time exponent (soft-saturation) due to the reduction of stress field with time, when the transistor studied in Fig. 5.4 is stressed at different $V_{G\text{-}STR}$ and at $T = 125$ °C. Here, the *dashed lines* consider the effect of soft-saturation, whereas *solid lines* do not [20]

5.3 RD Formalism for NBTI Relaxation Phase

One important feature of NBTI-induced generation of interfacial defects is their reverse-annealing (or relaxation), once NBTI stress voltage is removed [24–31]. This recovery makes NBTI distinct compared to other transistor reliability concerns, such as TDDB [32–38] and HCI [39–41], where generated defects show negligible recovery. When NBTI stress is removed after a time t_{STR} (i.e., in the NBTI relaxation phase), the Hydrogen species generated during the stress phase diffuse back towards the Si/oxide interface and passivates interface traps. Note that a similar Hydrogen diffusion and dangling Si- bond passivation has been the key for operation of a CMOS transistor [1–4, 8–10]. The start of interface trap passivation during the NBTI relaxation phase, therefore, depends on the time required for backward diffusion of Hydrogen species, which is relatively slow compared to the other components of NBTI (Fig. 5.1 and Sect. 6.3). In the following discussion, the physics of this slow start and continuous recovery of interface traps will be elaborated based on RD formalism.

If $V_{G\text{-}STR}$ is completely removed during NBTI relaxation (the typical case), the first term in (5.2) drops out, which allows annealing of interface traps or Si- bonds via backward diffusion of the Hydrogen species. For partial reduction in $V_{G\text{-}STR}$, some of the interface traps are also passivated. The RD simulation of such relaxation with a non-zero stress voltage (Sect. 6.3) requires a change in k_F as per Sect. 5. 6 in (5.2). A subtraction of this back-diffused Hydrogen component running from t_{STR} till time t ($>t_{STR}$), one can write the following expression for N_{IT} in the typical relaxation phase [12–25]:

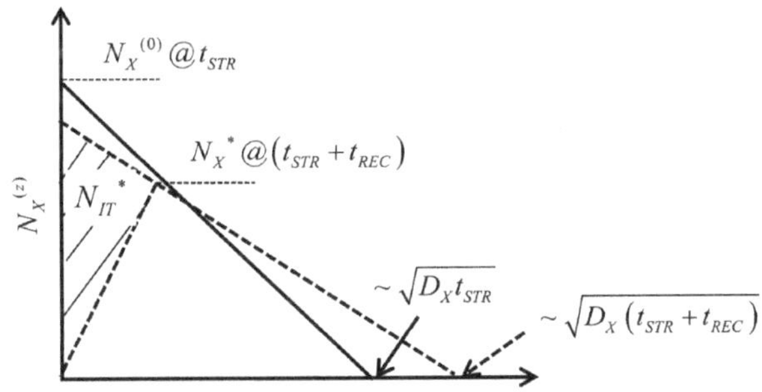

Fig. 5.9 Approximate Hydrogen profile at the start of NBTI relaxation (*solid line*; $t = t_{STR}$) and after a recovery time of t_{REC} (*dashed line*; $t = t_{STR} + t_{REC}$). See text for the definition of other parameters

$$N_{IT}(t > t_{STR}) = N_{IT}(t_{STR}) - N_{IT}^*(t) \tag{5.28}$$

The calculation of $N_{IT}^*(t)$ in (5.28) involves consideration of back-diffused Hydrogen species near the interface (schematically illustrated in Fig. 5.9) and takes the form [25]:

$$N_{IT}^*(t) \approx N_X^* \sqrt{\delta D_X (t - t_{STR})} \tag{5.29}$$

where ξ is a constant indicating the asymmetry in diffusion and equals 0.5 for symmetric diffusion; use of $\xi \sim 0.58$ fits detailed simulation of interface defect passivation [25]. Using Fig. 5.9, a second approximation of $N_{IT}(t)$ takes the following form:

$$N_{IT}(t) \approx N_X^* \sqrt{D_X t}. \tag{5.30}$$

Eliminating $N_X^*(t)$ from (5.28) to (5.30), one obtains the following expression for N_{IT} in the relaxation phase:

$$N_{IT}(t_{REC}) = N_{IT}(t_{STS}) \Big/ \left[1 + \sqrt{\frac{\xi t_{REC}}{t_{STS} + t_{REC}}}\right] \tag{5.31}$$

where t_{REC} ($=t - t_{STR}$) indicates the duration of NBTI relaxation.

The calculations for the relaxation of interface defects presented in (5.28)–(5.31) is valid in general for any neutral diffusing species (H or H_2). Note that, as discussed in Sect. 5.2, these diffusing species are more relevant for NBTI stress. For the H–H_2RD model with direct H to H_2 dimerization, RD simulation predicts a 5 % trap passivation (i.e., $1 - N_{IT}(t_{REC})/N_{IT}(t_{STR}) = 5$ %) of defects at $t_{REC} \sim 0.005 t_{STR}$ and a complete passivation at $t_{REC} \rightarrow \infty$. To understand this onset of passivation, we analyze the H_2 profile for a transistor stressed at a certain $V_{G\text{-}STR}$, T and for $t_{STR} = 1000$ s. Figure 5.10 plots the percentage of Hydrogen species (mostly H_2 for

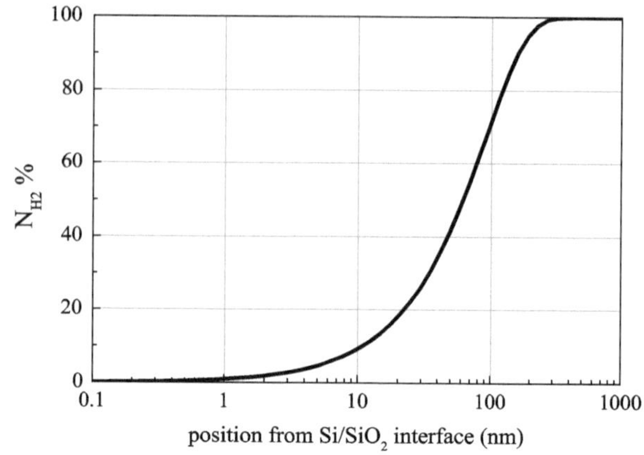

Fig. 5.10 Percentage of H_2 as a function of distance from the interface (z) at the end of stress ($V_{G\text{-}STR} = -2.3$ V, $T = 125$ °C) for a transistor having 1.4 nm gate dielectric after $t_{STR} = 1000$ s

this RD model) at different distances away from the interface and suggests the presence of ~ 5 % H_2 within 4 nm away from the interface. To achieve the recovery of interface traps from this 5 % H_2, one needs $t_{REC} \sim z^2/4D_{H2} \sim 1$ s (for $D_{H2} \sim 10^{-14}$ cm^2 s^{-1} [23]) that supports the analytical calculation using (5.31).

For the second-interface assisted H to H_2 dimerization, the faster diffusion of Hydrogen species inside the oxide compared to that in the poly-Si gate results faster start of interface trap passivation compared to the single-interface model [23], as confirmed in Fig. 5.11 using a single set of parameters for the single-interface and double-interface models.

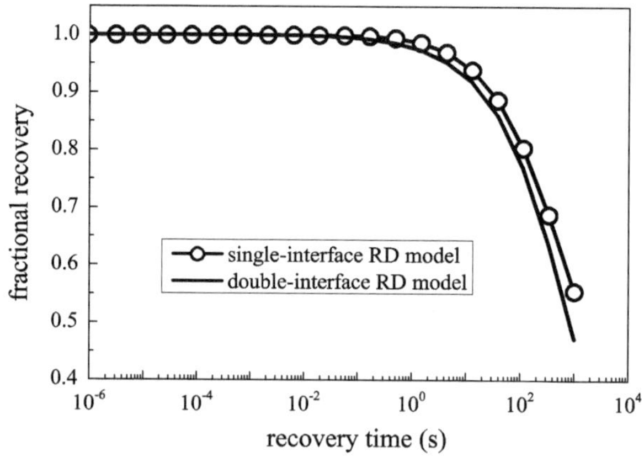

Fig. 5.11 Comparison of interface trap passivation during the NBTI relaxation phase for the single interface RD model: direct H to H_2 dimerization (*line with symbols*) and second interface assisted H to H_2 dimerization (*solid line*). The simulation uses a transistor with 1.4 nm gate dielectric stressed using $V_{G\text{-}STR} = -2.3$ V, $T = 125$ °C for $t_{STR} = 1000$ s and then relaxed at 0 V

5.4 RD Formalism for AC NBTI Stress

The passivation of interface traps due to the removal of stress generates some of the interesting features for the AC NBTI stress, where the stress is turned ON and OFF periodically at a certain frequency and duty cycle. As diffusion of Hydrogen species dictates the NBTI dynamics, for the ON state of the first AC cycle (t_{ON}), we can write:

$$N_{IT}(t_{ON}) = At_{ON}^{n} \tag{5.32}$$

where t_{ON} is the duration of ON state during AC NBTI stress, A is the pre-factor containing reaction-rate and diffusion coefficient related terms, and values of n depends on the diffusing species (see Sect. 5.2). Similarly, we can rewrite (5.31) to obtain the following expression:

$$R_2 = \frac{N_{IT}(P)}{N_{IT}(t_{ON})} = \left(1 + \sqrt{\frac{\xi t_{OFF}}{P}}\right)^{-1} = \frac{R_1}{1 + \sqrt{(1-d)\xi}} \tag{5.33}$$

where P is the period and t_{OFF} is the duration of OFF state during AC NBTI stress, d (=t_{ON}/P) is the duty cycle, R_i is the ratio of N_{IT} at the end of i-th step of AC stress (e.g., $i = 2$ for one complete AC cycle) to N_{IT} at the end of 1st step; hence, $R_1 = 1$. Note that (5.33) is an exclusive universal function of time and is independent of voltage, temperature and power-law time exponent. As discussed below, this universality leads to one of the characteristic feature of RD formalism, namely the frequency independence of NBTI response. Specifically, consider two transistors subjected to a long series of NBTI stress-relaxation sequences at two different frequencies f_1 and f_2. The theory suggests that if integrated operation time remains the same, i.e., $P_{max} = N_1 * f_2 = N_2 * f_2$, the total degradation also remains the same, i.e., $N_{IT}(f_1) = N_{IT}(f_2)$.

The proof of this ‘frequency-independence’ is elementary, but involves a tedious derivation [17]. If the transistor is re-stressed after the first stress-relaxation cycle, the forward reaction will resume and new traps will be generated. The history of the traps generated in the complex interplay of stress and relaxation in the first cycle can be captured in the notion of an ‘equivalent time’, as defined in Fig. 5.12 and calculated as $R_2 A t_{ON}{}^{n} = A t_e{}^{n}$, or equivalently, $t_e/t_{ON} = (R_2)^{1/n}$. With this definition of t_e, the number of interface traps generated in the stress phase of second AC cycle is $N_{IT}(P + t) = N_{IT}(t_e + t) = A(t_e + t)^n$, so that we obtain:

$$R_3 = \frac{N_{IT}(P + t_{ON})}{N_{IT}(t_{ON})} = \frac{A(t_e + t_{ON})^n}{At_{ON}^{n_{IT}}} = \left(1 + R_2^{\frac{1}{n}}\right)^n; R_3^{\frac{1}{n}} = 1 + R_2^{\frac{1}{n}} \tag{5.34}$$

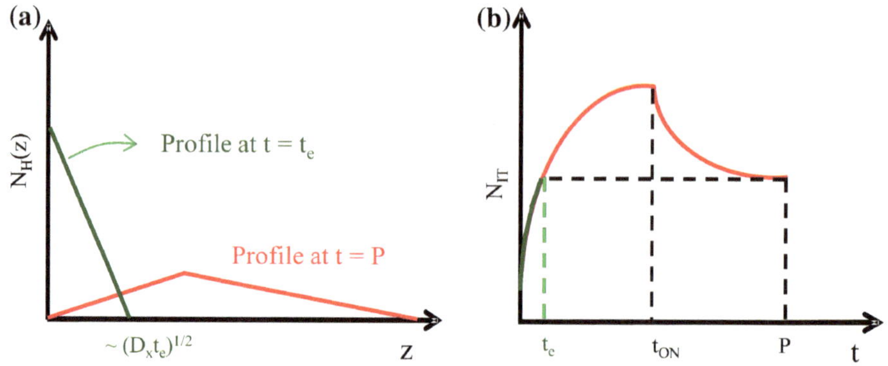

Fig. 5.12 **a** Hydrogen profiles at T and at effective time t_e. Here, t_e is computed such that areas under both the profiles are same. **b** One can also compute t_e by projecting the residual interface trap density at the end of an AC stress cycle (P) to the degradation curve and then by finding the corresponding degradation time

In general, R_i at the end of ON-state of kth AC cycle is related to that at the end of $(k-1)$th AC cycle is given by:

$$R_{2k-1}^{\frac{1}{n}} = 1 + R_{2k-2}^{\frac{1}{n}} \tag{5.35}$$

Following steps similar to the first AC cycle as in (5.28)–(5.30), for the relaxation phase of the second AC cycle, we can write: $N_{IT}^{*}(2P) \approx \sqrt{\xi D_X t_{OFF}} N_X^{*}$, $N_{IT}(2P) - N_{IT}(P) \approx \sqrt{D_X P} N_X^{*}$ and $N_{IT}(2P) = N_{IT}(P + t_{ON}) - N_{IT}^{*}(2P)$. Eliminating N_X^{*} and N_{IT}^{*} from these equations, we obtain:

$$N_{IT}(2P) = \frac{N_{IT}(P + t_{ON})}{1 + \sqrt{\frac{\xi t_{OFF}}{t_{OFF}+t_{ON}}}} + \frac{\sqrt{\frac{\xi t_{OFF}}{t_{OFF}+t_{ON}}}}{1 + \sqrt{\frac{\xi t_{OFF}}{t_{OFF}+t_{ON}}}} N_{IT}(P), \tag{5.36}$$

$$R_4 = \frac{1}{1 + \sqrt{(1-d)\xi}} R_3 + \frac{\sqrt{(1-d)\xi}}{1 + \sqrt{(1-d)\xi}} R_2. \tag{5.37}$$

A similar argument can be applied for any even numbered AC cycle and hence (5.37) can be generalized as:

$$R_{2k-2} = \frac{1}{1 + \sqrt{(1-d)\xi}} R_{2k-3} + \frac{\sqrt{(1-d)\xi}}{1 + \sqrt{(1-d)\xi}} R_{2k-4}. \tag{5.38}$$

By combining (5.35) and (5.38), and also by considering $R_{2k-3}^{1/n} = 1 + R_{2k-4}^{1/n}$ as an extension of (5.35) and $R_{2k-3} \gg 1$ (for large number of AC cycles), we find the following recursion relationship:

$$R_{2k-1}^{\frac{1}{n}} = 1 + R_{2k-3}^{\frac{1}{n}}\left[1 - n\frac{\sqrt{(1-d)\xi}}{1+\sqrt{(1-d)\xi}}\frac{1}{R_{2k-3}^{\frac{1}{n}}}\right] \cong \frac{1}{1+\sqrt{(1-d)\xi}} + R_{2k-3}^{\frac{1}{n}}, \tag{5.39}$$

which has the following general form:

$$R_{2k}^{\frac{1}{n}} \approx \frac{k}{1+\sqrt{(1-d)\xi}}. \tag{5.40}$$

5.4.1 Frequency Independence

Let us now evaluate (5.40) for two AC waveforms with time periods P_1 (=1/f_1) and P_2 (=1/f_2) and number of cycles k_1 and k_2 (≫1). If the duty cycle and the duration of AC NBTI stress for these two waveforms are same, i.e., $k_1P_1 = k_2P_2$, the ratio of generated interface trap can be expressed as:

$$\mathrm{AC/DC-Ratio} = \frac{N_{\mathrm{IT1}}(f_1, d)}{N_{\mathrm{IT2}}(f_2, d)} \approx \left(\frac{k_1}{k_2}\right)^n \left(\frac{t_{\mathrm{ON},1}}{t_{\mathrm{ON},2}}\right)^n = 1. \tag{5.41}$$

Therefore, (5.41) establishes that interface trap density for any frequency will be same, provided the transistor is stressed with same d, in other words, AC NBTI is frequency independent irrespective of the voltage and temperature of operation. Anyone familiar with traditional linear systems based on RC circuits would immediately appreciate the highly unusual nature of the proposition. It is as if one has defined a broadband filter with same gain starting from a few Hz to 1 GHz with no variation in the gain!

5.4.2 Duty Cycle Dependence

In the discussion above, we have seen that AC degradation is frequency independent for a fixed duty cycle of operation, however, that does not mean that the AC/DC ratio is fixed for any duty cycle. Obviously, if the duty cycle is close to 100 %, NBTI degradation approaches values as obtained for the DC stress and degradation disappears for zero duty cycle. To derive this duty cycle dependence, we use (5.40) and calculate the ratio of the number of interface traps generated at the end of kth AC cycle and at the end of same duration DC stress, as given below:

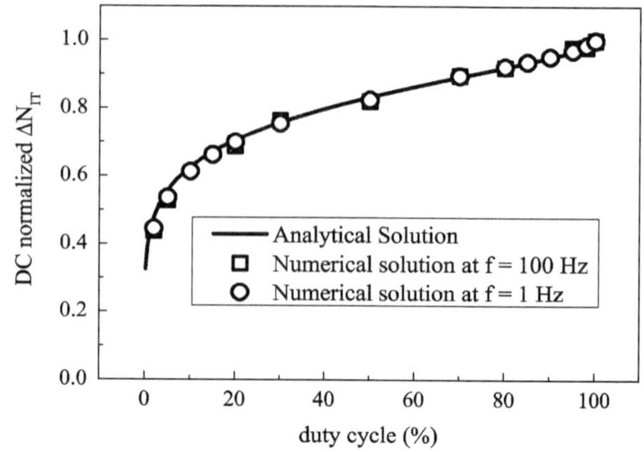

Fig. 5.13 Numerical solution of duty cycle dependence at different frequency compares well with the analytical solution of (5.42)

$$\mathrm{AC/DC-Ratio}(d) = \frac{N_{\mathrm{IT,AC}}(kP)}{N_{\mathrm{IT,DC}}(kP)} \approx \left(d \Big/ \left(1+\sqrt{\frac{1-d}{2}}\right)\right)^n. \tag{5.42}$$

As seen in Fig. 5.13, (5.42) matches the numerical solution and establishes the duty cycle dependence of AC/DC-Ratio as a universal observation for any frequency.

5.5 Key Features of RD Theory

RD theory makes the following four unique and self-consistent predictions for NBTI without any adjustable parameters:

(a) At long stress time, interface trap generation is described by a power-law, i.e., $N_{\mathrm{IT}}(t) = At^n$
(b) Once the stress is removed, the relaxation given by (5.31) is only a function of $t_{\mathrm{REC}}/t_{\mathrm{STR}}$
(c) For a given duty cycle, the ratio of degradation due to DC stress and AC stress is frequency independent
(d) AC/DC-Ratio for any frequency is a unique function of the duty cycle.

These features of RD theory are critical for NBTI reliability projection and equivalent circuit modeling. For example, if instead of a power-law, NBTI degradation had been described by $N_{\mathrm{IT}}(t) = B\ln(t)$, and if the AC/DC ratio decreased with frequency, NBTI would have become an unimportant reliability phenomenon at high-frequency and for long period of usage.

5.6 Mechanism of Si–H Bond Dissociation

The mechanism of Si–H bond dissociation defines the factor k_F in (5.7), (5.15), (5.27) and the factor $k_{F(1)}$ in (5.18), thus critically controls interface trap generation at different $V_{G\text{-}STR}$. Though first-principle calculations [42, 43] suggest large energy barrier $\sim$2.5 eV (which in presence of hole reduces to $\sim$2.3 eV) and endothermic reaction (with reaction energy $\sim$2 eV) for the Si–H bond dissociation, application of electric field under NBTI stress reduces the energy barrier and makes the reaction exothermic [20, 27].[2] The field dependence of interface trap generation has been considered using several empirical field models, e.g., exponential model with a dissociation factor $\propto \exp(\gamma E)$ [44, 45], power-law model with the factor $\propto (E)^P$ [47], and mixed model with the factor $\propto E\exp(\gamma E)$ [14, 21, 47, 48], where γ is the field acceleration factor. As shown below, a physical consideration of Si–H bond stretch and dissociation [14, 47] can validate the model with $E\exp(\gamma \mathrm{E})$ dissociation factor for proper estimation of NBTI lifetime at operating conditions.

5.6.1 *Hole-Assisted, Field-Enhanced, Thermal Dissociation of Si–H Bond*

The model for Si–H bond dissociation is presented in Fig. 5.14. Under NBTI stress condition, there are holes near the silicon-oxide interface with a concentration, $p_h \sim E_c$; where E_c represents the electric field due to mobile carriers and its magnitude equals the total electric field E_{OX} in accumulation and equals E_{OX} minus the depletion charge contribution in inversion. The holes from accumulation or inversion layer tunnel towards the Si–H bond (bond length $\sim$1.5 Å [10]) with a tunneling probability T_H. This capture of holes by the Si–H bond reduces the energy barrier for its dissociation to $E_{A,kF}$. In addition to this hole capture, the polarity of electric field during NBTI stress stretches the Si–H bond and further reduces its barrier by aE_{OX}, where a is the effective dipole moment for the Si–H bond [20, 49]. This results a factor $B \propto \exp[-(E_{A,kF} - aE_{OX})/k_BT]$ for the thermal dissociation of Si–H bond. Therefore, the Si–H bond under NBTI stress breaks at a rate of:

$$k_F \sim p_h T_H B \sim E_c T_H \exp\left[-\left(E_{A,kF} - aE_{ox}\right)/k_B T\right] \tag{5.43}$$

The tunneling probability T_H in (5.43) can be estimated using the following Wentzel–Kramers–Brillouin (WKB) approximation:

---

[2]An alternate dissociation mechanism of Si–H via passivated dopants like phosphorous-Hydrogen (P–H) complex though have a lower dissociation energy and a exothermic reaction, P–H assisted dissociation of Si–H results $n \sim 1/4$ [12, 26] and $E_{A,IT} \sim 0.36$ eV, which is not supported by recent NBTI measurements, see Chap. 3.

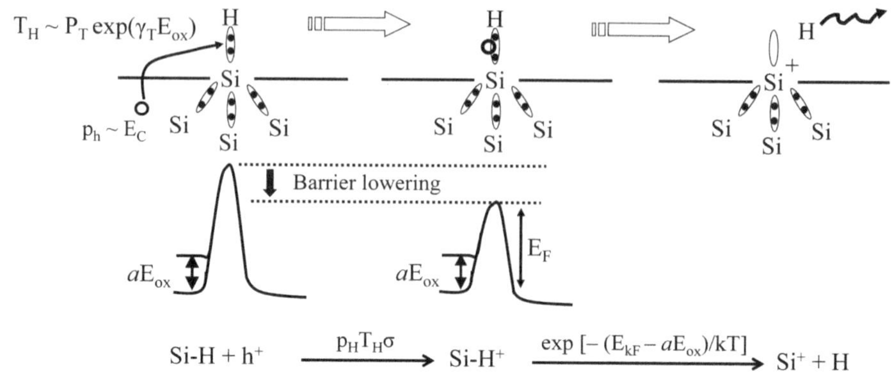

Fig. 5.14 The model for Si–H bond dissociation considers holes from the channel of a transistor (concentration $p_h \sim E_c$ in inversion and $p_h \sim E_{OX}$ in accumulation) near the Si/dielectric interface tunnel into and is captured by the Si–H bond, thus leading to a hole-assisted, field-enhanced, thermal dissociation of the Si–H bond and generation of interface defects (Si-) at the silicon-oxide interface

$$T_H \sim \exp\left[-2\int_0^{t_{int}} dx \frac{\sqrt{2qm_{ox}(\varphi_{bh} - xE_{ox})}}{\hbar}\right], \tag{5.44}$$

where t_{int} is the interfacial layer thickness ($\sim$ Si–H bond length of 1.5 Å), m_{ox} is the effective mass of hole in the oxide, and φ_{bh} is the potential energy barrier for hole tunneling at the silicon-oxide interface. The completion of integration in (5.44) and a subsequent Taylor series expansion gives:

$$\ln T_H \sim -2\frac{\sqrt{2qm_{ox}\varphi_{bh}}}{\hbar} t_{int} + \sqrt{\frac{m_{ox}}{2q\varphi_{bh}}} \frac{qt_{int}^2 E_{ox}}{\hbar} - \cdots \tag{5.45}$$

Ignoring the higher order terms in (5.45), we can approximate the tunneling probability using $T_H \sim P_T \exp(\gamma_T E_{OX})$, where,

$$P_T \sim \exp\left[-2\frac{\sqrt{2qm_{ox}\varphi_{bh}}}{\hbar} t_{int}\right] \quad \text{and} \quad \gamma_T \sim \sqrt{\frac{m_{ox}}{2q\varphi_{bh}}} \frac{qt_{int}^2}{\hbar} \tag{5.46}$$

Overall, the hole-assisted, field enhanced, thermal dissociation of Si–H bond presented in Fig. 5.14 suggests

$$k_F \sim E_C P_T \exp\left[\gamma_T E_{ox} - \left(E_{A,kF} - aE_{ox}\right)/k_B T\right] \tag{5.47}$$

Among different parameters in (5.47), the magnitude of applied NBTI stress defines E_C and E_{OX}. Activation energy for hole-assisted Si–H bond dissociation $E_{A,F}$ are obtained from first-principle calculation [50, 51] and the remaining parameters

γ_T, P_T, and a can be theoretically calculated as shown below. Proper estimation of these three parameters are critical to analyze field dependence of interface trap generation, as it effects the prediction of NBTI lifetime in the operation condition of a device from the measured lifetime at accelerated stress conditions.

5.6.2 *Theoretical Estimation of Tunneling Factors*

Among the tunneling parameters in (5.47), γ_T can be calculated from the T_H vs E_{OX} plot obtained using detailed quantum-mechanical model [52] or modified WKB approximation [53]; see Fig. 5.15a [47, 54]. The straight-line trend in the semilog-*y* plot for T_H as a function of E_{OX} confirms the insignificance of higher order terms in (5.45). The calculated γ_T from the slope of $\ln T_H$ vs E_{OX} plot varies negligibly with nitridation in the gate dielectric (Fig. 5.15a) and with uniaxial (Fig. 5.15b) or

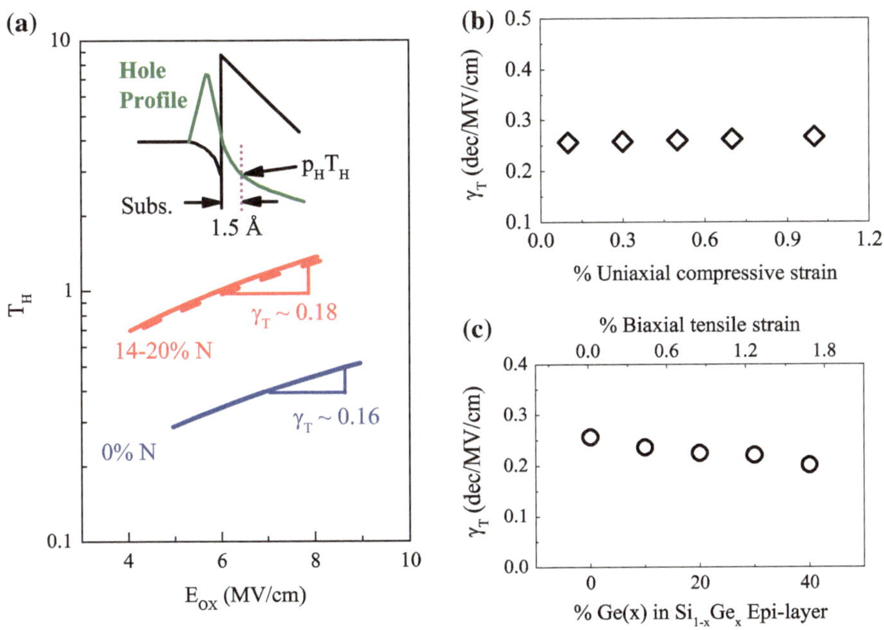

Fig. 5.15 **a** Tunneling probability (T_H) at 1.5 Å away from the transistor channel calculated at different E_{OX}. *Inset* shows the quantum-mechanical penetration of hole profile used to calculate T_H; here, the valence band-diagram is drawn upside-down for clarity. One can estimate γ_T from the slope of $\ln T_H$ vs E_{OX} plot for different nitrided gate dielectric. Calculated γ_T for strained transistors, where **b** uniaxial compressive strain in the channel is applied using SiGe source and drain, and **c** biaxial, tensile strain in the channel is applied using SiGe epitaxial layer under the channel. Theoretically extracted values of γ_T show negligible variation with any process technology, especially for strained transistors

biaxial (Fig. 5.15c) channel strain [55]. The calculation details and device parameters for nitrided dielectric and strained channel are provided in [20, 47, 54–56].

Similar to γ_T, tunneling simulation also allows calculation P_T. Among different parameters that define P_T (see (5.46)), m_{ox} and φ_{bh} are the ones that changes with gate dielectric and channel strain. For example, nitridation of gate oxide reduces both m_{ox} and φ_{bh} [47, 54] and hence reduces P_T. Similarly, compressive strain increases φ_{bh}, while tensile strain reduces φ_{bh} without effecting m_{ox} [55] and hence also changes P_T accordingly. Note that variation of P_T with channel strain and a fixed γ_T universally explains the field dependence of NBTI measured over a broad range of uniaxial/biaxial, compressive/tensile strained transistors [55]. For the analysis of NBTI in nitrided transistors, variations in γ_T (though small, see Fig. 5.15a) and P_T require appropriate consideration [47, 54].

5.6.3 *Theoretical Estimation of Effective Dipole Moment*

One can evaluate the effective dipole moment for Si–H bond [the parameter a in (5.47)] by considering the electronegativity difference between the Si (x_{Si} = 1.8) and H (x_H = 2.1) atoms, as per Pauling's scale [57].[3] The corresponding ionic bond energy, $U_{H\text{-}Si} = 1.3(x_{Si} - x_H)^2 = 0.1$ eV leads to an estimation for the effective charge transfer within Si–H dipole as, $\sqrt{(4\pi\varepsilon_0 r U_{H-Si})}$ [49, 57], where q is the electron charge and r is the Si–H bond length. The resultant dipole moment for the Si–H bond, $p = \sqrt{(4\pi\varepsilon_0 r U_{H-Si})}\ r = 0.15q$ Å. Considering that a Si–H dipole resembles a dipolar orientation normal to a thin slab (Fig. 5.16), the local electric field [59] within the dipole can be calculated as $E_{loc} = (1 + L\chi_{int})E_{OX} = \varepsilon_{int}E_{OX}$; where, $L = 1$ is the depolarization factor [59], χ_{int} is the dielectric susceptibility at the interface and ε_{int} is the relative dielectric permittivity at the interface with values ~3.9–8 [60]. Under NBTI stress, the polarization vector of the Si–H bond is opposite to the electric field (Fig. 5.16). The potential energy of Si–H bond will therefore be increased by $pE_{loc} = p\varepsilon_{int}E_{OX} = aE_{OX}$ (Fig. 5.14). Using the parameters provided above, we can estimate the effective dipole moment as, $a = p\varepsilon_{int} = 0.6\text{–}1.2q$ Å.

5.6.4 *Procedure for Estimating Model Parameters*

The estimation of γ_T and a using measured data primarily requires a separation of interface trap component of DC NBTI degradation (ΔV_{IT} of Fig. 5.1) and then an

[3] Similar calculation can also be performed using Sanderson's scale [58] that leads to similar values.

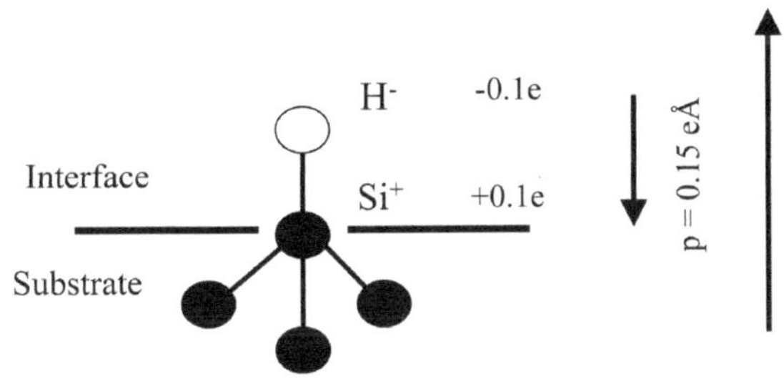

Fig. 5.16 Polarization properties of Si−H bond provides an estimation for $a = p\varepsilon_{\text{int}}$. Here, the polarization vector is opposite to the direction of the applied field under NBTI stress

implementation of the following equations. As H_2 diffusion dictates ΔV_{IT} at long stress time, one can replace k_F from (5.47) in (5.22) and write:

$$\Delta V_{\text{IT}} = \frac{q\Delta N_{\text{IT}}}{C_{\text{OX}}} = A_{\text{IT}} * \text{EOT} * (E_c)^{2/3} \exp\left(\frac{2\gamma E_{\text{ox}}}{3}\right) \exp(-nE_{\text{D1}}/k_B T) t^n \quad (5.48)$$

where

$$\gamma = \gamma_T + a/k_B T \quad (5.49)$$

$$nE_{\text{D1}} = nE_{\text{A,DH}_2} + \frac{2}{3}\left(E_{\text{A,kF}} - E_{\text{A,kR}}\right) \quad (5.50)$$

and the pre-factor

$$A_{\text{IT}} \sim (N_0 P_T)^{2/3} \sim \left[N_0 \exp\left(-\sqrt{m_{\text{ox}}\varphi_{\text{bh}}}\right)\right]^{2/3} \quad (5.51)$$

The overall activation energy for ΔV_{IT} from (5.48) can be written as,

$$E_{\text{A,IT}} \equiv nE_{\text{A,DH}_2} + \frac{2}{3}\left(E_{\text{A,kF}} - E_{\text{A,kR}} - aE_{\text{ox}}\right) = nE_{D1} - 2/3aE_{\text{ox}} \quad (5.52)$$

For the neutral Hydrogen species [14, 47], k_R, D_{H2}, k_H, and k_{H2} are field independent therefore the field-dependence of k_F dictates the field-dependence of N_{IT} and hence allows extraction of γ_T and a. To perform this extraction [20, 47], the measured ΔV_{IT} versus t data is fitted with (5.48) at different $V_{\text{G-STR}}$ and T. This fit allows extraction of A_{IT}, γ, n at different T (here, the manner of fitting includes the variation in the field independent E_{D1} within the parameter A_{IT}). A plot of extracted γ versus $1/k_B T$ gives tunneling parameter γ_T as the intercept at $T \to \infty$ and the effective dipole moment a as the slope; see (5.49). Alternately, one can also estimate a using (5.52), hence using the extracted $E_{\text{A,IT}}$ from the temperature dependent ΔV_{IT} versus t data at different voltage and plotting it as a function of E_{OX}. However, presence of noise in the extracted $E_{\text{A,IT}}$ incorporates inconsistency for the later technique.

5.7 Summary

The unique formulation of the RD model presented in this chapter in the context of NBTI allows interpretation of the kinetics of interface defects during NBTI stress and relaxation. For the NBTI stress phase, RD theory captures the bond dissociation at a particular stress bias and temperature and predicts power-law time evolution ($\sim t^n$) for the interface trap generation. The magnitude of this power-law exponent at long stress time depends only on the diffusing species. The density of generated interface defects at a particular stress time, however, depends on how Si−H bonds are broken at different stress conditions like bias, temperature and frequency. A polarized nature of the Si−H bond results their field-enhanced, hole-assisted dissociation with a field dependence of $E\,\exp(\gamma E)$. The temperature dependence of Si−H bond dissociation, as well the diffusion of Hydrogen species, takes the Arrhenius form and defines the Arrhenius activation energy for the interface defect generation. For the NBTI relaxation phase, the time evolution of interface trap repassivation after a fixed duration of stress t_{STR} takes the form $\sim 1/\left[1+\sqrt{(\xi/(1+t_{\mathrm{SRT}}/t_{\mathrm{REC}}))}\right]$. This unique nature of the time evolution for stress and relaxation and its dependence only on the diffusing species leads to frequency independence for interface trap generation during AC NBTI stress. Finally, it is important to understand that the RD model is a general formulation independent of many details of the physical process. The predictions of the model do not distinguish between Si−H bonds present at the interface (P_b centers [6, 9, 61, 62]) or Hydrogen-terminated near-interface E′ centers [62, 63], as long as the released species diffuses back and forth within a diffusing medium. It is precisely this generality of the RD framework that allows capture of versatile ranges of robust phenomena like power-law exponent, frequency independence which is universality measured over a wide range of transistors.

Acknowledgments The authors acknowledge Narendra Parihar for his editorial support.

References

1. A.S. Grove, *Physics and Technology of Semiconductor Devices* (Wiley, London, 1971)
2. B.E. Deal, M. Sklar, A.S. Grove, E.H. Snow, Characteristics of the surface-state charge (Qss) of thermally oxidized silicon. J. Electrochem. Soc. **114**, 266 (1967)
3. P.V. Gray, D.M. Brown, Density of SiO_2–Si interface states. App. Phys. Lett. **8**, 31 (1966)
4. Y. Nishi, Study of silicon-silicon dioxide structure by electron spin resonance 1. Japanese J. App. Phys. **10**, 52 (1971)
5. P.J. Caplan, E.H. Poindexter, B.E. Deal, R.R. Razouk, ESR centers, interface states, and oxide fixed charge in thermally oxidized silicon wafers. J. App. Phys. **50**, 5847 (1979)
6. K.L. Brower, Structural features at the Si–SiO_2 interface. Zeitschrift Fur Physikalische Chemie Neue Folge **151**, 177 (1987)
7. E. Cartier, J.H. Stathis, D.A. Buchanan, Passivation and depassivation of silicon dangling bonds at the Si/SiO_2 interface by atomic-hydrogen. App. Phys. Lett. **63**, 1510 (1993)

8. L.D. Thanh, P. Balk, Elimination and generation of Si–SiO_2 interface traps by low-temperature hydrogen annealing. J. Electrochem. Soc. **135**, 1797 (1988)
9. A.H. Edwards, Theory of P_b center at ⟨111⟩ Si/SiO_2 interface. P. Rev. **36**, 9638 (1987)
10. R. Helms, E.H. Poindexter, The silicon silicon-dioxide system—Its microstructure and imperfections. Rep. Prog. Phys. **57**, 791 (1994)
11. D.K. Schroder, J.A. Babcock, Negative bias temperature instability: road to cross in deep submicron silicon semiconductor manufacturing. J. App. Phys. **94**, 1 (2003)
12. K.O. Jeppson, C.M. Svensson, Negative bias stress of mos devices at high electric-fields and degradation of Mnos devices. J. App. Phys. **48**, 2004 (1977)
13. S. Ogawa, N. Shiono, Generalized diffusion-reaction model for the low- field charge-buildup instability at the Si–SiO_2 interface. P. Rev. **51**, 4218 (1995)
14. M.A. Alam, S. Mahapatra, A comprehensive model of PMOS NBTI degradation. Microelectron. Reliab. **45**, 71 (2005)
15. D.K. Schroder, Negative bias temperature instability: what do we understand? Microelectron. Reliab. **47**, 841 (2007)
16. M.A. Alam, H. Kufluoglu, D. Varghese, S. Mahapatra, A comprehensive model for PMOS NBTI degradation: recent progress. Microelectron. Reliab. **47**, 853 (2007)
17. S. Kumar, C.H. Kim, S.S. Sapatnekar, An analytical model for negative bias temperature instability, in *International Conference on Computer-Aided Design*, *6D.1*, 2006
18. H. Kufluoglu, M.A. Alam, A generalized reaction-diffusion model with explicit H–H_2 Dynamics for negative bias temperature instability (NBTI) degradation. IEEE Trans. Electron Devices **54**, 1101 (2007)
19. A.E. Islam, H. Kufluoglu, D. Varghese, M.A. Alam, Critical analysis of short-term negative bias temperature instability measurements: Explaining the effect of time-zero delay for on-the-fly measurements. App. Phys. Lett. **90**, 083505 (2007)
20. A.E. Islam, H. Kufluoglu, D. Varghese, S. Mahapatra, M.A. Alam, Recent issues in negative bias temperature instability: initial degradation, field-dependence of interface trap generation, hole trapping effects, and relaxation. IEEE Trans. Electron Devices **54**, 2143 (2007)
21. A.T. Krishnan, S. Chakravarthi, P. Nicollian, V. Reddy, S. Krishnan, Negative bias temperature instability mechanism: The role of molecular hydrogen. App. Phys. Lett. **88**, 153518 (2006)
22. J.B. Yang, T.P. Chen, S.S. Tan, L. Chan, Analytical reaction-diffusion model and the modeling of nitrogen-enhanced negative bias temperature instability. App. Phys. Lett. **88**, 172109 (2006)
23. A.T. Krishnan, C. Chancellor, S. Chakravarthi, P.E. Nicollian, V. Reddy, A. Varghese, R.B. Khamankar, S. Krishnan, L. Levitov, Material dependence of hydrogen diffusion: implications for NBTI degradation, IEEE Int. Electron Devices Meet. Tech. Digest, **688** (2005)
24. S. Chakravarthi, A.T. Krishnan, V. Reddy, S. Krishnan, Probing negative bias temperature instability using a continuum numerical framework: physics to real world operation. Microelectron. Reliab. **47**, 863 (2007)
25. M.A. Alam, A critical examination of the mechanics of dynamic NBTI for PMOSFETs, IEEE Int. Electron Devices Meet. Tech. Digest, **345** (2003)
26. J. Crank, *The Mathematics of Diffusion*, 2nd ed. (Oxford University Press, Oxford, 1980)
27. T. Grasser, P.J. Wagner, P. Hehenberger, W. Goes, B. Kaczer, A rigorous study of measurement techniques for negative bias temperature instability. IEEE Trans. Device Mater. Reliab. **8**, 526 (2008)
28. S. Rangan, N. Mielke, E.C.C. Yeh, Universal recovery behavior of negative bias temperature instability [PMOSFETs], IEEE Int. Electron Devices Meet. Tech. Digest, **341** (2003)
29. H. Reisinger, O. Blank, W. Heinrigs, A. Muhlhoff, W. Gustin, C. Schlunder, Analysis of NBTI degradation- and recovery-behavior based on ultra fast V_T measurements, in *IEEE International Reliability Physics Symposium Proceedings*, (2006), p. 448
30. T. Grasser, W. Gos, V. Sverdlov, and B. Kaczer, The universality of NBTI relaxation and its implications for modeling and characterization, in *IEEE International Reliability Physics Symposium Proceedings*, (2007), p. 268

31. C.R. Parthasarathy, M. Denais, V. Huard, G. Ribes, E. Vincent, A. Bravaix, New insights into recovery characteristics post NBTI stress, in *IEEE International Reliability Physics Symposium Proceedings*, (2006), p. 471
32. N. Klein, Mechanism of self-healing electrical breakdown in MOS structures. IEEE Trans. Electron Devices **13**, 788 (1966)
33. P. Solomon, Breakdown in Silicon-Oxide. J. Vac. Sci. Technol. **14**, 1122 (1977)
34. K.F. Schuegraf, C.M. Hu, Hole injection SiO_2 breakdown model for very-low voltage lifetime extrapolation. IEEE Trans. Electron Devices **41**, 761 (1994)
35. R. Degraeve, G. Groeseneken, R. Bellens, M. Depas, H.E. Maes, A consistent model for the thickness dependence of intrinsic breakdown in ultra-thin oxides. IEEE Int. Electron Devices Meet. Tech. Digest, **863** (1995)
36. D.J. DiMaria, J.H. Stathis, Explanation for the oxide thickness dependence of breakdown characteristics of metal-oxide-semiconductor structures. App. Phys. Lett. **70**, 2708 (1997)
37. J.H. Stathis, Percolation models for gate oxide breakdown. J. App. Phys. **86**, 5757 (1999)
38. M. Alam, B. Weir, P. Silverman, A future of function or failure? IEEE Circuits Devices **18**, 42 (2002)
39. D. Varghese, S. Mahapatra, M.A. Alam, Hole energy dependent interface trap generation in MOSFET Si/SiO_2 interface. IEEE Electron Device Lett. **26**, 572 (2005)
40. S. Mahapatra, D. Saha, D. Varghese, P.B. Kumar, On the generation and recovery of interface traps in MOSFETs subjected to NBTI, FN, and HCI stress. IEEE Trans. Electron Devices **53**, 1583 (2006)
41. D. Varghese, H. Kufluoglu, V. Reddy, H. Shichijo, D. Mosher, S. Krishnan, M.A. Alam, OFF-State degradation in drain-extended NMOS transistors: Interface damage and correlation to dielectric breakdown. IEEE Trans. Electron Devices **54**, 2669 (2007)
42. L. Tsetseris, X.J. Zhou, D.M. Fleetwood, R.D. Schrimpf, S.T. Pantelides, Physical mechanisms of negative-bias temperature instability. App. Phys. Lett. **86**, 142103 (2005)
43. L. Tsetseris, X.J. Zhou, D.M. Fleetwood, R.D. Schrimpf, S.T. Pantelides, Hydrogen-related instabilities in MOS devices under bias temperature stress. IEEE Trans. Device Mater. Reliab. **7**, 502 (2007)
44. C.L. Chen, Y.M. Lin, C.J. Wang, K. Wu, A new finding on NBTI lifetime model and an investigation on NBTI degradation characteristic for 1.2 nm ultra thin oxide, in *IEEE International Reliability Physics Symposium Proceedings*, (2005), p. 704
45. N. Kimizuka, T. Yamamoto, T. Mogami, K. Yamaguchi, K. Imai, T. Horiuchi, The impact of bias temperature instability for direct-tunneling ultrathin gate oxide on MOSFET scaling, in *Symposium on VLSI Technology*, (1999), p. 73
46. H. Aono, E. Murakami, K. Okuyama, A. Nishida, M. Minami, Y. Ooji, K. Kubota, Modeling of NBTI degradation and its impact on electric field dependence of the lifetime, in *IEEE International Reliability Physics Symposium Proceedings*, (2004), p. 23
47. A.E. Islam, G. Gupta, S. Mahapatra, A. Krishnan, K. Ahmed, F. Nouri, A. Oates, M.A. Alam, Gate leakage vs. NBTI in plasma nitrided oxides: characterization, physical principles, and optimization. IEEE Int. Electron Devices Meet. Tech. Digest, **329** (2006)
48. D. Varghese, D. Saha, S. Mahapatra, K. Ahmed, F. Nouri, M.A. Alam, On the dispersive versus Arrhenius temperature activation of NBTI time evolution in plasma nitrided gate oxides: measurements, theory, and implications. IEEE Int. Electron Devices Meet. Tech. Digest, **684** (2005)
49. J.W. McPherson, C.H. Mogul, Underlying physics of the thermochemical E model in describing low-field time-dependent dielectric breakdown in SiO_2 thin films. J. App. Phys. **84**, 1513 (1998)
50. B. Tuttle, Energetics and diffusion of hydrogen in SiO_2. Phys. Rev. **61**, 4417 (2000)
51. C.G. Van de Walle, R.A. Street, Structure, energetics, and dissociation of Si−H bonds at dangling bonds in silicon. Phys. Rev. **49**, 14766 (1994)

52. A. Ghetti, A. Hamad, P.J. Silverman, H. Vaidya, N. Zhao, Self-consistent simulation of quantization effects and tunneling current in ultra-thin gate oxide MOS devices, in *International Conference on Simulation of Semiconductor Processes and Devices (SISPAD)*, (1999), p. 239
53. L.F. Register, E. Rosenbaum, K. Yang, Analytic model for direct tunneling current in polycrystalline silicon-gate metal-oxide-semiconductor devices. App. Phys. Lett. **74**, 457 (1999)
54. A.E. Islam, G. Gupta, K.Z. Ahmed, S. Mahapatra, M.A. Alam, Optimization of gate leakage and NBTI for plasma-nitrided gate oxides by numerical and analytical models. IEEE Trans. Electron Devices **55**, 1143 (2008)
55. A.E. Islam, J.H. Lee, W.H. Wu, A. Oates, M.A. Alam, Universality of interface trap generation and its impact on i_d degradation in strained/unstrained PMOS devices during NBTI stress. IEEE Int. Electron Devices Meet. Tech. Digest, **107** (2008)
56. A.E. Islam, Ph.D. Dissertation, Theory and characterization of random defect formation and its implication in variability of nanoscale transistors, Electrical and Computer Engineering, Purdue University, 2010
57. L. Pauling, *The Nature of The Chemical Bond*, **3rd** edn. (Cornell University Press, New York, 1960)
58. R.T. Sanderson, *Chemical Bonds and Bond Energy* (Academic Press, New York, 1971)
59. C. Kittel, *Introduction to Solid State Physics* (Wiley, London, 1996), p. 381
60. D.A. Muller, T. Sorsch, S. Moccio, F.H. Baumann, K. Evans-Lutterodt, G. Timp, The electronic structure at the atomic scale of ultrathin gate oxides. Nature **399**, 758 (1999)
61. G.J. Gerardi, E.H. Poindexter, P.J. Caplan, N.M. Johnson, Interface traps and P_b centers in oxidized (100) silicon-wafers. App. Phys. Lett. **49**, 348 (1986)
62. P.M. Lenahan, Dominating defects in the MOS system: P_b and E′ centers, in *Defects in Microelectronic Materials and Devices*, ed. by S.P.D. Fleetwood, R.D. Schrimpf (CRC Press, Boca Raton, 2008), p. 163
63. R.A. Weeks, The many varieties of E′ centers—A review. J. Non Crystalline Solids **179**, 1 (1994)

Chapter 6
Modeling of DC and AC NBTI Degradation and Recovery for SiON and HKMG MOSFETs

Nilesh Goel and Souvik Mahapatra

Abstract In this chapter, a comprehensive model is proposed to describe NBTI degradation in SiON and HKMG p-MOSFETs. The model is based on the physical mechanism of NBTI established in earlier chapters, has mutually uncorrelated trap generation and trapping subcomponents, and can predict ultra-fast measured stress and recovery data under DC and AC stress. Time evolution of NBTI degradation and recovery during and after DC stress as well as during multiple DC stress and recovery cycles for different stress bias, temperature and recovery bias can be successfully explained. The model can explain time evolution of AC degradation for different pulse frequency, duty cycle, and pulse low bias, and can explain measurements after last half or full cycle of AC pulse. The model is consistent with the compact model presented in Chap. 4, and can successfully explain the gate insulator process and material dependence for both SiON and HKMG devices, and can also predict long-time DC and AC degradation to determine NBTI lifetime.

6.1 Introduction

Negative bias temperature instability (NBTI) is an important reliability concern for p-channel metal oxide semiconductor field effect transistors (MOSFETs). As shown in Chap. 1, NBTI was discovered more than 40 years ago [1], however, it became a reliability concern with the introduction of Silicon Oxynitride (SiON) devices [2–7] and continues to threaten the reliability of latest planar and FinFET devices having High-K Metal Gate (HKMG) gate insulator [8–17]. NBTI results in gradual buildup of positive charges in the gate insulator, and causes degradation of device parameters such as threshold voltage (ΔV_{T}) transconductance (Δg_m), linear drain current (ΔI_{DLIN}), saturation drain current (ΔI_{DSAT}), subthreshold slope (ΔSS),

N. Goel · S. Mahapatra (✉)
Department of Electrical Engineering, Indian Institute of Technology Bombay, Mumbai 400076, India
e-mail: souvik@ee.iitb.ac.in; mahapatra.souvik@gmail.com

S. Mahapatra (ed.), *Fundamentals of Bias Temperature Instability in MOS Transistors*, Springer Series in Advanced Microelectronics 139,
DOI 10.1007/978-81-322-2508-9_6

gate-to-drain capacitance (ΔC_{GD}) [18] etc., and hence, it impacts the performance of digital, analog and memory circuits [19–24]. NBTI degradation demonstrates several universal features for production quality devices and circuits, for different technology nodes and across different industries, refer to [24] for further details.

A plethora of reports have been published in the last 15 years on NBTI characterization techniques, process and materials impact, and physical mechanism. The characterization methods for measuring NBTI degradation and the underlying defects responsible for the degradation have been discussed in Chap. 2. Gate insulator process impact on NBTI degradation and its underlying defects is discussed in Chaps. 1, 3 and 4, and is explained using a physics-based compact model in Chap. 4. NBTI physical mechanism and model has been a subject of great debate and is recently reviewed in [25], and is also briefly mentioned in Chaps. 3 and 4. The following NBTI experimental signatures have been universally observed and reported by various groups:

(a) NBTI degradation substantially recovers after removal of stress [26–29], results are discussed in Chap. 1 and also later in this chapter, and therefore, several ultra-fast characterization techniques have been proposed to make "recovery free" measurements [26, 27, 30–32], refer to Chap. 2 for implementation details of different characterization methods.

(b) NBTI degradation characterized by ultra-fast method shows rapid increase at the initiation of stress and power-law time dependence at longer stress time (t_{STR}) for both SiON and HKMG devices [7, 10, 32], results are discussed in Chap. 1 and later in this chapter.

(c) Long-time power-law time exponent n depends on measurement delay [30, 31], and is independent of stress bias ($V_{G\text{-}STR}$) and temperature (T), when experiments are performed in the absence of different stress and measurement artifacts, such as stress saturation [33], recovery impact [34], impact due to extraneous trap generation effects [35], etc., refer to Chap. 1 for detailed discussion on these experimental artifacts. However, the exponent n depends strongly on the gate insulator process for both SiON and HKMG devices [7, 15], as shown in Chaps. 1 and 4.

(d) Measured degradation in production quality devices shows power-law dependence with $n \sim 1/6$ at very long t_{STR} [36, 37], and similar behavior is observed in actual circuit operation [20], refer to Chap. 4, Fig. 4.27 and [24] for additional details.

(e) The degradation magnitude depends on inversion hole density as well as on gate oxide field (E_{OX}) but not on $V_{G\text{-}STR}$ [3, 4, 35]; physics-based calculation suggests $E_{OX} * \exp(\Gamma_E \cdot E_{OX})$ dependence where Γ_E is the field acceleration factor [33], although for simplicity and without loss of much accuracy, power-law E_{OX} or $V_{G\text{-}STR}$ dependence is often used in the industry [38, 39], refer to Chap. 4, Fig. 4.35 for relative comparison of power-law E_{OX} acceleration factor Γ_E across different technology nodes. The factor Γ_E is a strong function of gate insulator process as well [7, 15]; refer to Chaps. 1 and 4 for additional details.

(f) NBTI degradation, when properly measured, shows Arrhenius T activation [4–7, 10, 15, 34, 40] with energy E_A that depends on the gate insulator processes [7, 15]; reported non-Arrhenius behavior [41] with time exponent n being a function of stress T has been shown to be an artifact of measurement delay [34]. Moreover, T activation of time to reach a particular degradation shows universality across SiON and HKMG devices as shown in Chap. 4, Fig. 4.31 and also discussed in this chapter.

(g) Field acceleration factor Γ_E and T activation energy E_A extracted at fixed t_{STR} are found to be independent of stress T and E_{OX}, respectively [42], and obtained Γ_E and E_A values are independent of t_{STR}, when stress and measurements remain free from artifacts mentioned in (c). This aspect is discussed in detail in Chap. 1.

(h) Gate insulator and other processes impact NBTI degradation; in particular, the presence of Nitrogen (N) in the gate insulator stack increases degradation magnitude but reduces time exponent n, field acceleration Γ_E and T activation E_A for both SiON and HKMG devices [4–7, 10, 15, 42], relevant results are shown in Chaps. 1, 3 and 4, and also in this chapter.

(i) For HKMG devices, NBTI degradation depends on the quality of the interlayer (IL) separating MOSFET channel and the High-K layer [15]; reduction in interlayer (IL) thickness increases ΔV_T but reduces parameters n, Γ_E and E_A [13, 15, 42]; results are presented in Chaps. 1, 3 and 4, and also in this chapter.

(j) Recovery of NBTI degradation after stress implies lower degradation for AC compared to DC stress. Similar to DC, ultra-fast techniques have been developed and used to characterize AC degradation, which can be measured after the end of last half cycle (Mode-A) or after end of last full cycle (Mode-B), refer to Chap. 2, Fig. 2.14 [32, 43]. AC degradation shows frequency (f) independence for Mode-B stress, but measured ΔV_T depends on f, especially at lower f, for Mode-A stress [42]. Higher ΔV_T is observed for Mode-A compared to Mode-B stress at lower f, and Mode-A ΔV_T reduces with increase in f and merges with Mode-B ΔV_T at higher f. Measured results are discussed in Chaps. 1 and 2 and also later in this chapter. Different groups have reported f independence for AC NBTI measurements till GHz [21, 44].

(k) NBTI degradation increases with increase in pulse duty cycle (PDC) of AC stress and shows a characteristic "S" shaped PDC dependence with a large kink or jump between high PDC AC and DC data [21, 32, 42, 45–48]. Different published data show different AC to DC ratio when normalized to their corresponding DC value as per usual practice. However, a remarkably universal PDC dependence is observed for up to ~85 % PDC when published data are normalized to their corresponding 50 % AC value [48], and this universality holds for differently processed devices and for different AC pulse bias conditions [42], and is discussed later in this chapter. Ultra-fast measured PDC dependent data are shown in Chaps. 1 and 2 and also in this chapter.

As discussed in detail in Chaps. 3 and 4, it is now universally accepted that positive charge buildup in the gate insulator during NBTI stress predominantly arises out of generation of new traps at the interface between device channel and gate insulator (ΔN_{IT}) and in certain situations inside the gate insulator bulk (ΔN_{OT}), as well as hole trapping in process related pre-existing bulk gate insulator traps (ΔN_{HT}) [25]. As shown in Chaps. 2 and 3, trap generation during NBTI stress has been independently estimated using Charge Pumping (CP) [4–6, 11, 35, 48–51], Gated Diode (or DCIV) [25, 48, 50, 52–55] and low voltage stress induced leakage current (LV-SILC) [2, 40, 53] techniques, while pre-existing gate insulator traps are independently estimated using flicker noise technique [15, 48, 51, 56–58]. SILC has been used to independently characterize generated bulk traps [35].

In spite of such overwhelming experimental evidence of trap generation during NBTI stress, some authors have proposed ΔN_{HT} to be the sole contributor to measured ΔV_T [59–63], which is definitely not physically justifiable. On the other hand, although ΔN_{IT} contribution is dominant and determines end-of-life degradation for production quality devices and circuits [24], it alone cannot explain the short-time degradation measured using ultra-fast methods, and gate insulator process dependence of NBTI. Therefore, ΔN_{IT} only framework cannot provide a complete physical picture of NBTI [34, 40], and needs to be augmented by additional contribution from ΔN_{HT} [25]. Although some report suggests strong coupling between the two components [64], most evidences suggest mutually uncoupled trap generation and trapping [4–6, 15, 25, 27, 33, 42, 46, 48, 51, 65–69], which has now become the acceptable physical mechanism of NBTI. In situations such as thick gate insulators and large $V_{G\text{-}STR}$ and/or high stress T, uncoupled contribution from ΔN_{OT} also needs to be taken into account to compute overall ΔV_T [25, 35, 48, 70, 71]. As discussed in Chap. 4, compact model based on mutually uncoupled trap generation and trapping components can explain gate insulator process dependence of NBTI degradation in different SiON and HKMG p-MOSFETs [15, 25, 48, 67, 72, 73].

Many reports have suggested the involvement of Hydrogen (H) species in generation and passivation of gate insulator traps (sometimes referred to as trap volatility) [2–6, 25, 27, 33–35, 40, 42, 46, 48, 50, 51, 55, 74–87]. The breaking and annealing processes associated with H passivated defects, respectively, responsible for trap generation and anneal, have been formulated using the Reaction-Diffusion (RD) model, first proposed in [77], and subsequently reformulated in [78–81] for macroscopic differential equation based framework for average behavior in large area devices, and in [82–84] for Kinetic Monte Carlo based framework for stochastic behavior in small area devices. Although basic one-dimensional (1D) RD model formulation described in Chap. 5 can successfully explain trap generation during stress, especially the universal power-law time dependence with exponent $n \sim 1/6$ at long t_{STR} and AC frequency independence [78–81], it cannot capture the exact physics associated with stochastic hopping of H especially during long recovery experiments after stress, which is explained in [25, 48, 83] and later in this chapter. Therefore, the basic 1D macroscopic RD formulation has been suitably modified to effectively capture delayed recovery of generated traps due to H hopping effects [25, 48].

Moreover, although the reformulated RD model can successfully explain both generation and recovery of ΔN_{IT} during and after DC and AC NBTI stress, it must be augmented by models associated with calculating the occupancy of generated traps, as well as with calculation of generation and recovery for ΔN_{HT} and ΔN_{OT}, to explain the generation and recovery of measured ΔV_T, as demonstrated in [25, 42, 48, 85–87] and also discussed in this chapter. However, some reports have failed to realize these important aspects, and especially highlighted the incompatibility of ΔN_{IT} recovery calculated using the basic RD model to measured ΔV_T recovery [27, 45, 47, 61–64, 69, 75, 76, 82, 84], without accounting for stochastic nature of H hopping during trap recovery, occupancy of generated traps during recovery, as well as recovery of generated ΔN_{HT} and ΔN_{OT}, and unfairly criticized the capability of RD model. As a consequence, this has lead to development of several alternative NBTI models that focused exclusively on ΔV_T recovery and AC PDC and *f* dependence [45, 61–64], but unfortunately, these alternative models fail to predict basic features associated with measured time evolution of ΔV_T during stress, and hence is not pursued further; refer to [25, 48, 88] for additional details.

In this chapter, a comprehensive RD model based framework will be discussed, which can predict the time evolution of ΔV_T during and after DC stress and during multiple stress and recovery sequences for different $V_{G\text{-}STR}$, stress *T* and recovery bias ($V_{G\text{-}REC}$), and during AC stress at different PDC, *f* and pulse low bias ($V_{G\text{-}LOW}$) stress condition [42, 86, 87]. The framework is consistent with model proposed in Chap. 4, and can explain data measured in SiON and HKMG devices with different gate insulator processes. As mentioned before, the framework calculates generation and recovery of interface traps using RD model and augments with calculation of occupancy of generated traps, and also calculates trapping and detrapping of holes in pre-existing traps and generation and recovery of bulk insulator traps. Finally, a simple compact model is discussed and compared with the full-feature model to aid circuit analysis under DC and AC NBTI stress.

6.2 Degradation During DC Stress

Figure 6.1 illustrates the cross section of a HKMG gate insulator stack and shows different defect precursors. The stack has SiON based IL and HfO_2 based High-K layers, with H passivated P_b centers (Si–H bonds) at Si/IL interface. Other H passivated defects in the bulk of the gate insulator are denoted as X–H, where X can be Silicon (Si), Nitrogen (N) or Oxygen (O) atoms, and different Oxygen vacancy (Ov) related defects, located in SiON IL, Hafnium Oxide (HfO_2) High-K, and also in the HfSiON transition region between IL and High-K layers. Note that the presence of N in the High-K and IL layers can be intentional, e.g., resulting from post High-K nitridation, or unintentional, due to N diffusion from the Titanium Nitride (TiN) metal gate and Silicon Nitride (SiN) spacer of the MOSFET.

As explained later in this chapter, dissociation of Si–H and X–H bonds results in creation of Si– and X–traps. In addition, hole trapping take place in pre-existing IL

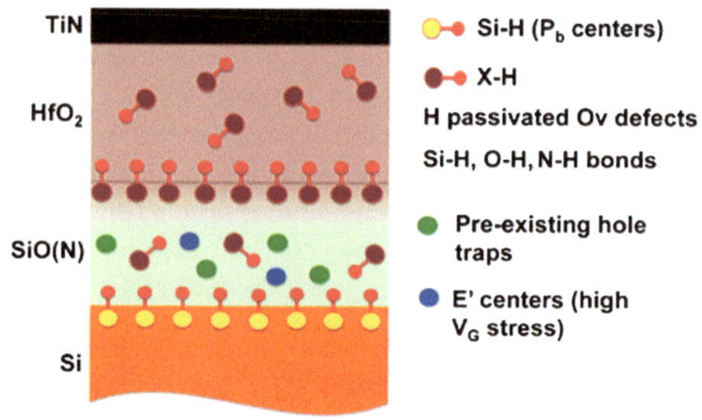

Fig. 6.1 Schematic of HKMG gate insulator stack and different defect precursors for NBTI degradation

defects, and E′ centers are created especially at higher $V_{G\text{-}STR}$ by breaking of Si–O bonds in the IL bulk. As mentioned before, both trap generation and trapping contributes to overall NBTI degradation. Note that identical trap generation and trapping mechanisms describe the scenario for devices having SiON gate insulators, which does not contain the HfO_2 High-K layer and Hf related defects.

Figure 6.2 shows atomic configurations of different Ov related defects (top) and associated energy levels obtained using Density Functional Theory (DFT) calculations (bottom) [86]. Ov is attributed to a missing O atom between two Si atoms resulting in Si–Si dimer, Ov (Si–Si) in the IL, missing O between two Hf and one Si atoms, Ov (HHS), and between one Hf and two Si atoms, Ov (HSS), in the HfSiON transition layer, and missing O between three and four Hf atoms in the HfO_2 High-K layer, resulting in 3- and 4-coordinated vacancies, Ov (3c) and Ov (4c), respectively. The energy levels of traps are shown without and with one or two occupied electrons, and trap energy levels relax after electron occupation due to structural relaxation effects. Note that the energy levels do not exist as long as Ov defects remain passivated with H, and appear only after Ov–H bonds are broken and H atoms are released. This aspect is discussed later in the following section.

Figure 6.3 shows the atomic configurations, corresponding density of states and energy levels of traps associated with the presence of interstitial Hf or N atoms inside the IL of a HKMG gate insulator stack obtained using DFT calculations [57]. The reference case of clean IL without any impurity is also shown. Note that Hf and N atoms in the HfSiON transition layer result in mid-Si gap defects. However, when Hf and N atoms are located in the IL bulk, defect levels appear near the valence band of Si substrate, and these defects can appear as hole traps. Both SiON and HKMG gate insulators show larger pre-existing traps due to the presence of N when measured using flicker noise [15, 56, 57]; results are shown in Chaps. 2 and 4. Moreover, HKMG stacks with IL grown using the conventional Chemical Oxide (Chem-Ox) process [89] shows larger pre-existing trap density compared to thermal IL process [90] when measured using flicker noise, which is due to larger IL and High-K intermixing and higher Hf density in the IL for the Chem-Ox process, refer to [15, 57] for details. Therefore, the quality of gate insulator influences the magnitude of hole trapping, as described and analyzed in detail in Chap. 4.

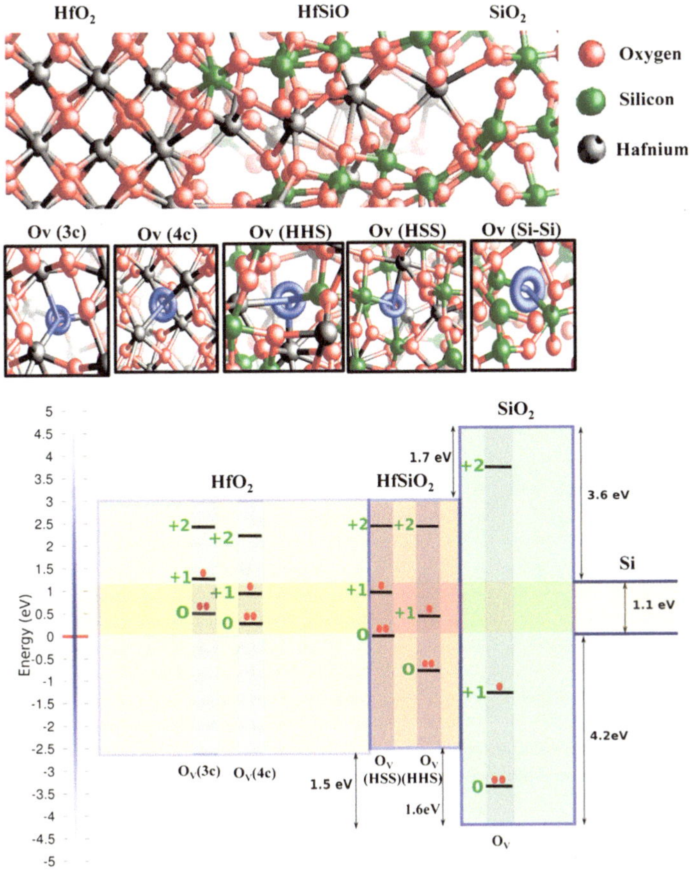

Fig. 6.2 *Top* Illustration of atomic configuration and different oxygen vacancies in HKMG gate insulator stack. *Bottom* Energy levels corresponding to oxygen-vacancy defects from DFT simulations, for different electron occupancy levels (*shown as dots*)

6.2.1 Reaction-Diffusion (RD) Model for Interface Trap Generation

Interface trap generation is an important component of NBTI degradation and can be directly estimated using CP, DCIV and LV-SILC measurements as mentioned earlier in this chapter and also explained in detail in Chap. 2. DCIV measurements with suitable delay and band gap corrections have been used in Chap. 3 to perform a detailed analysis of ΔN_{IT} time evolution under different DC and AC NBTI stress conditions in differently processed HKMG p-MOSFETs. A remarkable universality of power-law time evolution of ΔN_{IT} with exponent $n \sim 1/6$ has been observed for longer t_{STR}, for both DC and AC stress and also across different HKMG processes, which is a unique signature of H/H_2 RD model as discussed in detail in Chap. 5.

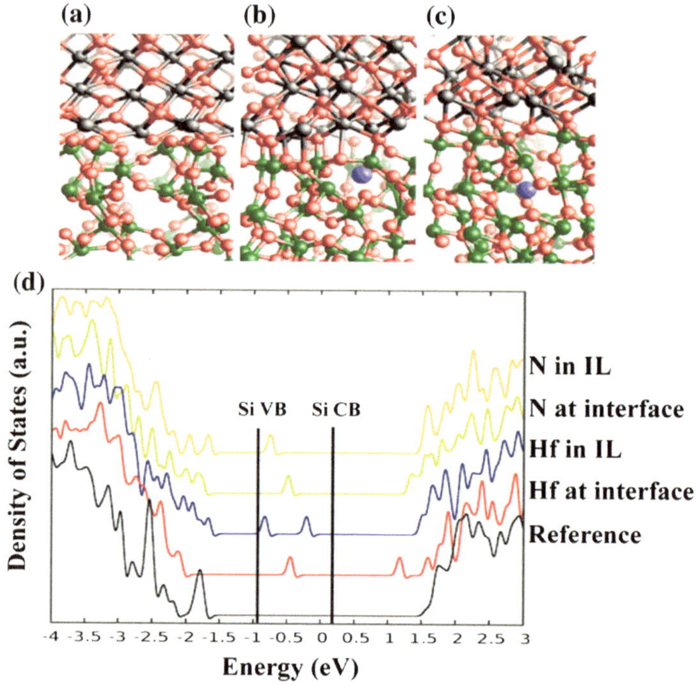

Fig. 6.3 Illustration of atomic configuration of HKMG gate insulator stack, showing **a** reference defect free configuration, and interstitial Hf or N atom at **b** the interface between HfO_2 and IL and **c** inside the IL. The corresponding density of states is shown below in **d**

Note that $n \sim 1/6$ power-law time evolution is also observed for ΔV_T time evolution in production quality devices and in circuit performance drift for very long-time NBTI stress, refer to Chap. 4, Fig. 4.27.

Figure 6.4 shows the schematic of double interface H/H_2RD model for generation of interface traps; refer to Chap. 5 for implementation details. Si–H bonds are located at the Si/SiON interface, while all bulk X–H bonds are effectively assumed to be at a "second interface", which can be presumed to be at the center of the gate dielectric for SiON insulator, or at the IL/High-K interface for HKMG gate stacks. When $V_{G\text{-}STR}$ is applied, inversion layer holes tunnel into and get captured in Si–H bonds at the Si/SiON interface. The covalent bonds get weakened by hole capture, and can be easily broken by thermal process to form Si or interface traps at first interface (ΔN_{IT1}) [37]. Release H atoms from broken Si–H bonds move away from the Si/SiON interface and diffuse towards the gate due to concentration gradient. Diffusing H atoms react with X–H bonds at second interface to form H_2 molecules. Resulting X– bonds are generated traps at the second interface (ΔN_{IT2}). Note that trap generation at the first interface is primary and at the second interface is a secondary or related phenomenon. Generated H_2 molecules eventually diffuse out and control the time evolution of ΔN_{IT} at longer t_{STR}, resulting in characteristic power-law time dependence with exponent $n = 1/6$ [25, 40].

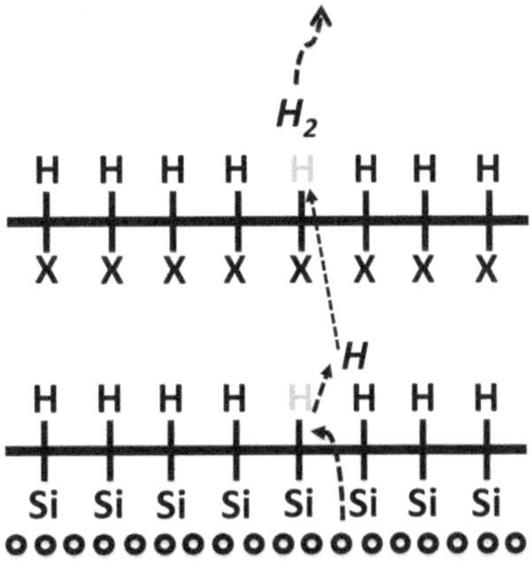

Fig. 6.4 Schematic of double interface H/H_2 Reaction-Diffusion model for interface trap generation during stress

When gate stress bias is removed or reduced, recovery of generated traps is initiated by back diffusion of H_2 molecules and subsequent reaction of H_2 molecules with broken X– bonds at the second interface and formation of X–H bonds and H atoms. The released H atoms diffuse towards the first interface and passivate Si–dangling bonds. Recovery of generated traps is discussed later in this chapter.

Table 6.1 lists model equations and parameters for double interface H/H_2 RD model [40]; detailed explanation of the model equations is provided in Chap. 5 and is briefly reviewed here for completeness. Equations (6.1) and (6.4) represent the forward and reverse reactions associated with trap generation and recovery at first and second interfaces, respectively. Equations (6.2) and (6.3) are the detailed balance between reaction and diffusion as per Fick's first law of diffusion. Equations (6.5) and (6.6) represent diffusion of H and H_2 in gate dielectric and beyond as stated by Fick's second law of diffusion. Subscript (1) and (2) in equations represent first and second interface, respectively. N_O, N_{IT}, N_H, N_{H2}, D_H, and D_{H2}, respectively, denote densities of Si–H bonds and interface traps, concentration of hydrogen atoms and molecules and their corresponding diffusivities; k_F and k_R are forward and reverse reaction rates; δ is interfacial thickness ($\sim$1.5 Å) assumed for dimensionality considerations.

Although the model equations are solved in a self-consistent manner to obtain time evolution of ΔN_{IT}, as shown in Chap. 5 and also later in this chapter, a simplified analytical expression for long-time behavior is shown in (6.7). Note that the model suggests $n = 1/6$ power-law time dependence at long t_{STR} without using any adjustable parameters. This is a unique feature of the model, and is consistent with experimental results shown in Chap. 3.

Arrhenius T activation is used for reaction parameters k_F and k_R at both interfaces and diffusion parameters D_H and D_{H2} as shown (6.8), with corresponding activation energies E_{AKF}, E_{AKR}, E_{ADH}, and E_{ADH2}, respectively. The T activation energy E_{AIT} of ΔN_{IT} at long t_{STR} is shown in (6.9), which depends on E_{AKF}, E_{AKR} at both interfaces and also on E_{ADH2}. It is important to remark that under the

Table 6.1 Equations for double interface H/H_2 Reaction-Diffusion model. Fixed parameters are universal across devices. Adjustable parameters depend on gate insulator process

$\frac{dN_{IT(1)}}{d_t} = k_{F(1)}(N_{O(1)} - N_{IT(1)}) - k_{R(1)}N_{IT(1)}N_H^{(1)}$ (6.1) [1st interface]
$\frac{\delta}{2}\frac{dN_H^{(1)}}{d_t} = D_H\frac{dN_H^{(1)}}{d_x} + \frac{dN_{IT}}{d_t}$ (6.2)
$\frac{\delta}{2}\frac{dN_{H2}^{(1)}}{d_t} = D_{H2}\frac{dN_{H2}^{(1)}}{d_t}$ (6.3)
$\frac{dN_{IT(2)}}{d_t} = k_{F(2)}(N_{O(2)} - N_{IT(2)})N_H^{(2)} - k_{R(2)}N_{IT(2)}N_{H2}^{(2)}$ (6.4) [2nd interface]
$\frac{dN_H}{d_t} = D_H\frac{d^2N_H}{dx^2}$ (6.5);
$\frac{dN_{H2}}{d_t} = D_{H2}\frac{d^2N_{H2}}{dx^2}$ (6.6)
$N_{IT(1)} = [D_{H2}t]^{\frac{1}{6}}\frac{q}{C_{OX}}\left[\frac{k_{F0(1)}k_{F0(2)}N_{O(1)}N_{O(2)}}{k_{R(S)}k_{R(P)}}\right]^{\frac{1}{3}}$ (6.7)
$k_{F(1)} = k_{F0(1)}f(E_{ox})e^{-\frac{E_{AKF(1)}}{kT}}$; $k_{R(1)} = k_{R0(1)}e^{-\frac{E_{AKR(1)}}{kT}}$ $k_{F(2)} = k_{F0(2)}e^{-\frac{E_{AKF(2)}}{kT}}$; $k_{R(2)} = k_{R0(2)}e^{-\frac{E_{AKR(2)}}{kT}}$ (6.8) $D_H = D_{H0}e^{-\frac{E_{ADH}}{kT}}$; $D_{H2} = D_{H20}e^{-\frac{E_{ADH2}}{kT}}$
$E_{AIT} = \frac{1}{3}[E_{AKF(1)} + E_{AKF(2)} - E_{AKR(1)} - E_{AKR(2)}] + \frac{1}{6}E_{ADH2}$ (6.9)
$f(E_{OX}) = (E_{OX})^3e^{(3E_{OX}\Gamma_{IT})}$ (6.10)

Adjustable parameters: $k_{F0(1)}$, $k_{F0(2)}$, Γ_{IT} and $E_{AKF(1)}(=E_{AKF(2)})$
Fixed parameters: $k_{R0(1)}$; $k_{R0(2)}$; D_H; D_{H20}; $E_{AKR(1)}$; $E_{AKR(2)}$; E_{ADH} and E_{ADH2}

assumption of detailed balance of forward and reverse reactions such that $E_{AKF} \sim E_{AKR}$ at both interfaces, the T activation energy E_{AIT} of ΔN_{IT} at longer t_{STR} is equal to the T activation energy associated with molecular H_2 diffusion, E_{ADH2} [25, 48]. This aspect is verified in the following section.

Oxide field (E_{OX}) dependence is only incorporated in the forward reaction parameter k_{F1} at the first interface with exponent Γ_{IT} as shown in (6.10). This particular form of E_{OX} dependence of k_{F1} is used to obtain identical Γ_{IT} values for this model and the compact model listed in Chap. 4, Table 4.1.[1] Refer to Chap. 5 for a detailed physics-based calculation of the forward reaction process. Note that this model has mutually uncorrelated E_{OX} and T dependence of ΔN_{IT} during stress, which is consistent with results shown in Chap. 3, Fig. 3.5.

The fixed and process dependent adjustable parameters are listed in Table 6.1. Most parameters are fixed across differently processed gate stacks for both SiON and HKMG devices. The model has only four process dependent parameters, and the field acceleration factor Γ_{IT} and T activation energy E_{AIT} are consistent with the corresponding compact model parameters shown in Chap. 4.

[1]When (6.10) is inserted in (6.7), the resultant equation shows $\Delta N_{IT} \sim E_{OX} * \exp(\Gamma_{IT} \cdot E_{OX})$ for long stress time, i.e., identical E_{OX} dependence and Γ_{IT} as used in Table 4.1.

6.2.2 Experimental Validation of RD Model

Figure 6.5 plots the time evolution of DCIV measured ΔN_{IT} after measurement delay and band gap correction, see Chap. 2, in HKMG p-MOSFETs having different gate insulator processes. As described in Chap. 3, Fig. 3.2, D2 and D4 devices, respectively, have rapid thermal process (RTP) grown non-nitrided and nitrided IL layers but identical HfO_2 High-K layer deposited using Atomic Layer Deposition (ALD) method. Experiments are performed at three different stress T, and for each T, three different $V_{G\text{-}STR}$ is used for stress. Note that DCIV is a slow measurement method and only provides meaningful results at longer time as shown in Chap. 2.

Time evolution of ΔN_{IT} obtained using numerical simulation of RD model is shown from short to long t_{STR}. As explained in detail in Chap. 5, the initial degradation phase shows reaction-limited $n = 1$ power-law time dependence, with transforms into H_2 diffusion limited $n = 1/6$ dependence at long stress time. The model can predict measured data for different stress conditions and across different devices with only four device dependent adjustable parameters listed in Table 6.1; once fixed for a particular device, the parameters are not changed for variations in $V_{G\text{-}STR}$ and T. Note that ΔN_{IT} is converted to ΔV_{IT} $(=q * \Delta N_{IT}/C_{OX})$, q and C_{OX} are

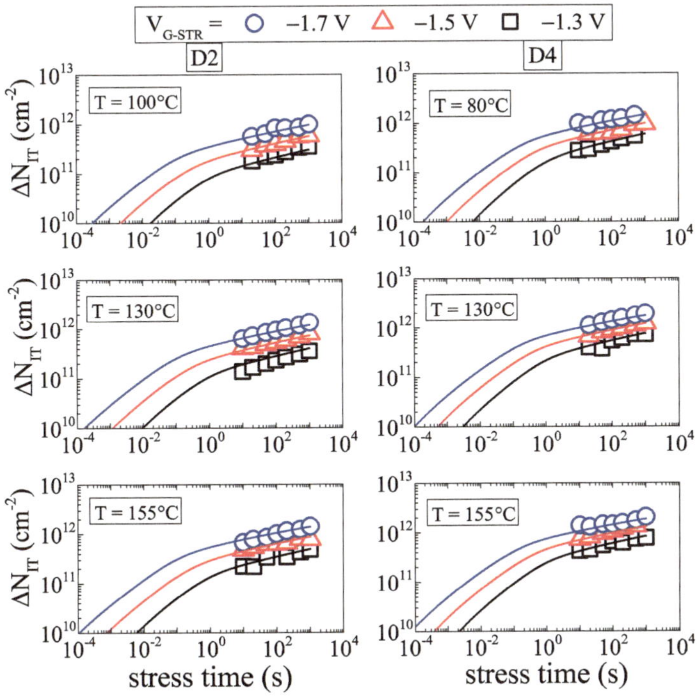

Fig. 6.5 Time evolution of DCIV measured ΔN_{IT} after delay and band gap correction (*symbols*), for NBTI stress in different HKMG devices for different $V_{G\text{-}STR}$ and T. *Lines* represent RD model simulation

the electronic charge and gate insulator capacitance, respectively, to compute contribution by trap generation to overall ΔV_T during stress. Time evolution of ΔV_{IT} obtained from the calibrated RD model simulation is used to calculate time evolution of overall ΔV_T shown later.

Although ΔN_{IT} dominates overall ΔV_T, in general, contributions from ΔN_{HT} and ΔN_{OT} cannot be ignored. However, ΔN_{HT} contribution can be minimal for certain specific situations, e.g., for NBTI stress in plasma nitrided oxide (PNO) based SiON gate insulators having lower N content and proper post nitridation anneal (PNA) [91, 92], and especially when ΔV_T is measured using relatively slower on-the-fly (OTF) method [5]. Moreover, ΔN_{OT} contribution remains negligible when NBTI stress is done using low $V_{G\text{-}STR}$ as shown in Chap. 4. In these situations, time evolution of ΔV_T can be considered to be primarily due to time evolution of ΔN_{IT}.

Longer-time stress experiments are performed at different stress T in different PNO with proper PNA based SiON devices [5], and ΔV_T is measured using OTF method with time-zero (t_0) delay of 1 ms, which captures negligible ΔN_{HT}. Refer to Chap. 2 for OTF measurement details. As mentioned before, low $V_{G\text{-}STR}$ is used to ensure negligible ΔN_{OT} contribution. Measured time evolution of ΔV_T has power-law time dependence with $n \sim 1/6$, and is plotted in a log–log scale versus t_{STR} for different stress T. Since ΔV_T versus t_{STR} curves are parallel to each other at different T, they are scaled by a factor t_{SCALE} along the time axis (X-axis) to overlap with each other. Figure 6.6 illustrates the scaling method mentioned above, and also plots the T activation of $1/t_{SCALE}$ for different PNO-SiON devices.

Similar T dependent stress experiments are performed for longer t_{STR} in pure Silicon Dioxide (SiO_2) devices, and the time evolution of ΔV_T and ΔN_{IT} are measured, respectively, using slow I–V and CP methods [35]. Once again, precautions are used such that measured ΔV_T is dominated by ΔN_{IT}. In addition, DCIV measurements are done to obtain time evolution of ΔN_{IT} at different stress T in HKMG devices [55]. The X-axis scaling exercise is performed on measured data, and the corresponding T activation of $1/t_{SCALE}$ for these devices are also shown in Fig. 6.6. Note that the scale factors for different devices are arbitrarily shifted in the

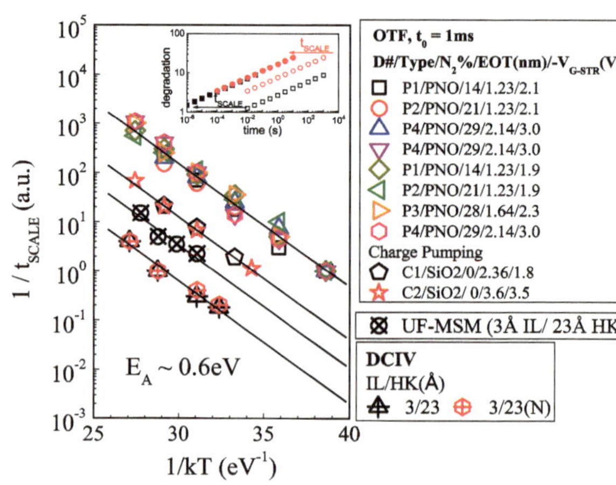

Fig. 6.6 T activation of $1/t_{SCALE}$, where t_{SCALE} is time to reach a particular degradation obtained from X-axis scaling of T dependent NBTI data. *Inset* shows the scaling procedure

plot for viewing clarity, and the relative values have no real significance. Of significance is the T activation energy of $1/t_{SCALE}$, which remarkably shows universality across different types of devices and measurement conditions.

Note that under the detailed balance assumption discussed before, (6.7) and (6.9) of Table 6.1 would suggest $\Delta N_{IT} \sim (D_{H2} * t_{STR})^n$ at longer t_{STR}, where $n = 1/6$ is governed by molecular H_2 diffusion and D_{H2} being the corresponding diffusivity. Therefore, t_{SCALE} ($=t_{STR}$) to achieve identical ΔN_{IT} at different stress T would be inversely proportional to D_{H2}, and the T activation of $1/t_{SCALE}$ should be equal to the T activation of D_{H2} [25]. Remarkably, note that measured T activation values listed in Fig. 6.6 yields $E_A \sim 0.6$ eV, which is similar to the T activation energy reported for molecular H_2 diffusion [93]. Therefore, both the time and T dependence of ΔN_{IT} at long stress time suggests degradation is governed by molecular H_2 diffusion, and verifies the H/H_2 RD model. Moreover, the above relation suggests T activation energy $E_{AIT} \sim 0.1$ eV for ΔN_{IT} at long but constant t_{STR}, since $E_{ADH2} \sim 0.6$ eV and $n = 1/6$. Indeed, a similar value has been used for ΔN_{IT} component for different SiON and HKMG devices, refer to Chap. 4 for details.

6.2.3 *Hole Trapping and Bulk Trap Generation*

As mentioned earlier in this chapter and discussed in detail in Chap. 4, although ΔN_{IT} plays an important role, the contribution from ΔN_{HT} and ΔN_{OT} must be considered to compute overall ΔV_T. Note that ΔN_{HT} is a faster process and significantly depends on the quality of the gate insulator, while ΔN_{OT} gets triggered under high $V_{G\text{-}STR}$, is a slower process and builds up over time. Table 6.2 shows the empirical relations used for ΔV_{HT} ($=q * \Delta N_{HT}/C_{OX}$) and ΔV_{OT} ($=q * \Delta N_{OT}/C_{OX}$) components during stress; refer to (6.11) and (6.12), respectively.

As discussed in Chap. 4, ΔV_{HT} is assumed to saturate at longer t_{STR}. The ΔV_{HT} expression shown in Table 6.2 uses a stretched exponential form, which becomes identical to the expression used in Table 4.1 at long t_{STR}. As mentioned in Chap. 4, ΔV_{HT} has identical E_{OX} acceleration factor as that of ΔV_{IT} ($\Gamma_{HT} = \Gamma_{IT}$), but has much lower T activation energy (E_{AHT}) than the T activation of ΔV_{IT}; identical values of Γ_{HT} and E_{AHT} are used in this model and the compact model of Chap. 4. The parameters τ_{STR} and β_{STR} determine the shape and time-to-saturation for time

Table 6.2 Equations for hole trapping and bulk trap generation

$\Delta V_{HT}(t) = \mathrm{B}E_{OX} e^{\Gamma_{HT} E_{OX}} e^{-\frac{E_{AHT}}{kT}} \left(1 - e^{-\left(\frac{t}{\tau_{STR}}\right)^{\beta_{STR}}}\right)$	(6.11)
$\Delta V_{OT}(t) = \mathrm{C}\left(1 - e^{-\left(\frac{t}{m}\right)^{\beta_{OT}}}\right)$; where $m = \eta(V_{G,STR} - V_{T0} - \Delta V_T)^{-\frac{\Gamma_{OT}}{\beta_{OT}}} e^{\left(\frac{E_{AOT}}{kT\beta_{OT}}\right)}$	(6.12)

Adjustable parameters: B, Γ_{HT}, τ_{STR}, β_{STR}; C
Fixed parameters: E_{AHT}, E_{AOT}, Γ_{OT}; β_{OT}; η

evolution of ΔV_{HT}. Note that τ_{STR} is T activated with energy E_A–τ_S while β_{STR} remains independent of stress T. The model for ΔV_{OT} uses the same empirical equation and parameters as shown in Chap. 4. Unlike ΔV_{IT} and ΔV_{HT}, ΔV_{OT} depends on $V_{G\text{-}STR}$ and not E_{OX}, has voltage acceleration factor (Γ_{OT}) and Arrhenius T activation energy (E_{AOT}). The fixed and device dependent adjustable parameters for ΔV_{HT} and ΔV_{OT} are listed in Table 6.2. Although ΔV_{OT} contribution can be significant for thicker SiON devices especially at higher $V_{G\text{-}STR}$, it is found to be much lower for HKMG devices, refer to Chap. 4 for details.

6.2.4 SiON Process Dependence

In this section, the composite framework will be used to predict measured data in SiON p-MOSFETs, refer to Chaps. 1 and 4 for device details. Mobility corrected on-the-fly (OTF) method is used for measurements with default time-zero (t_0) delay of 1 μs unless mentioned otherwise; refer to Chap. 2 for detailed description of OTF measurement and mobility correction methods.

Figure 6.7 shows the time evolution of ΔV_T from very short to long t_{STR}, measured using OTF method with t_0 delay of 1 μs and 1 ms, respectively, for SiON p-MOSFETs having (a) lower and (b) much higher N content in the gate insulator. The devices have different pre-stress $V_{T0}$and equivalent oxide thickness (EOT), and $V_{G\text{-}STR}$ is suitably adjusted to have similar stress E_{OX} for both devices. Note that the device having higher N content shows larger overall ΔV_T and much larger ΔV_T in the sub 1 ms time scale when measured using t_0 = 1 μs method, as well as larger difference between 1 μs and 1 ms measurements when compared to the device having low N content in the gate insulator. As mentioned in Chap. 2, the first data

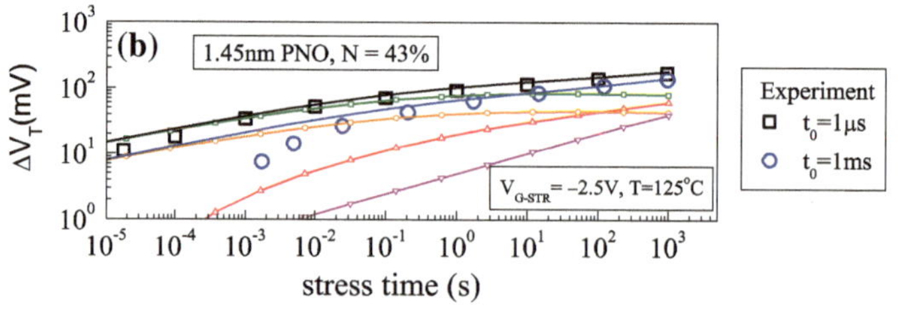

Fig. 6.7 Time evolution of ΔV_T, obtained using mobility corrected OTF measurements for different time-zero delay (*symbols*), for NBTI stress in different SiON devices. Overall composite model solution and its underlying trap generation and trapping subcomponents are shown as *lines*

point is considered as unstressed in OTF method, and therefore, short-time measured data gets corrupted especially for slower measurements. As a consequence, the actual difference between the 1 μs and 1 ms measurements is somewhat lower than that shown in Fig. 6.7, as lower than actual ΔV_T is measured by the 1 ms OTF method especially at shorter stress time. Nevertheless, the gap in ΔV_T magnitude between 1 μs and 1 ms measurements is much larger for the device having higher N content as shown.

Figure 6.7 also plots the calculated time evolution of ΔV_T and underlying ΔV_{IT}, ΔV_{HT} and ΔV_{OT} components using the composite framework. Note that time evolution of ΔV_{IT} is calculated by using numerical simulation of RD model listed in Table 6.1, while ΔV_{HT} and ΔV_{OT} are calculated using analytical expressions listed in Table 6.2. The model can explain measured data quite well for both devices; data for different t_0 delay is predicted by different ΔV_{HT} but identical ΔV_{IT} and ΔV_{OT} components, refer to Chap. 4 for further details. Note that the difference between model calculation and measurement, i.e., lower measured ΔV_T than calculated, for t_0 = 1 ms data especially at shorter t_{STR} is due to measurement inaccuracy associated with slower OTF method as discussed above.

Since ΔV_{HT} is a fast component and its magnitude increases with increase in N as shown in Chap. 4, higher ΔV_{HT} contribution can explain large sub 1 ms degradation observed for the device having larger N in the gate insulator stack. Moreover, as slower measurement fails to capture the early part of degradation, and since the device having larger N starts to degrade rapidly right after the initiation of stress due to higher ΔV_{HT} contribution, a large difference is observed between 1 μs and 1 ms measurements for this device. However, due to lower ΔV_{HT} contribution, the device having low N shows negligible sub 1 ms degradation and much smaller difference is observed between 1 μs and 1 ms data. Such calculations are done for different devices and for NBTI stress using different $V_{G\text{-}STR}$ and T, and the results are shown below.

Figure 6.8 shows the time evolution of ΔV_T from short to long t_{STR}, measured in SiON devices having different N content and EOT, refer to Chaps. 1 and 4 for device details. Experiments have been performed using t_0 = 1 μs OTF measurements with mobility correction, for NBTI stress at different $V_{G\text{-}STR}$ and fixed stress T and at different stress T and fixed $V_{G\text{-}STR}$. Calculated time evolution of ΔV_T using the composite modeling framework is also shown. The model can accurately explain measured data from very short to long t_{STR} under such diverse set of devices and experimental conditions with fixed and device dependent adjustable parameters listed in Table 6.1 for ΔV_{IT} and Table 6.2 for ΔV_{HT} and ΔV_{OT}. Identical fixed and device dependent model parameters are used here and in Chap. 4. Once the parameters are fixed for a particular device, they are not adjusted with variation in $V_{G\text{-}STR}$ and T. Although not shown here for brevity, t_0 = 1 ms OTF measurements for these devices and stress conditions can also be successfully explained with exactly identical values for all parameters except one; only exception being the parameter B determining the saturated ΔV_{HT} component. This is fully consistent with modeling results discussed in Chap. 4.

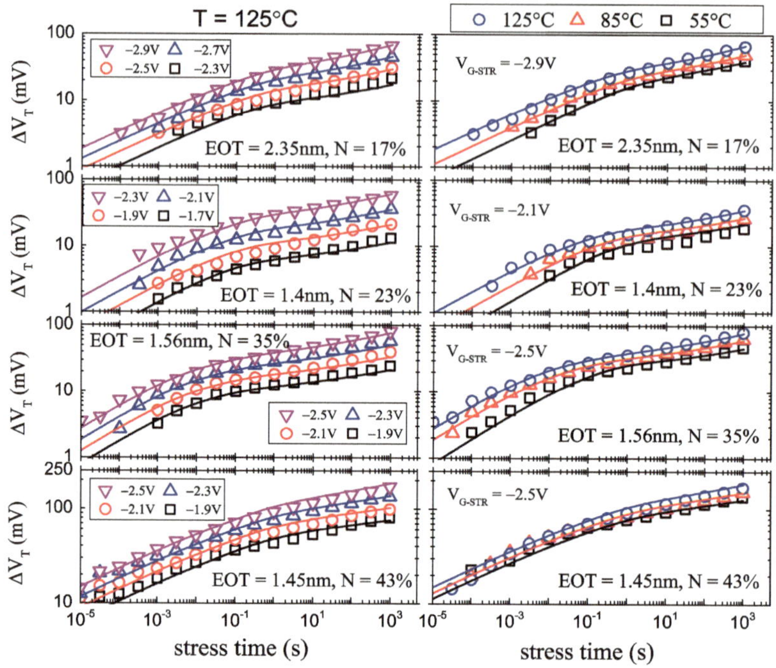

Fig. 6.8 Time evolution of ΔV_T, obtained using mobility corrected OTF measurements (*symbols*), for NBTI stress using different $V_{G\text{-}STR}$ and T in different SiON devices. Overall composite model solution is shown as *lines*

Note that the ΔV_{HT} component increases for devices having higher N content in the gate insulator, as shown above in Fig. 6.6, and also discussed in much greater detail in Chap. 4. Also note that ΔV_{HT} has weaker T activation compared to ΔV_{IT} and ΔV_{OT} as shown in Chap. 4, Table 4.1, and moreover, unlike ΔV_{IT} and ΔV_{OT}, ΔV_{HT} saturates at longer t_{STR} as shown in Fig. 6.6. Since N has relatively lower impact on ΔV_{IT} and ΔV_{OT} components, measured ΔV_T for devices having larger N content has relatively higher contribution from ΔV_{HT} component, and such devices have relatively weaker T activation and lower long-time power-law time exponent n as shown. The process dependence of NBTI parameters is described in detail in Chap. 4.

6.2.5 HKMG Process Dependence

In this section, the composite framework will be used to predict measured data in HKMG p-MOSFETs, refer to Chap. 3, Fig. 3.2 for device details. Ultra-fast measure-stress-measure (UF-MSM) method is used for measurements with default

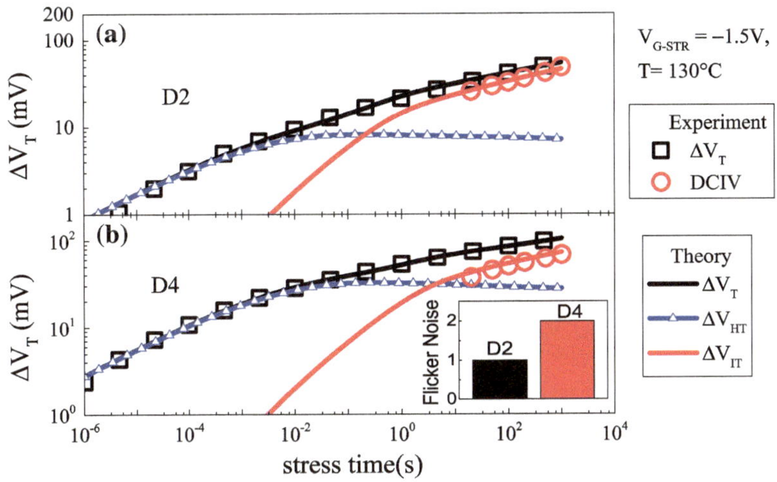

Fig. 6.9 Measured time evolution of ΔV_T from UF-MSM method and ΔV_{IT} from DCIV method after delay and band gap corrections (*symbols*), for NBTI stress in different HKMG devices (*symbols*). Overall composite model solution and its underlying trap generation and trapping subcomponents are shown as *lines*. *Inset* shows flicker noise data

delay of 10 μs unless mentioned otherwise; refer to Chap. 2 for detailed description of UF-MSM measurement technique.

Figure 6.9 shows the time evolution of UF-MSM measured ΔV_T from short to long t_{STR} and DCIV measured ΔV_{IT} only for longer t_{STR} in HKMG devices having RTP based (a) non-nitrided (D2) and (b) nitrided (D4) IL layers. Note that $V_{G\text{-}STR}$ is suitably adjusted to achieve identical stress E_{OX} for both devices, and the DCIV data are plotted after delay and band gap corrections; refer to Chap. 2 for additional details. Time evolution of model calculated ΔV_T and underlying ΔV_{IT} and ΔV_{HT} components are also plotted; calculated ΔV_T and ΔV_{IT} agrees well with measured data as shown. Note that the time evolution of ΔV_{IT} is calculated by using numerical simulation of RD model listed in Table 6.1, while ΔV_{HT} is calculated using analytical expression listed in Table 6.2. As mentioned earlier in this chapter and also analyzed in detail in Chap. 4, ΔV_{OT} component is negligible for HKMG devices and is not shown.

It is important to remark that similar to SiON devices, the D4 HKMG device having larger N content in the gate insulator shows higher overall ΔV_T as well as larger ΔV_T in the sub 1 ms time scale. It can be seen that ΔV_{HT} is larger for the D4 device due to larger N content, and can explain larger degradation observed in sub 1 ms time for this device. Both ΔV_{IT} and ΔV_{HT} components increase due to N, although the relative increase is much larger for ΔV_{HT} component, refer to Chap. 4 for additional details. Higher ΔV_{HT} component for the D4 device is consistent with flicker noise data shown in the inset.

Figure 6.10 shows the time evolution of measured ΔV_T from short to long t_{STR} in different HKMG devices having RTP based non-nitrided (D2) and nitrided (D4) IL

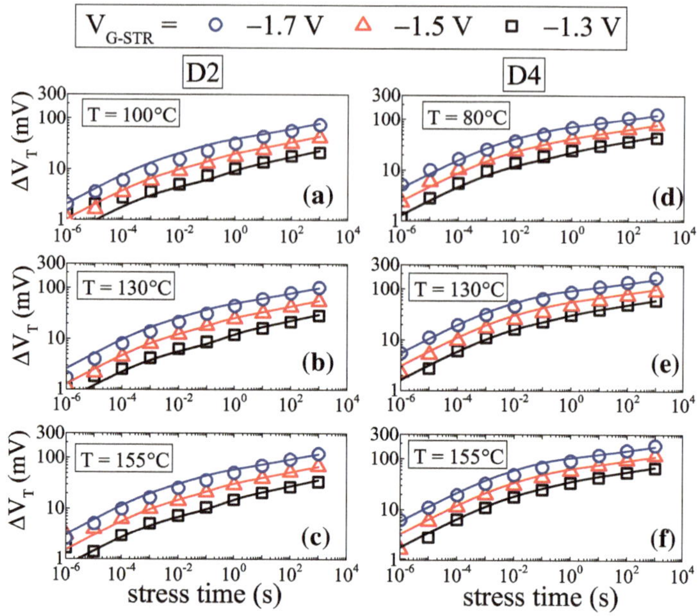

Fig. 6.10 Time evolution of UF-MSM measured ΔV_T for NBTI stress using different combinations of $V_{G\text{-}STR}$ and T in different HKMG devices. Overall composite model solution is shown as *lines*

layers. Experiments were done at three different T, and for each T, three different $V_{G\text{-}STR}$ values have been used for stress. The model calculated ΔV_T time evolution is also plotted, calculations agree well with measured data as shown. The corresponding DCIV measured and RD model calculated underlying time evolution of ΔV_{IT} are shown earlier in Fig. 6.5. The fixed and device dependent adjustable model parameters for ΔV_{IT} and ΔV_{HT} are listed, respectively, in Tables 6.1 and 6.2; similar fixed parameter values are used for SiON and HKMG devices. Moreover, the fixed and device dependent adjustable model parameters are identical to that used in Chap. 4.

Therefore, the composite modeling framework can readily explain time evolution of ΔV_T for differently processed SiON and HKMG p-MOSFETs, for different stress and measurement conditions as shown. Note that similar attempts were also made in [27, 65] to predict ΔV_T time evolution for SiON devices using RD model calculated ΔN_{IT} and fast saturating ΔN_{HT} components. However, these reports use RD model in its primitive form with only atomic H diffusion, which gives rise to power-law time dependence of ΔN_{IT} with time exponent $n = 1/4$, see Chap. 5, and is not consistent with experimental data shown in Chap. 3. Moreover, note that the alternative modeling approaches shown in [45, 61–64] fail to predict ΔV_T time evolution especially at longer t_{STR}, refer to [25, 48, 88], and therefore, these models have limited use and are not pursued.

6.3 Recovery After DC Stress

As mentioned earlier in this chapter, it is now well known that NBTI degradation substantially recovers after the removal of stress. The impact of recovery was discovered long ago [26], and has subsequently resulted in the development of different ultra-fast measurement methods [27, 30–32]. Recovery has been extensively studied in [28, 29, 45, 64, 69], and it was concluded, by highlighting the discrepancy between measured ΔV_T recovery and RD model calculated ΔN_{IT} recovery, that RD model is not suitable to explain the physical mechanism of NBTI. It was also opined that generated traps during NBTI stress are permanent, and it is only the trapped holes that get de-trapped and cause ΔV_T recovery. Therefore, ΔN_{IT} and ΔN_{HT} are, respectively, denoted as permanent (P) and recoverable (R) components [69]. There are several flaws in the above hypothesis since actual recovery mechanism is more complex than envisioned in these reports, and is due to the following reasons:

(a) As discussed earlier in this chapter and analyzed in detail in Chap. 4 for differently processed SiON and HKMG p-MOSFETs, NBTI stress results in ΔV_{IT}, ΔV_{HT} and in certain situations ΔV_{OT} subcomponents; overall ΔV_T is the uncorrelated sum of these subcomponents. Hence, it is naïve to assume ΔV_T recovery is due to the recovery of ΔV_{IT} only, without considering the recovery of ΔV_{HT} and ΔV_{OT}, as explained in [25, 48, 85] and discussed in this chapter. Therefore, it is improper to compare the RD model calculated ΔN_{IT} recovery to measured ΔV_T recovery.

(b) Moreover, direct experimental evidence of the recovery of generated ΔN_{IT} has been shown in Chaps. 2 and 3, and therefore, it is incorrect to denote ΔN_{IT} as permanent; however, it is possible that a fraction of ΔN_{IT} may never recover. Although the simple 1D macroscopic RD model framework described in Chap. 5 can predict time evolution of ΔN_{IT} during stress, the framework needs to be suitably modified to model stochastic hopping of H_2 during ΔN_{IT} recovery as explained in [25, 48], and is also discussed in this chapter.

(c) Finally, note that some fraction of the RD model generated traps, i.e., broken X– and Si– bonds, would go below the substrate Fermi level as $|V_G|$ is reduced from stress to recovery. These traps would capture electrons and neutralize before getting depassivated with returning H_2. Electron capture results in faster recovery of generated traps than predicted by RD model solution, which is explained in [42, 86, 87] and discussed in this chapter.

Therefore, the simplistic NBTI recovery picture depicted in [28, 29, 45, 64, 69] is inaccurate, and the exact physical process is explained hereinafter.

6.3.1 Modeling Framework

Figure 6.11 illustrates the schematic representation of the time evolution of different subcomponents during stress and recovery experiments. As mentioned before, ΔV_{IT} is the primary component during stress, is calculated using H/H_2 RD model solution, and shows power law time dependence with time exponent $n = 1$ at short and $n = 1/6$ at long t_{STR}. ΔV_{HT} is a fast saturation component and depends on the quality of the gate insulator, while ΔV_{OT} evolves at longer t_{STR}, shows power law time dependence with exponent $n \sim 1/3$, and makes significant contribution for thicker gate insulators and at higher $V_{G\text{-}STR}$ and stress T. As shown earlier in this chapter and in Chap. 4, ΔV_{OT} can be significantly large for SiON devices for certain stress conditions, while its contribution is much smaller for HKMG devices.

Figure 6.12 shows the schematic energy band diagrams drawn at the center of the channel for HKMG p-MOSFET biased under NBTI stress and recovery conditions, the recovery is shown for relatively higher and lower V_G. As $|V_G|$ is reduced from stress ($V_G = V_{G\text{-}STR}$) to recovery ($V_G = V_{G\text{-}REC}$), some of the RD model generated X- and Si- defects go below the Fermi level and capture electrons [94]. More defects would undergo this process when recovery is done at lower V_G. As explained using DFT simulation results for the particular case of different Ov related defects, see Fig. 6.2, the defect levels relaxes in energy after electron capture due to structural relaxation effects.

Therefore, as shown in Fig. 6.11, the generated ΔV_{IT} subcomponent is reduced by one of the two mechanisms during recovery. A fraction of traps ($\Delta V_{IT\text{-}1}$) that go below the Fermi level would recover by fast electron capture. The remaining fraction of traps ($\Delta V_{IT\text{-}2}$) would remain above the Fermi level, and would recover by back diffusion of H_2 and H and passivation of X- and Si-defects. Note that $\Delta V_{IT\text{-}1}$ includes traps from both first and second interfaces, and so is $\Delta V_{IT\text{-}2}$. Finally, some fraction of trapped holes in pre-existing (ΔV_{HT}) and generated (ΔV_{OT}) bulk traps would also go below the Fermi level and will detrap. Therefore, overall ΔV_Trecovery is uncorrelated sum of three subcomponents, fast electron capture in some fraction of RD model generated traps ($\Delta V_{IT\text{-}1}$), slow recovery of the remaining

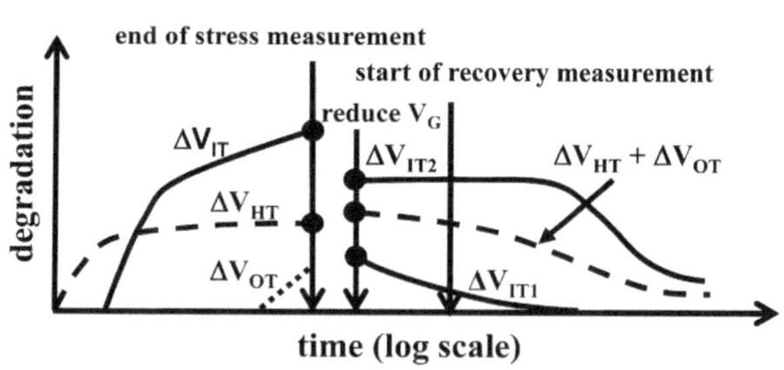

Fig. 6.11 Schematic representation of recovery of different NBTI subcomponents after stress. Recovery starts as soon as V_G is reduced, while the captured data depends on measurement speed (start of recovery measurement)

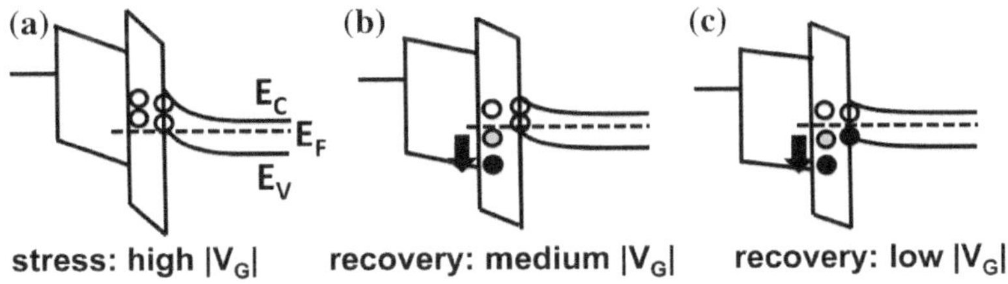

Fig. 6.12 Schematic energy band diagrams for **a** stress, as well as for recovery at **b** relatively larger and **c** small gate bias. The *circles* represent generated traps, their occupancy (empty of full) and energy relaxation of filled traps

RD model generated traps ($\Delta V_{IT\text{-}2}$), and fast hole detrapping from a fraction of pre-existing and generated bulk traps ($\Delta V_{HT} + \Delta V_{OT}$). As shown in Fig. 6.11, electron capture and hole detrapping are fast processes, and measured recovery would depend on the measurement speed.

Table 6.3 shows the stretched exponential analytical expressions used to calculate fast electron capture in RD model generated traps and hole detrapping from pre-existing and generated bulk insulator traps, see (6.13) and (6.14), respectively. Since electron capture reduces the positive gate insulator charge generated due to NBTI stress, both electron capture and hole detrapping expressions are similar, with T independent parameters τ_{EC}, β_{EC} and τ_{REC}, β_{REC}, respectively. These parameters determine the rate of electron capture and hole detrapping. The F_{FAST} parameter determines the fraction of RD model generated traps that go below the Fermi level and capture electrons. P_{HT} is the magnitude of pre-existing and generated bulk traps that remain above the Fermi level at a particular $V_{G\text{-}REC}$, and therefore, holes trapped in these traps do not detrap. The pre-factors D_1 and D_2 are calculated using $F_{FAST} * \Delta V_{IT}$ and ($\Delta V_{HT} + \Delta V_{OT}$) values obtained at the end of stress.

Figure 6.13 shows the schematic of double interface H/H_2 RD model during recovery after NBTI stress. H_2 molecules diffuse back towards the second interface and react with broken X– to form X–H bonds and H atoms. The resultant H atoms diffuse towards the first interface to passivate broken Si–bonds. The phenomenon is just opposite to stress, shown earlier in Fig. 6.4. Note that in reality, broken X-bonds are randomly located in 3D in the gate insulator, and the H_2 molecule, being a slow diffuser, needs to hop around till it gets captured and reacts with X–.

Table 6.3 Equations for electron capture in RD model generated traps and detrapping of holes from pre-existing and generated bulk traps

$\Delta V_{IT1}(t) = D_1\left(1 - e^{-\left(\frac{t}{\tau_{EC}}\right)^{\beta_{EC}}}\right)$	(6.13)
$(\Delta V_{HT} + \Delta V_{OT})(t) = D_2\, e^{-\left(\frac{t}{\tau_{REC}}\right)^{\beta_{REC}}} + P_{HT}$	(6.14)

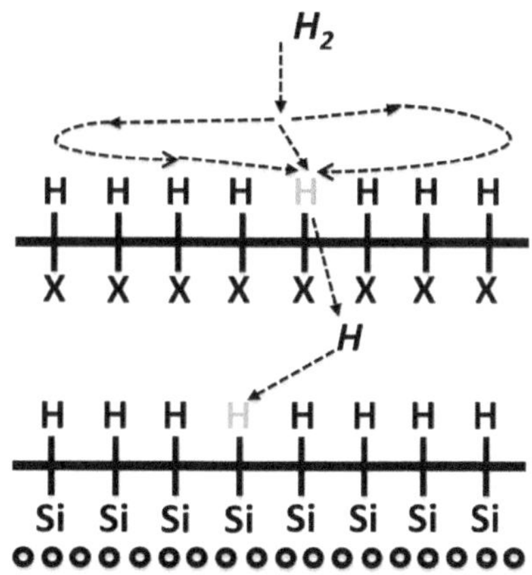

Fig. 6.13 Schematic of double interface H/H_2 Reaction-Diffusion model for interface trap passivation after stress

Therefore, when all broken X– bonds are assumed at the second interface, the H_2 molecule needs to hop around near the second interface till it finds a broken X– to react. Ideally, the same should hold for diffusing H atom during stress; however, being a fast diffuser, the atomic H can find X–H bonds in short-time scale and the H hopping is not a rate limiting step. Similarly during recovery, once H atom is released, it can very quickly find another broken Si– or X– bond and react, and is also not a rate limiting step.

The probability of H_2 finding an available X– is greater at the beginning of recovery. However, the hopping of H_2 prior to reaction with X– would be longer for longer recovery time, as most broken X– bonds would be consumed with passage of time. In a simple 1D RD model framework, this physical mechanism can be effectively captured by slowing down the H_2 diffusivity D_{H2} over time, using the relation $D_{H2}(t) = D_{H20}/[1 + B_D * (t/t_{STR})]$, where D_{H20} is the diffusivity of H_2 used during stress and t_{STR} is the total stress time before the start of recovery phase [25]. The parameter B_D determines the extent of reduction in diffusivity required in 1D simulation, $B_D = 0$ denotes basic RD model. Note that back diffused H_2 and H passivate all RD model generated X- and Si-defects, irrespective of whether they remain above or below the Fermi level for a particular $V_{G\text{-}REC}$. Of course, if a defect is already neutralized by electron capture, its subsequent passivation does not impact ΔV_T recovery. Moreover, some of the released H species during stress can become trapped or otherwise lost in the system. These H species can be considered as locked and will not be available during recovery [26], and therefore, some of the RD model generated defects will remain as permanent.

Figure 6.14 shows the time evolution of UF-MSM measured ΔV_T after NBTI stress in HKMG p-MOSFETs for identical $V_{G\text{-}STR}$, t_{STR} and stress T but relatively (a) higher and (b) lower $|V_{G\text{-}REC}|$. Refer to Chap. 2 for details of UF-MSM method for recovery characterization. Note that for identical stress condition, ΔV_T at the start of recovery is lower and the rate of recovery is much faster at lower $|V_{G\text{-}REC}|$ when compared to measured data at higher $|V_{G\text{-}REC}|$. Calculated ΔV_T and its underlying subcomponents obtained by using the composite modeling framework

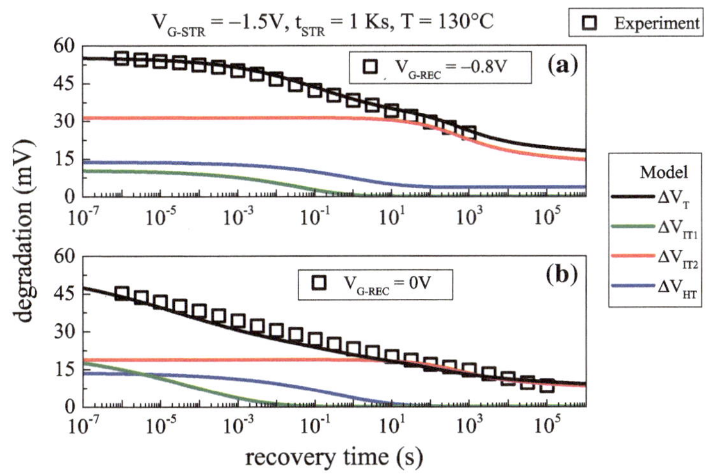

Fig. 6.14 Time evolution of UF-MSM measured ΔV_T recovery after NBTI stress in HKMG devices (*symbols*), for different $V_{G\text{-}REC}$. Overall composite model solution and its underlying trap passivation, electron capture and hole de-trapping subcomponents are shown as *lines*

are also shown. The time-dependent shape of hole detrapping ($\Delta V_{HT} + \Delta V_{OT}$) component and the associated permanent component P_{HT} depends on $V_{G\text{-}REC}$ as shown, faster rate of detrapping and smaller residual P_{HT} is observed for lower $|V_{G\text{-}REC}|$. As explained before, ΔV_{IT} recovers by fast electron capture ($\Delta V_{IT\text{-}1}$) and by slower H_2 diffusion related passivation ($\Delta V_{IT\text{-}2}$) processes, a larger ΔV_{IT} fraction recovers by electron capture at lower $|V_{G\text{-}REC}|$ as expected. The recovery of $\Delta V_{IT\text{-}2}$ is calculated using the RD model solution with reduced D_{H2} as discussed above, and a fraction of $\Delta V_{IT\text{-}2}$ remains unrecovered, denoted as P_{TG}, due to locking effect mentioned above. The ratio of $\Delta V_{IT\text{-}2}$ at the beginning of stress to P_{TG} at the end of stress remains independent of $|V_{G\text{-}REC}|$ as neutral H species are involved.

The H_2 hopping related slowing down of $\Delta V_{IT\text{-}2}$ recovery and the permanent P_{TG} component associated with H_2 locking effect determine recovery at long time and need further attention. Figure 6.15 plots the time evolution of RD model calculated $\Delta V_{IT\text{-}2}$ recovery without and with D_{H2} slow down effect. Reduced rate of recovery is observed when D_{H2} reduction is invoked. The permanent P_{TG} component is also shown, the ratio of $\Delta V_{IT\text{-}2}$ at start of recovery to P_{TG} at end of recovery is kept identical to that used in Fig. 6.14. Time evolution of total $\Delta V_{IT\text{-}2}$ recovery including the reduced D_{H2} solution and P_{TG} component is shown, which can be effectively modeled using a single reduced D_{H2} RD model simulation, but with larger value of parameter B_D that results in larger reduction at longer time, which is akin to simulating the H_2 lock-in effect. Note that for simplicity, all recovery simulations done using the H/H_2 RD model to predict the time evolution of $\Delta V_{IT\text{-}2}$ component of overall ΔV_T use an effective but larger reduced diffusivity, which automatically takes into account the permanent component. Simulated results are shown in the following section.

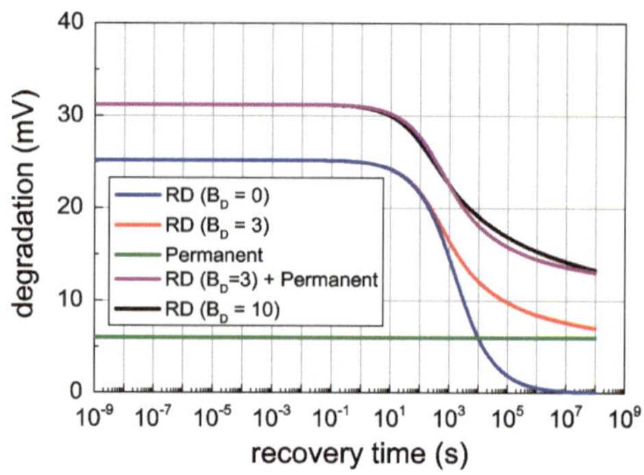

Fig. 6.15 Macroscopic 1D implementation of 3D stochastic hopping effect for trap passivation: time evolution of RD model calculated recovery without and with delayed diffusivity. The sum of delayed diffusivity and non-recoverable component is shown, which can be modeled using an effective RD model solution with larger reduction in diffusivity

6.3.2 Analysis of SiON and HKMG Devices

Figure 6.16 plots the time evolution of ΔV_T recovery measured using $t_0 = 1$ µs OTF method in SiON p-MOSFETs having different gate insulator processes. Recovery is measured after devices are stressed at fixed $V_{G\text{-STR}}$ and t_{STR} but with different T (left panels), and also at fixed $V_{G\text{-STR}}$ and T but for different t_{STR} (right panels). The OTF method for recovery measurements is explained in Chap. 2. The OTF measured time evolution of ΔV_T during stress for these devices are shown and modeled in Fig. 6.8.

Calculated ΔV_{IT}, ΔV_{HT} and ΔV_{OT} subcomponents are obtained at end-of-stress, and used to model time evolution of recovery using the framework discussed earlier. Note that due to the use of relatively high $|V_{G\text{-REC}}|$, recovery of RD model generated traps occurs mostly via trap passivation due to H_2 back diffusion and not so much via electron capture. The time evolution of ΔV_T recovery calculated using the composite modeling framework is also plotted in Fig. 6.16, the model agrees reasonably well with measured data across different devices and experimental conditions. Consistent model parameters have been used for stress and recovery simulations. Once the device dependent parameters are adjusted for a particular device, they are not tweaked for variations in experimental conditions.

Figure 6.17 plots the time evolution of measured ΔV_T recovery following NBTI stress in SiON devices at different t_{STR}, obtained from published report [69]. Experiments have been performed at fixed $V_{G\text{-STR}}$ and T while t_{STR} is varied over a wide range, and recovery is measured over long recovery time (t_{REC}). Two different combinations of $V_{G\text{-STR}}$ and T are used for stress. The corresponding time evolution of ΔV_T during stress for these stress conditions, also obtained from [69], is modeled to determine the end-of-stress ΔV_{IT}, ΔV_{HT} and ΔV_{OT} subcomponents (not plotted here for brevity). The composite modeling framework is used to calculate the time evolution of recovery of these subcomponents and hence the recovery of overall

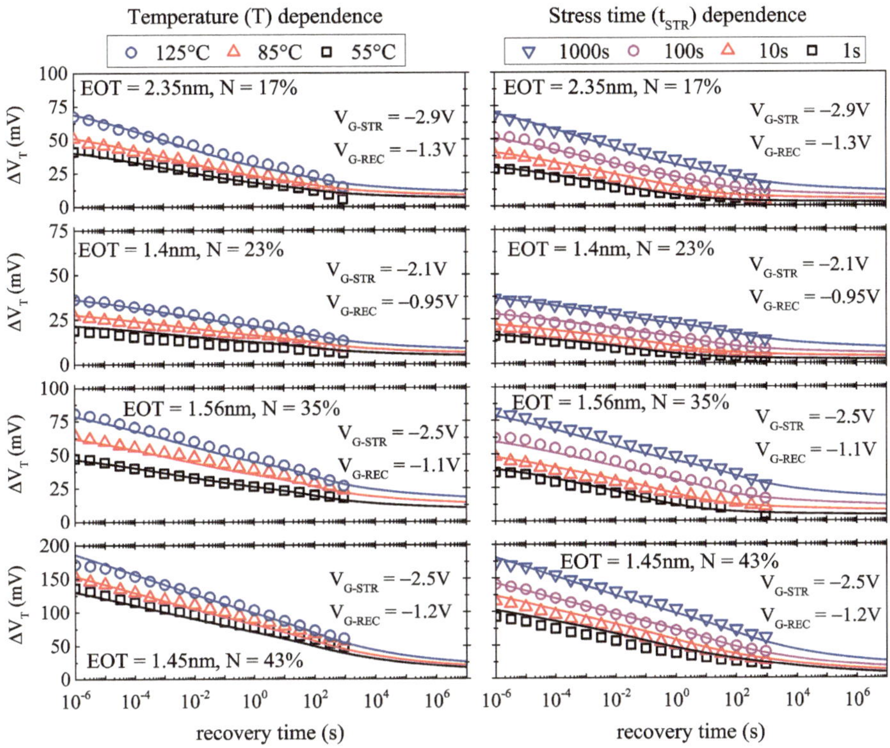

Fig. 6.16 Time evolution of ΔV_T recovery measured using mobility corrected UF-OTF method (*symbols*), after NBTI stress in different SiON devices for different stress T and stress time. Overall composite model solution is shown as *lines*

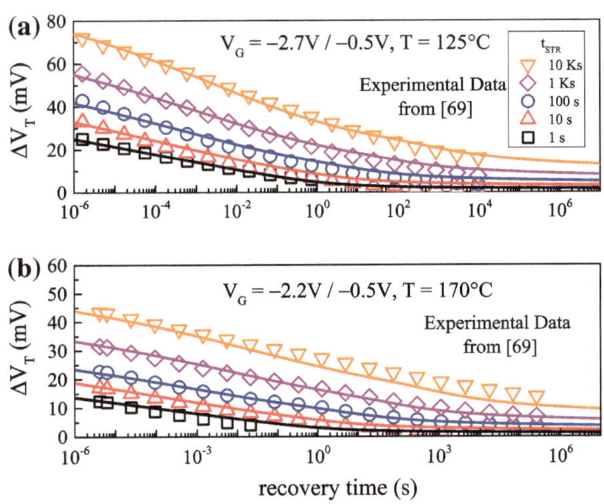

Fig. 6.17 Time evolution of ΔV_T recovery following NBTI stress in SiON devices for very long stress and recovery time. Measured data (*symbols*) and overall composite model solution (*lines*) are shown

ΔV_T. The time evolution of calculated ΔV_T recovery is also shown, which agrees reasonably well with measured data as shown. Consistent model parameters are used to calculate ΔV_T recovery in Figs. 6.16 and 6.17, as very similar SiON devices have been used [48, 69].

Figure 6.18 plots the time evolution of ΔV_T recovery measured using UF-MSM method in HKMG devices. NBTI stress is done using different $V_{G\text{-}STR}$ (top panels), stress T (middle panels) and t_{STR} (bottom panels), and recovery is done using high $|V_{G\text{-}REC}|$ (left panels) and using $|V_{G\text{-}REC}| = 0$ V (right panels). Recovery measurement using the UF-MSM method is discussed in Chap. 2. As mentioned before, for identical $V_{G\text{-}STR}$, stress T and t_{STR}, the magnitude of ΔV_T at start of recovery is lower and the rate of recovery is faster for lower $|V_{G\text{-}REC}|$, and this remains valid for different type of stress scenarios shown in Fig. 6.18. The time evolution of UF-MSM measured ΔV_T during stress using different $V_{G\text{-}STR}$ and T is plotted and

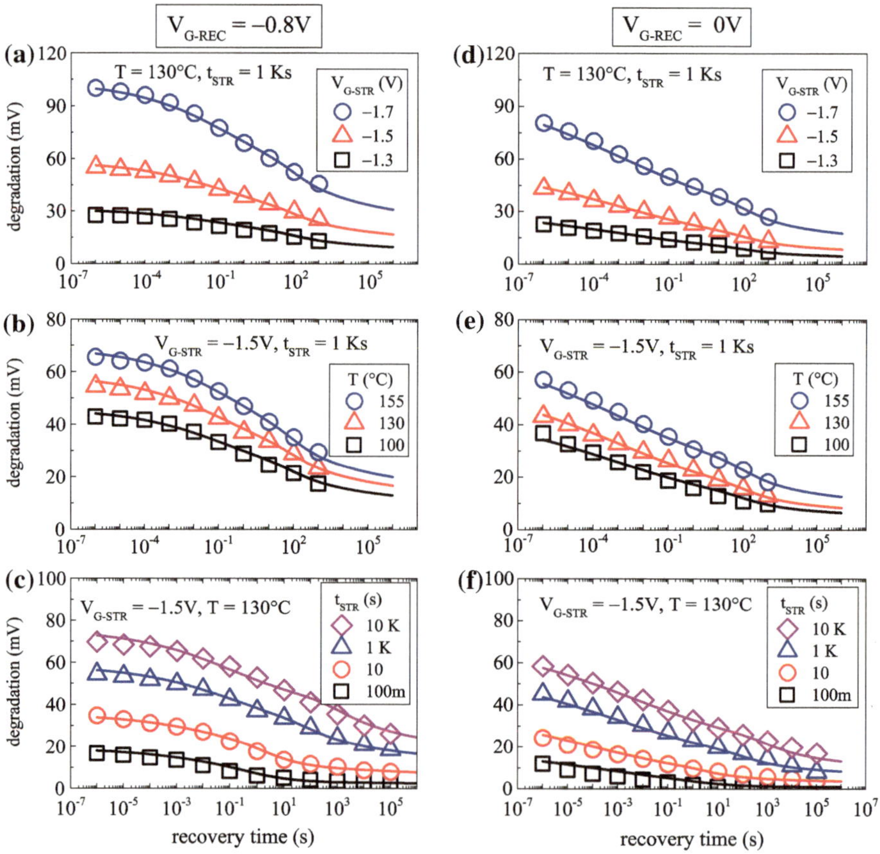

Fig. 6.18 Time evolution of UF-MSM measured ΔV_T recovery (*symbols*), after NBTI stress in HKMG devices for different stress bias, stress T and stress time, and measured using relatively high as well as 0 V $V_{G\text{-}REC}$. Overall composite model solution is shown as *lines*

analyzed in Fig. 6.10. The composite modeling framework is used to calculate ΔV_{IT}, ΔV_{HT} and ΔV_{OT} subcomponents at end-of-stress, and to calculate their time evolution during recovery after stress. The calculated time evolution of overall ΔV_T during recovery for different experimental conditions is plotted in Fig. 6.18; the model is in good agreement with measured data for such diverse experimental conditions. Note that consistent model parameters are used during stress and recovery calculations.

Figure 6.19 plots the time evolution of ΔV_T recovery measured using UF-MSM method at different $|V_{G\text{-}REC}|$, following NBTI stress in HKMG devices performed at fixed $V_{G\text{-}STR}$ and T but for (a) long and (b) short stress duration. Note that the magnitude of ΔV_T at the start of recovery and the rate of recovery is a strong function of $|V_{G\text{-}REC}|$, a significant reduction in the starting magnitude is observed when recovery is measured at lower $|V_{G\text{-}REC}|$. Also note that ΔV_T continues to recover in time for all values of $|V_{G\text{-}REC}|$, although at a reduced rate at higher $|V_{G\text{-}REC}|$, when measured after long t_{STR}. This also holds for recovery measured using low $|V_{G\text{-}REC}|$ after short t_{STR} stress. However, when recovery is measured at higher $|V_{G\text{-}REC}|$ after stress for short t_{STR}, ΔV_T does not continue to recover all the way, but turns around and subsequently increases at longer time; the turnaround kicks in earlier for larger $|V_{G\text{-}REC}|$. The composite framework has been used to calculate recovery of ΔV_T and its underlying subcomponents. The time evolution of calculated ΔV_T and ΔV_{IT} recovery are also plotted in Fig. 6.19. Remarkably, the composite model can successfully predict time evolution of ΔV_T recovery for different experimental conditions, and particularly, can also predict the turn around and increase in ΔV_T observed at longer time for certain experimental conditions. Note that the observed turnaround is due to delayed buildup of ΔV_{IT} because of trap generation when recovery measurements are done at higher $|V_{G\text{-}REC}|$, which can be successfully simulated using the RD model.

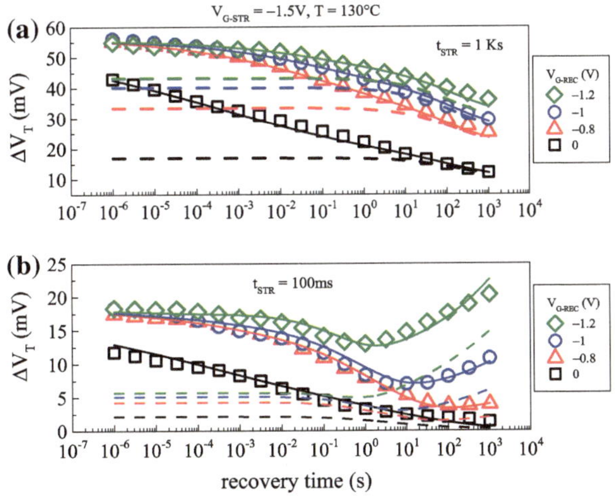

Fig. 6.19 Time evolution of UF-MSM measured ΔV_T recovery at different $V_{G\text{-}REC}$, after long and short-time NBTI stress in HKMG devices. Overall composite model solution and its underlying interface trap generation subcomponent are shown as *lines*

Therefore, contrary to some reports, the RD model based composite modeling framework can successfully explain diverse set of recovery measurements in differently processed SiON and HKMG devices, hitherto not achieved by any other modeling framework reported so far. The framework is used to predict AC degradation in the following section.

6.4 Degradation During AC Stress

As generated ΔV_T during DC stress recovers after the stress is removed, AC stress results in lower magnitude of ΔV_T than DC stress and is of interest from the viewpoint of digital switching logic circuits. Similar to DC stress, ΔV_T time evolution during AC stress also should be measured using ultra-fast method to obtain proper DC to AC ratio for a particular AC pulse frequency and duty cycle. As illustrated in Chap. 2, Fig. 2.14, ultra-fast AC measurements can be done using the UF-MSM technique after the end of last half cycle, denoted as Mode-A stress, or at the end of complete last cycle, denoted as Mode-B stress. Figure 2.15 in Chap. 2 shows the time evolution of UF-MSM measured ΔV_T, the impact of PDC, f and pulse low bias ($V_{G\text{-}LOW}$) on measured ΔV_T at fixed t_{STR}, and the impact of f on measured power-law time exponent n for AC Mode-A and Mode-B NBTI stress in HKMG devices. Additional AC stress results are shown in Chap. 1, Figs. 1.31 and 1.32. The following observations can be made:

(a) Mode-A measured ΔV_T magnitude is higher and the corresponding power-law time exponent n is lower compared to Mode-B measurements for low f AC stress; Mode-A ΔV_T reduces and time exponent n increases with f and merge with Mode-B measured data at higher f.
(b) Mode-B measured ΔV_T magnitude and time exponent n remains independent of f. Indeed, published data show f independence of ΔV_T for AC stress having f up to few GHz [21, 44]. The f independence of Mode-B stress is observed for different PDC, and the corresponding time exponent shows universal $n \sim 1/6$ value at different f and PDC.
(c) A characteristic “S” shaped PDC dependence of measured ΔV_T is observed for both Mode-A and Mode-B stress. A “kink” or jump in ΔV_T magnitude is observed between high PDC AC and DC measurements. The magnitude of this kink depends on pulse f and $V_{G\text{-}LOW}$ values, and also on the type, i.e., Mode-A or Mode-B, of AC stress.
(d) The DC to AC ratio at different PDC and f depends on $V_{G\text{-}LOW}$, although the shape of PDC and f dependence is maintained at different $V_{G\text{-}LOW}$ for both Mode-A and Mode-B stress. Lower ΔV_T magnitude is measured for lower $V_{G\text{-}LOW}$ for both Mode-A and Mode-B stress. This is consistent with $|V_{G\text{-}REC}|$ dependence of recovery discussed earlier, since for AC stress, degradation recovers during the pulse low phase.

In this section, the above experimental observations will be modeled using the composite framework discussed in previous sections. Note that AC stress experiments are performed in HKMG p-MOSFETs. Therefore, interface trap generation and hole trapping subcomponents are used to model measured data, and bulk trap generation is ignored.

6.4.1 RD Model for Generation of Interface Traps

Similar to DC stress, DCIV measurements have been performed to directly estimate interface trap generation during AC stress at different PDC and f. Measured results are shown in Chap. 3, and DCIV method is explained in Chap. 2. As DCIV is a slow measurement method, only Mode-B AC stress can be used, and the time evolution of measured ΔN_{IT} needs to be corrected for measurement delay and band gap as explained in Chap. 2. The following observations can be made regarding DCIV measured during AC stress, refer to from Chap. 3, Fig. 3.7 for details:

(a) Time evolution of ΔN_{IT} shows power law dependence with identical time exponent $n \sim 1/6$ for different PDC, f and $V_{G\text{-}LOW}$ values. Note the similarity of time exponent n between UF-MSM measured ΔV_T and DCIV measured ΔN_{IT} for Mode-B stress.
(b) Measured ΔN_{IT} shows f independence, and once again, UF-MSM measured ΔV_T and DCIV measured ΔN_{IT} show identical trend. However, the AC to DC ratio is much larger for ΔN_{IT} than ΔV_T. Moreover, unlike ΔV_T, the AC to DC ratio for ΔN_{IT} does not depend on $V_{G\text{-}LOW}$.
(c) Although ΔN_{IT}increases with increased PDC, the PDC dependent shape of ΔN_{IT} is somewhat different from that of ΔV_T. Moreover, unlike ΔV_T, measured ΔN_{IT} does not show a kink between high PDC AC and DC stress. It is important to remark that when normalized to 50 % AC, the PDC dependence of ΔN_{IT} and ΔV_T show universal shape up to ~85 % PDC for different devices and stress conditions, which is explained later in this section.

In previous sections, RD model has been used to predict the time evolution of ΔN_{IT} generated during DC stress and its subsequent recovery after stress. Note that contrary to some reports, ΔN_{IT} indeed recovers after stress, which is evident from direct DCIV measurements, results are shown in Chap. 2, Figs. 2.31 and 2.32. The RD model can be used to simulate time evolution of ΔN_{IT} during AC stress, refer to Chap. 5 for implementation details.

Figure 6.20 shows the time evolution of DCIV measured ΔN_{IT} after measurement delay and band gap correction, for AC NBTI stress in HKMG devices at different PDC; pulse $V_{G\text{-}HIGH}$ (=$V_{G\text{-}STR}$), $V_{G\text{-}LOW}$ (=$V_{G\text{-}REC}$), f and stress T are held constant. Time evolution of ΔN_{IT} calculated from RD model simulation for identical stress condition is also shown, identical model parameters are used for DC and AC stress. Note that simulated results agree well with measurements for AC stress at different PDC; the corresponding measured and simulated ΔN_{IT} time evolution

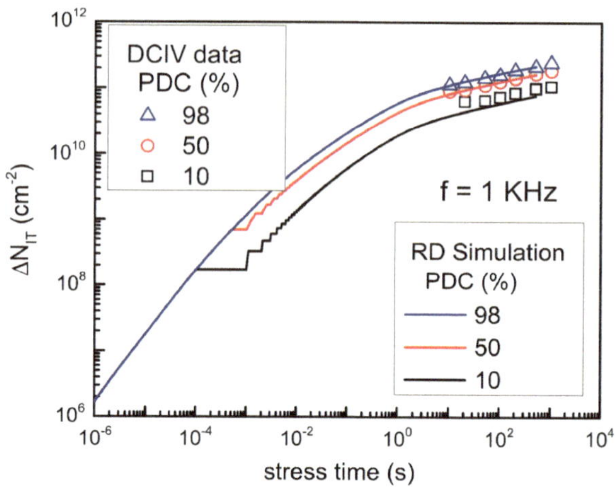

Fig. 6.20 Time evolution of DCIV measured ΔN_{IT} after delay and band gap correction (*symbols*), for AC NBTI stress in HKMG devices for different pulse duty cycle. *Lines* represent RD model simulation, shown up to 500 s stress, as AC simulation is computationally expensive

results for DC stress at different $V_{G\text{-}STR}$ and T are shown earlier in Fig. 6.5. It is important to remark that full AC simulation is computationally expensive, and therefore, AC simulations for very long t_{STR} and/or very high f become computationally challenging. However, time evolution of measured ΔN_{IT} demonstrates power law dependence with universal time exponent $n \sim 1/6$, and moreover, ΔN_{IT} magnitude is f independent as mentioned above. These features can be explained by RD model [78], refer to Chap. 5 for additional details. Therefore, simulated ΔN_{IT} time evolution data at relatively shorter t_{STR} and lower f can be extrapolated using simple analytical model to obtain ΔN_{IT} at longer t_{STR} and at higher f.

Figure 6.21 shows ΔN_{IT} at fixed t_{STR}, obtained from DCIV measured time evolution of ΔN_{IT} after measurement delay and band gap correction, for NBTI stress in HKMG devices for different (a) PDC and (b) f of the AC gate pulse. Measured AC stress data are normalized to the corresponding DC stress data at identical t_{STR}; therefore, the actual AC stress duration depends on PDC or the pulse on time. The PDC dependent experiments are done using different combinations of $V_{G\text{-}HIGH}$ and $V_{G\text{-}LOW}$; however, identical AC to DC ratio has been obtained for all combinations of stress as shown. Note that measured ΔN_{IT} shows f independence, and also note the absence of kink between high PDC AC and DC stress. Calculated PDC and f dependence of ΔN_{IT} obtained from RD model simulations are also shown, AC data are normalized to DC data at identical t_{STR} for fair comparison. Identical RD model parameters are used for DC and AC stress, and simulated results agree well with measurements as shown. Moreover, RD model suggests $V_{G\text{-}LOW}$ independence of ΔN_{IT}, which is consistent with recovery being governed by back diffusion of neutral H_2 species, and is also in agreement with measured data. The predicted f independence of ΔN_{IT} is also in agreement with measurements, and is a fundamental outcome of the RD model [78]. Note that RD model can predict $n = 1/6$ power-law time dependence of ΔN_{IT} for DC and AC stress and the f independence of ΔN_{IT} for AC stress with zero adjustable parameters, refer to Chap. 5 for additional details.

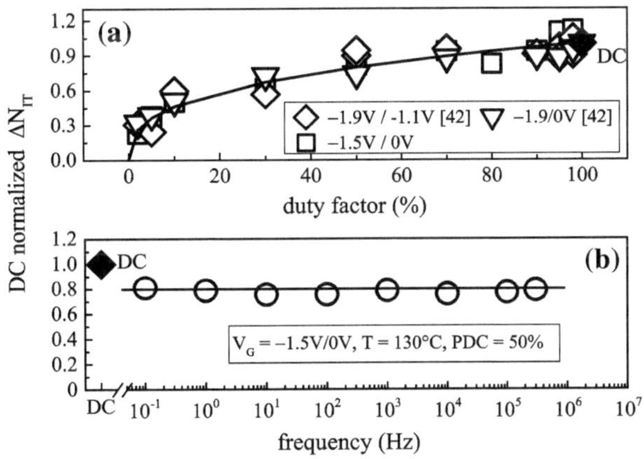

Fig. 6.21 DCIV measured ΔN_{IT} after delay and band gap correction (*symbols*) at fixed t_{STR}, for AC NBTI stress in HKMG devices for different pulse duty cycle and frequency. PDC dependent data measured for different pulse high and low bias combinations. *Lines* represent RD model solution. Simple empirical expression (see text) is used to fit actual simulated data at lower f, and the calibrated expression is used to predict data at higher f to save computation time

6.4.2 *Transient Trap Occupancy Calculation*

Although RD model can accurately predict the time evolution of DCIV measured ΔN_{IT} during and after DC stress and during AC stress for different experimental conditions, occupancy of generated traps needs to be considered for computing the charged state of generated traps and associated voltage shift. As explained earlier, the voltage shift ΔV_{IT} corresponding to generated traps equals $q * \Delta N_{IT}/C_{OX}$ during stress; ΔN_{IT} being the density of traps situated above the Fermi level during stress. However, ΔV_{IT} is divided into $\Delta V_{IT\text{-}1}$ and $\Delta V_{IT\text{-}2}$ components during recovery, depending on whether recovery occurs via fast electron capture in a part of ΔN_{IT} that goes below the Fermi level as $|V_G|$ is reduced, or by H_2 depassivation in the other part of ΔN_{IT} that remains above the Fermi level, as explained using Fig. 6.11. Since AC stress is equivalent to successive stress and recovery cycles, trap occupancy should be suitably calculated to convert RD model simulated time evolution of ΔN_{IT} to time evolution of ΔV_{IT}.

This conversion is performed using Transient Trap Occupancy Model (TTOM), which was proposed in [87] and explained using Fig. 6.22 and Table 6.4. First, AC simulation using the double interface H/H$_2$ RD model with reduced D_{H2} is done to obtain ΔN_{IT} (t) as a function of time (t) for successive on and off periods of gate stress pulse shown in Fig. 6.22. Table 6.4 lists the empirical equations used for trap occupancy calculation. To calculate ΔV_{IT} (t) at a given t during a particular on half cycle, see Fig. 6.22a, ΔN_{IT} needs to be calculated using the RD model at t_{EOPR} and t_{EOPS}, i.e., time corresponding to the end of last off half cycle and the end of previous on half cycle. In addition, ΔV_{IT} needs to be known at t_{EOPR}, which can be obtained form recovery calculation discussed below. At a time t_{ON} for the on half

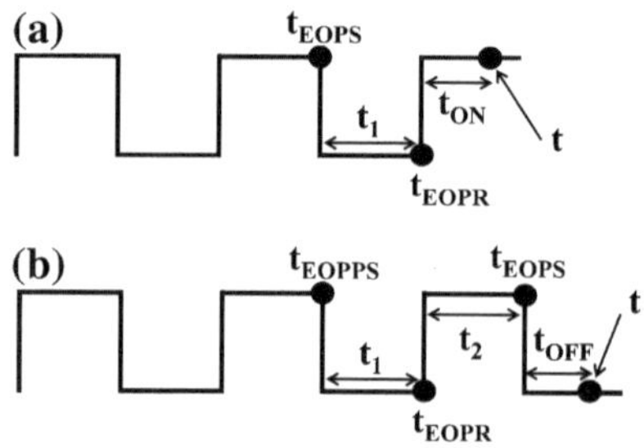

Fig. 6.22 Schematic of gate pulse showing the region of interest for **a** stress and **b** recovery for transient trap occupancy calculation

Table 6.4 Equations for transient trap occupancy calculation for RD model generated traps during multi-cycle DC or AC stress

$\Delta V_{IT}(t) = \Delta V_{IT}(t_{EOPR})$	(6.15)
$+ q/C_{OX}*[\Delta N_{IT}(t) - \Delta N_{IT}(t_{EOPR})]$	(6.16)
$+ q/C_{OX}*F_{FAST}*[\Delta N_{IT}(t_{EOPR}) - \Delta N_{IT}(t_{EOPS})*\exp\{-(t_1/\tau_{EC})^{\beta_{EC}}\}]$ $*[1 - \exp\{-(t_{ON}/\tau_{EE})^{\beta_{EE}}\}]$	(6.17)
$\Delta V_{IT}(t) = q/C_{OX}*(1\text{-}F_{FAST})*\Delta N_{IT}(t)$	(6.18)
$+ q/C_{OX}*F_{FAST}*[\Delta N_{IT}(t_{EOPS}) - \Delta N_{IT}(t_{EOPR})]*\exp\{-(t_{OFF}/\tau_{EC})^{\beta_{EC}}\}$	(6.19)
$+ q/C_{OX}*F_{FAST}*[\Delta N_{IT}(t_{EOPR}) - \Delta N_{IT}(t_{EOPPS})*\exp\{-(t_1/\tau_{EC})^{\beta_{EC}}\}]$ $*[1 - \exp\{-(t_2/\tau_{EE})^{\beta_{EE}}\}]*\exp\{-(t_{OFF}/\tau_{EC})^{\beta_{EC}}\}$	(6.20)
Parameters: Electron capture:- τ_{EC} and β_{EC}, Electron emission:- τ_{EE} and β_{EE}, Fast fraction:- F_{FAST}	

cycle of interest that corresponds to a total time simulation time t, ΔV_{IT} (t) is calculated by adding two equations to ΔV_{IT} (t_{EOPR}). The first (6.16) relates generation of new ΔN_{IT} in on half cycle of interest, which can be obtained from RD simulation. The second (6.17) calculates electron detrapping from the traps that have captured electrons in the previous off half cycle and is explained hereinafter.

Recall that F_{FAST} is the fraction of traps that go below Fermi level as $|V_G|$ is reduced for recovery. The pre-factor of (6.17) has two terms. The first term corresponds to traps that remain below Fermi level and are still un-passivated at t_{EOPR}. The second term corresponds to reduction in positively charged traps due to electron capture during previous off half cycle, and denotes residual charges due to traps that have gone below Fermi level but have not captured electrons at t_{EOPR}. Note that electron capture during off half cycle is analogous to emission of holes, i.e., reduction in positive charges, and is computed using stretched exponential equation with parameters τ_{EC} and β_{EC}. Therefore, the pre-factor of (6.17) denotes density of traps that remained occupied with electrons at the onset of on half cycle of interest, and these traps emit electrons during the current on cycle. The electron emission is analogous to hole capture and increase in positive charge, and a suitable stretched exponential equation is used with parameters τ_{EE} and β_{EE}. Electron capture is a fast process and uses very small τ_{EC}. However, traps relax in energy after electron capture as shown in Fig. 6.2, which increases the time constant τ_{EE} associated with subsequent electron emission. It can be shown that for first on half cycle, the TTOM model becomes the usual voltage shift expression $q * \Delta N_{IT}/C_{OX}$.

To calculate ΔV_{IT} (t) at a given t during a particular off half cycle, see Fig. 6.22b, ΔN_{IT} values need to be calculated at t_{EOPS}, t_{EOPR} and t_{EOPPS}, i.e., time corresponding to the end of last on half cycle, the end of the previous off half cycle, and also at the end of previous to previous on half cycle. At time t_{OFF} for the off half cycle of interest that corresponds to a total time simulation time t, ΔV_{IT} (t) is calculated by using three equations listed in Table 6.4. The first (6.18) corresponds to trap passivation and denotes residual charges in traps that remain above Fermi level and also remain un-passivated till time t_{OFF}. The pre-factor of the second (6.19) denotes generation of traps during the previous on half cycle, and hence the overall (6.18) corresponds to recovery due to electron capture in these generated traps, and denotes residual charges due to traps that have gone below Fermi level but have not captured electrons till time t_{OFF}. The third (6.20) has three terms. As explained above, the first term denotes density of traps that remained occupied with electrons at the onset of previous on half cycle, and the second term corresponds to electron emission from these traps during the previous on half cycle. Therefore, product of the first two terms corresponds to traps that got carried over from previous cycles, are not occupied by electrons, and go below Fermi level in the current off half cycle and become available to get neutralized by electron capture. The complete (6.20) denotes residual charges in these traps, which have still not captured electrons till t_{OFF}.

Therefore, generation and recovery of traps as calculated by the RD model and charge occupancy of these traps as calculated using the model described above can be used to calculate ΔV_{IT} for successive stress and recovery cycles. As an illustration, Fig. 6.23 plots voltage shift corresponding to trap generation and passivation calculated using the RD model, and also due to electron capture into these traps during recovery and emission from them during subsequent stress calculated using TTOM. The individual trap generation, passivation, electron capture and emission subcomponents for every on and off cycles are also shown. As mentioned before, electron capture is fast but subsequent re-emission is slow, and this can be visualized from Fig. 6.23. As illustrated, transient occupancy of generated traps plays a very important role in determining overall voltage shift contribution due to generated traps during multi-cycle DC or AC stress.

Figure 6.24 plots the time evolution of voltage shift corresponding to ΔN_{IT} and ΔV_{IT} calculated, respectively, using the RD model and TTOM with different F_{FAST}

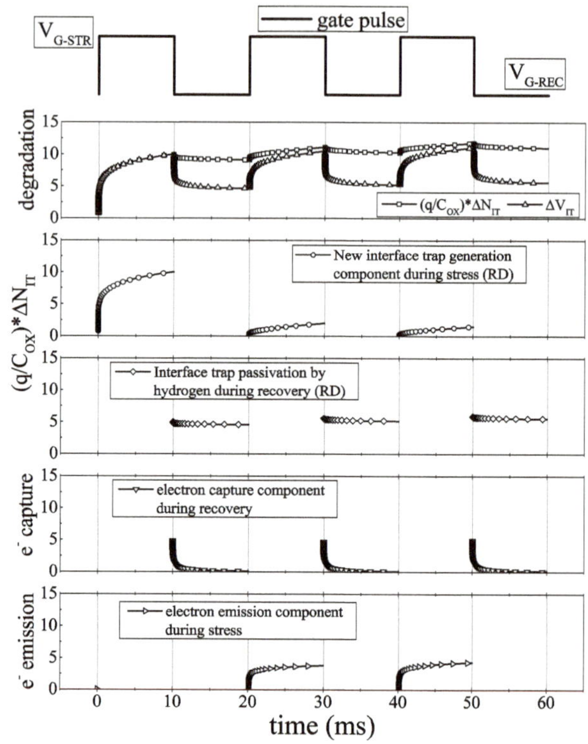

Fig. 6.23 TTOM enhanced RD model: time evolution of voltage shift due to trap generation and passivation from RD model and TTOM enabled RD model calculations for multiple stress and recovery cycles. Trap generation, passivation, electron capture and emission components are also shown separately for each cycle

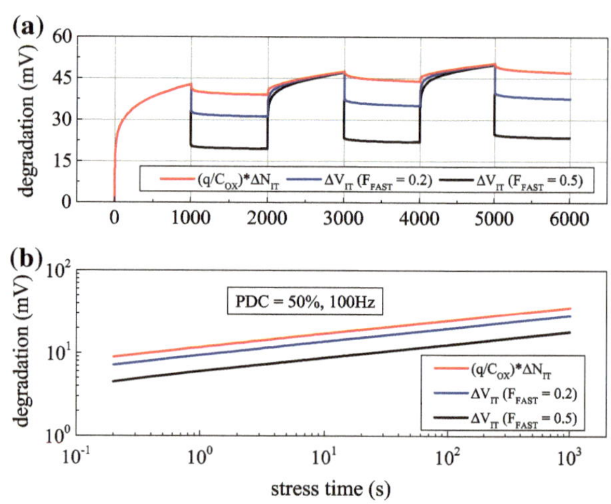

Fig. 6.24 Time evolution of voltage shift due to RD model calculated traps, and after transient occupancy calculation of these traps, for **a** multi-cycle DC stress and **b** AC stress. Calculations are done for different fast-fraction values corresponding to the electron capture process

parameters, for (a) multiple DC stress and recovery cycles as well as (b) higher f AC stress; Fig. 6.24a can be considered as AC stress with very low f. The DC data are also shown as reference. Figure 6.24a shows that $q * \Delta N_{IT}/C_{OX}$ and ΔV_{IT} are identical for the first on half cycle as expected, but they separate out from the first off half cycle due to electron capture as expected. ΔV_{IT} during recovery is lower than $q * \Delta N_{IT}/C_{OX}$; their separation depends on F_{FAST}, which is determined by $V_{G\text{-}REC}$. Figure 6.24b shows that the time evolution of $q * \Delta N_{IT}/C_{OX}$ and ΔV_{IT} show power-law dependence with identical exponent $n = 1/6$, and their separation depends on F_{FAST} or $V_{G\text{-}LOW}$ as expected. The DC to AC ratio is governed by the parameter F_{FAST} for both multi-cycle DC and high f AC stress as shown.

6.4.3 Hole Trapping and Detrapping

Although interface trap generation dominates NBTI, contribution due to trapping and detrapping of holes in pre-existing traps need to be considered to compute the time evolution of overall ΔV_T during multiple DC stress and recovery cycles as well as during high f AC stress. TDDB like bulk trap generation is ignored. Table 6.5 lists the cyclostationary stretched exponential (6.21) to calculate hole trapping into and detrapping from pre-existing bulk traps. Identical expressions as shown in Tables 6.2 and 6.3 have been used, with parameters τ_{STR}, β_{STR} and τ_{REC}, β_{REC}, respectively, for hole trapping and detrapping processes.

6.4.4 Analysis of Experimental Results

Figure 6.25 shows the time evolution of UF-MSM measured ΔV_T for subsequent DC stress and recovery cycles for NBTI stress in HKMG p-MOSFETs. Experiments are performed for symmetric stress and recovery periods for (a) high $V_{G\text{-}LOW}$ ($=V_{G\text{-}REC}$) and for (b) $V_{G\text{-}LOW} = 0$ V. Experiments are also performed for $V_{G\text{-}LOW} = 0$ V and for asymmetric stress and recovery periods, with (c) large on and short off and (d) short on and large off phases. Identical $V_{G\text{-}HIGH}$ ($=V_{G\text{-}STR}$) and stress T are used for all stress conditions.

Time evolution of calculated ΔV_T and its underlying ΔV_{IT} and ΔV_{HT} subcomponents are also shown. Excellent agreement has been obtained between measured and calculated degradation. Note that ΔV_{IT} is calculated using RD model and TTOM, while ΔV_{HT} is calculated using cyclostationary model, and all model parameters are kept exactly identical to that used for predicting single stress and recovery experiments shown earlier in this chapter.

Figure 6.26 shows the time evolution of UF-MSM measured ΔV_T for Mode-B AC NBTI stress in HKMG p-MOSFETs for (a) different $V_{G\text{-}LOW}$ but fixed PDC, and (b) different PDC but fixed $V_{G\text{-}LOW}$, while other stress parameters, i.e., $V_{G\text{-}HIGH}$,

Table 6.5 Equations for hole trapping and detrapping for multi-cycle DC and AC stress

$\Delta V_{HT,ON}(t) = \Delta V_{HT}(\mathrm{EOPR}) + [\Delta V_{HT,MAX} - \Delta V_{HT}(\mathrm{EOPR})] * [1 - \exp\{-(t_{ON}/\tau_{STR})^{\beta_{STR}}\}]$ $\Delta V_{HT,OFF}(t) = \Delta V_{HT}(\mathrm{EOPS}) * \exp\{-(t_{OFF}/\tau_{REC})^{\beta_{REC}}\}$	(6.21)

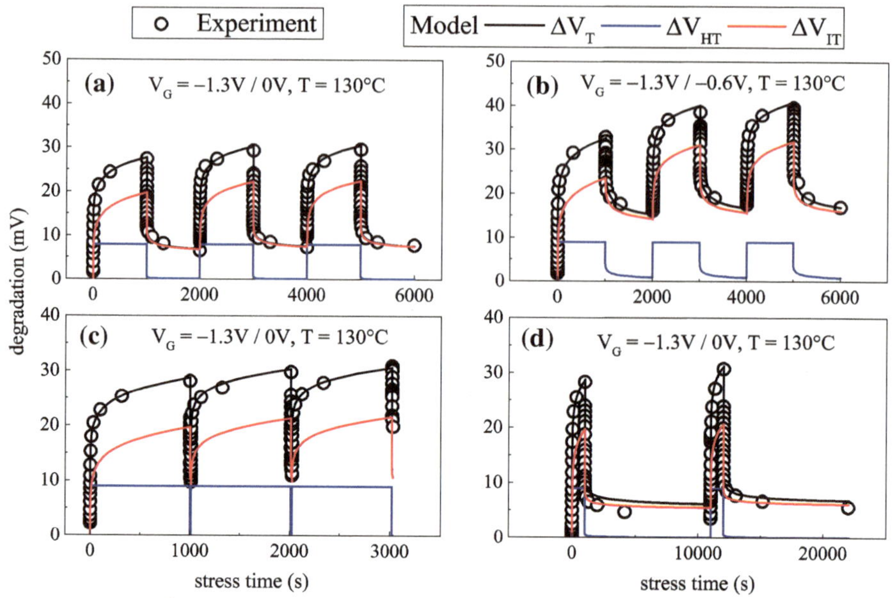

Fig. 6.25 Time evolution of UF-MSM measured ΔV_T during sequential DC NBTI stress and recovery cycles in HKMG devices (*symbols*), for (**a**, **b**) different $V_{G\text{-}REC}$ but symmetric stress and recovery cycles, and (**c**, **d**) for $V_{G\text{-}REC}$ = 0 V but asymmetric stress and recovery cycles involving long stress and short recovery and short stress and long recovery. Overall composite model solution and its underlying trap generation and trapping subcomponent are shown as *lines*

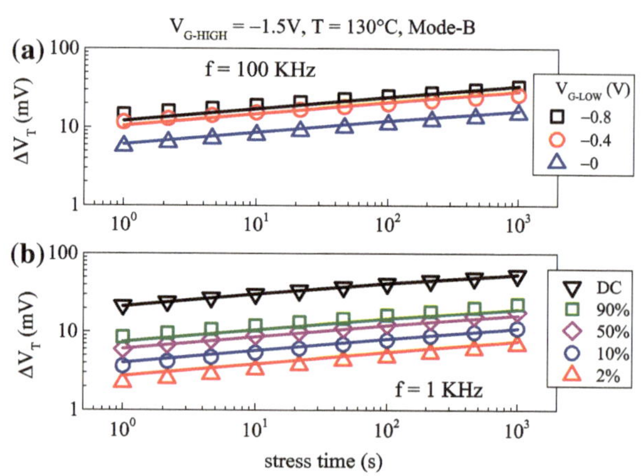

Fig. 6.26 Time evolution of UF-MSM measured ΔV_T during Mode-B AC NBTI stress in HKMG devices (*symbols*), for **a** different $V_{G\text{-}LOW}$ and **b** different PDC. DC data are shown as reference. *Lines* represent composite model solution

f and stress *T* are kept constant. The DC data are shown as reference. Note that power-law time dependence is observed for different $V_{G\text{-}LOW}$ and PDC with identical exponent $n \sim 1/6$ that is characteristics of RD model. It will be shown later that hole-trapping component is negligible for Mode-B stress, and therefore, time evolution of measured ΔV_T can be explained using ΔV_{IT} obtained from RD model and TTOM calculations across different PDC and $V_{G\text{-}LOW}$ as shown. The model agrees well with measured data, and calculations are done with fixed model parameters to predict PDC dependence, only F_{FAST} is varied to predict $V_{G\text{-}LOW}$ dependence that governs the ratio of generated traps that undergo fast electron capture. Note that although not encountered in digital circuits, the predictive capability of the model for different $V_{G\text{-}LOW}$ dependent data suggests accuracy of TTOM calculations.

Figure 6.27 plots the PDC dependence of UF-MSM measured ΔV_T at fixed t_{STR}, for Mode-B AC NBTI stress in HKMG p-MOSFETs for different $V_{G\text{-}LOW}$; all other stress parameters are kept constant. The DC value is shown as reference. As explained earlier in this chapter, ΔV_T for DC stress consists of ΔV_{IT} and ΔV_{HT}, while ΔV_{OT} remained negligible for HKMG devices. As mentioned before and will also be explained later in this chapter, ΔV_{HT} is negligible for Mode-B AC stress. Therefore, measured ΔV_T for Mode-B stress is solely due to ΔV_{IT}. The dotted line represents ΔV_{IT} for AC stress calculated using RD model without considering electron trapping. The difference between calculated ΔV_{IT} and measured ΔV_T at DC is due to ΔV_{HT}. However, the magnitude of ΔV_{HT} is small and its absence for Mode-B AC stress is not sufficient to explain the large AC to DC ratio observed for different PDC and $V_{G\text{-}LOW}$. The solid lines represent calculated ΔV_{IT} with trap occupancy obtained using RD model and TTOM based framework. The model agrees well with measured data across different PDC and $V_{G\text{-}LOW}$. As mentioned before, PDC dependence is calculated using identical model parameters, while only the F_{FAST} parameter is varied to explain $V_{G\text{-}LOW}$ dependence.

Figure 6.28 shows the PDC dependence of ΔV_T for AC Mode-B NBTI stress in HKMG p-MOSFETs, measured using UF-MSM technique having different

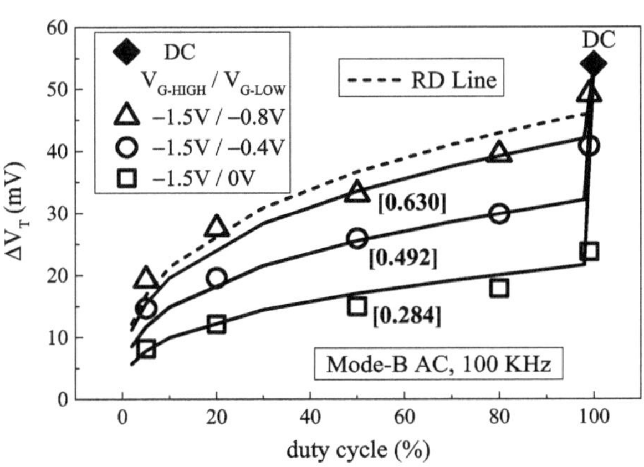

Fig. 6.27 PDC dependence of UF-MSM measured ΔV_T at fixed t_{STR}, for Mode-B AC NBTI stress in HKMG devices at different $V_{G\text{-}LOW}$ (*symbols*). DC data shown as reference. *Dotted line* represents RD model solution. *Solid lines* represent RD model solution together with transient trap occupancy. Hole trapping contribution is negligible

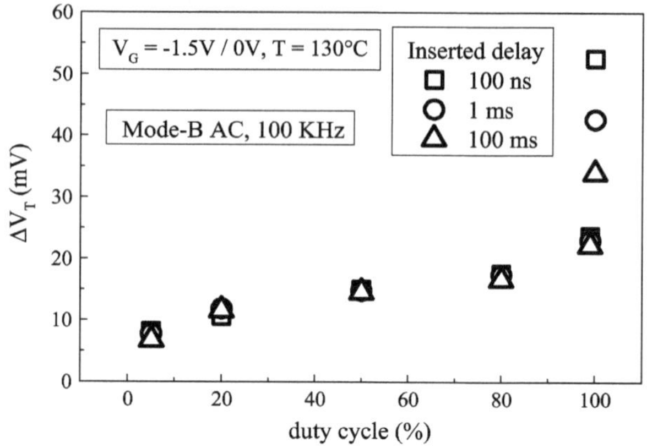

Fig. 6.28 PDC dependence of ΔV_T at fixed t_{STR} for Mode-B AC stress in HKMG devices, measured using UF-MSM method for different measurement delay. DC data shown as reference

measurement delay. Refer to Chap. 2 for a detailed discussion on the impact of measurement delay on UF-MSM measurements for AC stress. The DC data measured using identical measurement delay are also shown as reference. Except PDC, other stress conditions, e.g., $V_{G\text{-}HIGH}$, $V_{G\text{-}LOW}$, f and T are kept constant. Note that measurement delay impacts ΔV_T for DC but not for Mode-B AC stress.

Recall that ΔV_T for Mode-B AC stress is dominated by ΔV_{IT}. AC stress is done at $V_{G\text{-}LOW} = 0$ V that ensures electron capture induced trap neutralization during the last off half cycle. As electron capture is fast, measurements with additional delay do not impact resultant ΔV_T, since generated traps in the last on half cycle gets neutralized in the last off half cycle. Moreover, as ΔV_{HT} is negligible for Mode-B stress, ΔV_T is also not influenced by hole detrapping time constant during delayed measurements. However, the same is not true for DC stress, which results in both ΔV_{IT} and ΔV_{HT}. Therefore, electron capture into generated traps and hole detrapping takes place as $|V_G|$ is reduced after stress for measurement, and their magnitude depends on measurement delay. Finally, note that the passivation of interface traps is a slower process and does not play a big role unless measurement delay is made artificially large.

Figure 6.29 plots ΔV_T as a function of AC pulse f for Mode-B NBTI stress; DC data are also shown as reference. Data are obtained from various sources and are normalized to $f = 1$ kHz AC data.[2] Note that ΔV_T for AC Mode-B stress is governed by ΔV_{IT}, and its magnitude does not depend on measurement delay. Therefore, data from different sources show universal f independence; the f independence being the characteristics of RD model that governs ΔV_{IT}, as discussed earlier in this section. It is important to remark that the occupancy of RD model generated traps determine the magnitude of ΔV_{IT}. However, trap occupancy calculation results in a f independent constant scaling factor that depends only on $V_{G\text{-}LOW}$, for the range of f studied in this work. Therefore, measured data demonstrate the characteristic f independence of RD model as shown.

[2]Since measured data show f independence, the normalization can be done using any f.

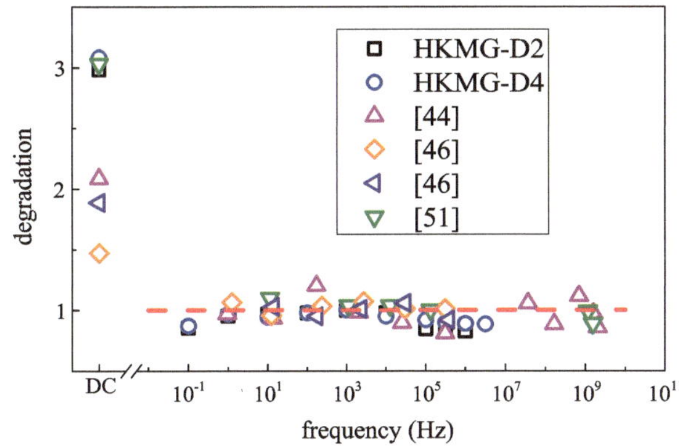

Fig. 6.29 Frequency dependence of ΔV_T at fixed t_{STR} for Mode-B AC stress at 50 % PDC. Data obtained from SiON devices (slow MSM), HKMG devices (UF-MSM) and also from difference published reports. DC data shown as reference. All f dependent dataset are normalized to their respective AC data at 1 kHz. *Line* represents RD model solution

However, the DC data show a large spread, and is due to differences in measurement speed used by different groups. As discussed above, measurement speed determines the extent of recovery due to electron capture and hole detrapping as $|V_G|$ is reduced from stress. Moreover, the spread in DC data also depends on the quality of the gate insulator used by different groups, which determines the magnitude of trapped holes during stress. The AC Mode-B data are free from hole trapping as discussed below, and hence is not impacted by gate insulator quality.

Figure 6.30 plots the PDC dependence of ΔV_T for Mode-B NBTI stress in different SiON and HKMG p-MOSFETs. SiON data are obtained using slow MSM method for different $V_{G\text{-}LOW}$. HKMG-A data are obtained using UF-MSM method on devices having different IL quality, for different f and $V_{G\text{-}LOW}$. HKMG-B data are obtained using slow MSM method for different $V_{G\text{-}LOW}$. HKMG-C data are also obtained using slow MSM method on devices having different IL quality. Additional data are plotted from published reports. Such diverse set of data show a remarkable universality when normalized to their respective 50 % PDC value, and this universality is seen for AC stress up to ~85 % PDC. However, a large spread in normalized ΔV_T is observed for higher PDC AC and DC (PDC = 100 %) stress.

As mentioned before and also explained in the next section, measured ΔV_T for Mode-B AC stress is predominantly due to ΔV_{IT} and contribution from ΔV_{HT} is negligible unless PDC becomes very high. Hence, ΔV_T is not influenced by reasonable measurement delay as discussed in Fig. 6.28. Note that PDC dependence of ΔV_{IT} is governed by RD model. Although the actual magnitude of ΔV_{IT} depends on occupancy of generated traps, trap occupancy calculation introduces a constant scaling factor, which is independent of f and PDC and only depends on $V_{G\text{-}LOW}$ of the gate pulse as shown earlier. Therefore, the RD model dictated PDC dependent shape, represented by the dotted line, is maintained for different type of measured data up to moderately high PDC stress as shown in Fig. 6.30.

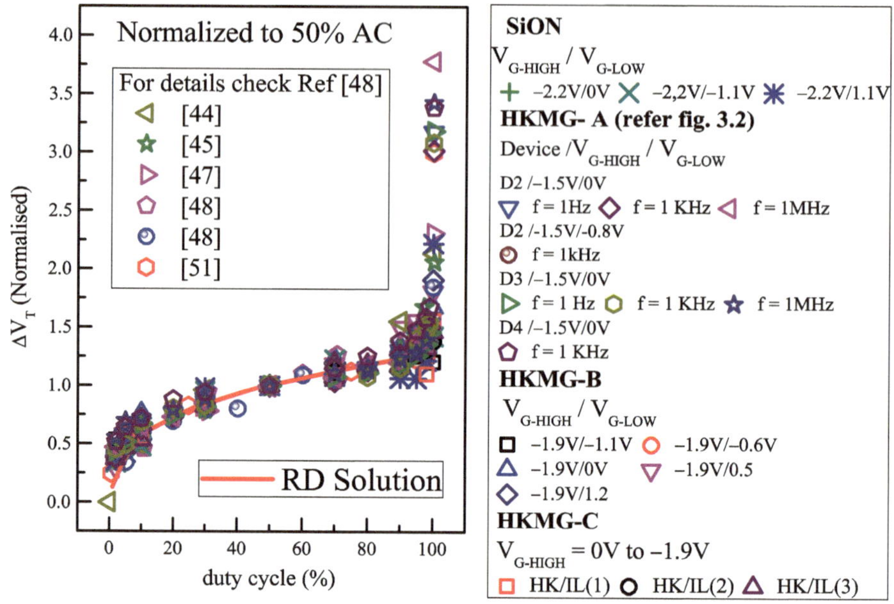

Fig. 6.30 PDC dependence of ΔV_T at fixed t_{STR} for Mode-B AC stress for different devices and experimental conditions. SiON, HKMG-B and different HKMG-C devices are measured using slow MSM for different $V_{G\text{-HIGH}}$ and $V_{G\text{-LOW}}$ combinations. Different HKMG-A devices are measured using UF-MSM for different values of $V_{G\text{-LOW}}$. Data also obtained from published reports. DC data shown as reference. All PDC dependent dataset are normalized to their respective 50 % PDC AC data. *Line* represents RD model solution

A scatter in measured ΔV_T is observed for very high PDC AC stress. Note that the pulse off half cycle becomes smaller at larger PDC. Therefore, measured ΔV_T would depend on both ΔV_{IT} and to some extent also by ΔV_{HT}, especially for devices with inferior quality of gate insulator, as holes trapped during on half cycle do not get sufficient time to detrap during off half cycle. As a consequence, measured ΔV_T is influenced by the quality of gate insulator that determines the magnitude of ΔV_{HT}. Moreover, ΔV_T will also be impacted by measurement delay that impacts the magnitude of hole detrapping. These aspects are responsible for the observed scatter at higher PDC. However, although small, the pulse off time is still good for fast electron capture in generated traps, and therefore, this phenomenon does not play any role, unless the *f* becomes too large and very high PDC data are analyzed.

The scenario becomes even more different for DC stress, which is governed by both ΔV_{IT} and ΔV_{HT}, and hence measured ΔV_T depends on the gate insulator quality and shows a spread as shown. However, more importantly, as $|V_G|$ is reduced for measurement, ΔV_T starts to recover not only by hole detrapping and reduction in ΔV_{HT}, but also due to fast electron capture and reduction in ΔV_{IT}, while trap passivation remains negligible for reasonably shorter measurement delay. Therefore, DC measurement is strongly influenced by measurement delay in addition to gate insulator quality, and different type of measurements with unspecified delay between stress and measurement on different types of devices result in a very large spread in ΔV_T for DC stress as shown.

6.4.5 Comparative Analysis of Mode-A and Mode-B Stress

Figure 6.31 shows the time evolution of UF-MSM measured ΔV_T for Mode-A and Mode-B AC stress in HKMG p-MOSFETs for (a) PDC = 50 % and (b) PDC = 99 %, all other stress conditions, i.e., $V_{G\text{-HIGH}}$, $V_{G\text{-LOW}}$ and f are kept constant. Note that ΔV_T shows power-law time dependence for both stress modes.

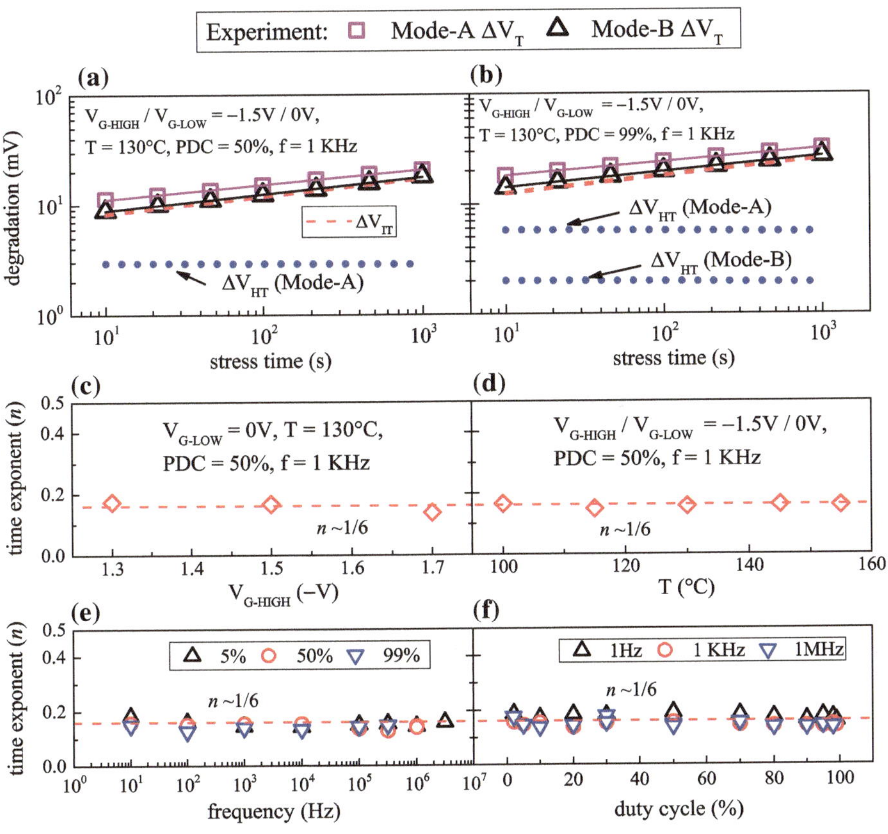

Fig. 6.31 Time evolution of UF-MSM measured ΔV_T for Mode-A and Mode-B AC NBTI stress in HKMG devices (*symbols*) at **a** low PDC and **b** high PDC. Overall model prediction and its underlying subcomponents are shown as *lines*. Power-law time exponent *n* for ΔV_T time evolution at longer t_{STR} for Mode-B AC stress, versus **c** pulse high ($V_{G\text{-HIGH}}$), **d** stress *T*, **e** frequency for different PDC and **f** duty cycle for different *f*

The Mode-B stress shows universal time exponent $n \sim 1/6$ for both PDC, while the magnitude of ΔV_{T} increases with PDC as expected. However, the Mode-A stress shows higher ΔV_{T} and lower n than Mode-B, and the difference increases for higher PDC stress.

Time evolution of ΔV_{IT} calculated using RD model and TTOM are also shown, and calculated ΔV_{IT} magnitude is quite close to measured ΔV_{T} for Mode-B stress at both PDC, and verifies that Mode-B stress is dominated by ΔV_{IT}. Note that due to high f, calculated ΔV_{IT} values for Mode-A stress is closer to Mode-B value and not shown. The difference between calculated ΔV_{IT} and measured ΔV_{T}, i.e., extracted ΔV_{HT} component is shown for Mode-A and Mode-B stress for both PDC. Note that ΔV_{HT} is non negligible for Mode-A stress and increases significantly for higher PDC. On the other hand, ΔV_{HT} is negligible for lower PDC and becomes visible only at higher PDC Mode-B stress, although, ΔV_{T} is always dominated by ΔV_{IT} as shown.

Time evolution of ΔV_{T} is measured using UF-MSM for different AC Mode-B stress conditions and power-law time exponent is extracted. Figure 6.31 also plots extracted time exponent n as a function of (c) $V_{G\text{-HIGH}}$, (d) stress T, (e) frequency and (f) PDC; the f dependence is studied at difference PDC and similarly, the PDC dependence is studied at different f. Note that the universal time exponent $n \sim 1/6$ is obtained for all stress conditions, which also matches well with that predicted using the TTOM augmented RD model solution for AC NBTI stress. Therefore, it is unequivocally established that ΔV_{IT} dominates ΔV_{T} for Mode-B AC stress for diverse experimental conditions.

Figure 6.32 shows UF-MSM measured ΔV_{T} at fixed t_{STR}, for AC Mode-A and Mode-B NBTI stress in HKMG p-MOSFETs as a function of (a, b) PDC and (c, d) frequency. Experiments are performed for different $V_{G\text{-LOW}}$, as shown in (a) and (c) for PDC and f dependence, respectively. Moreover, PDC dependence is studied at different f and the frequency dependence at different PDC, as shown in (b) and (d), respectively. Other stress parameters are kept constant. Note that PDC dependence of ΔV_{T} shows S shaped characteristics and a kink or jump between high PDC AC and DC stress. The AC to DC ratio at different PDC and the magnitude of the kink depends on f and $V_{G\text{-LOW}}$. Mode-B stress shows f independent ΔV_{T} for different PDC and $V_{G\text{-LOW}}$. Mode-A ΔV_{T} is larger compared to Mode-B ΔV_{T} at lower f and this holds true for different PDC and $V_{G\text{-LOW}}$. However, Mode-A ΔV_{T} reduces and eventually merges with Mode-B ΔV_{T} at higher f and shows f independence.

The composite modeling framework consisting of ΔV_{IT} and ΔV_{HT} subcomponents is used to calculate PDC and f dependence of ΔV_{T} as also shown in Fig. 6.32. Modeled ΔV_{T} for AC stress is obtained using identical parameters as used for DC stress and recovery, and calculated ΔV_{T} agrees well with measurements for such diverse experimental conditions as shown. Figure 6.33 shows calculated (a) ΔV_{IT} and (b) ΔV_{HT} subcomponents as a function of frequency, for Mode-A and Mode-B AC stress for low and high PDC. Note that Mode-B ΔV_{T} is primarily due to ΔV_{IT} and is calculated using TTOM augmented RD model. Mode-A ΔV_{T} is due to both ΔV_{IT} and ΔV_{HT} especially at lower f, ΔV_{IT} is calculated as mentioned before, while

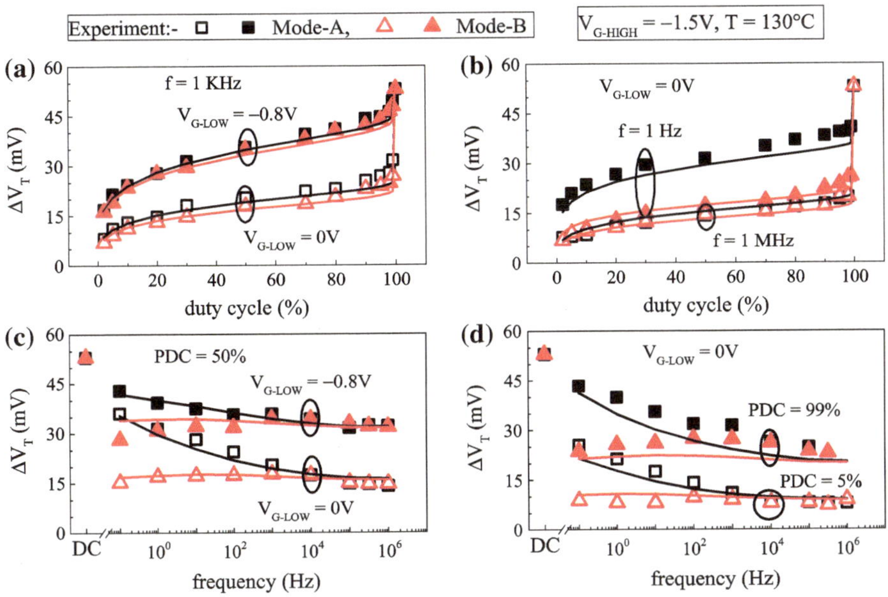

Fig. 6.32 UF-MSM measured ΔV_T for Mode-A and Mode-B AC stress in HKMG devices at fixed t_{STR} (*symbols*), versus (**a**, **b**) duty cycle and (**c**, **d**) frequency. (**a**, **c**) PDC and f dependence are studied for different $V_{G\text{-}LOW}$. **b** PDC dependence is studied for different f, and **d** frequency dependence is studied for different PDC. *Lines* represent composite model solution

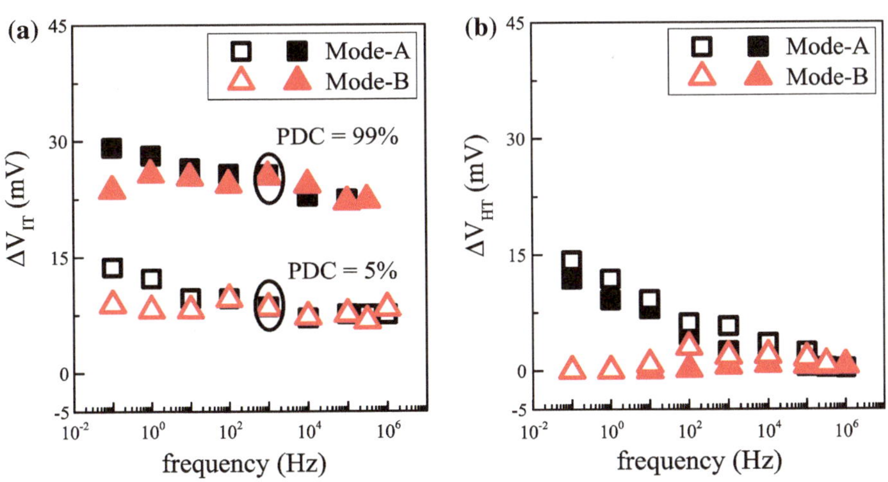

Fig. 6.33 Frequency dependence of model calculated **a** ΔV_{IT} and **b** ΔV_{HT} subcomponents at fixed t_{STR} for Mode-A and Mode-B AC NBTI stress in HKMG devices at different PDC

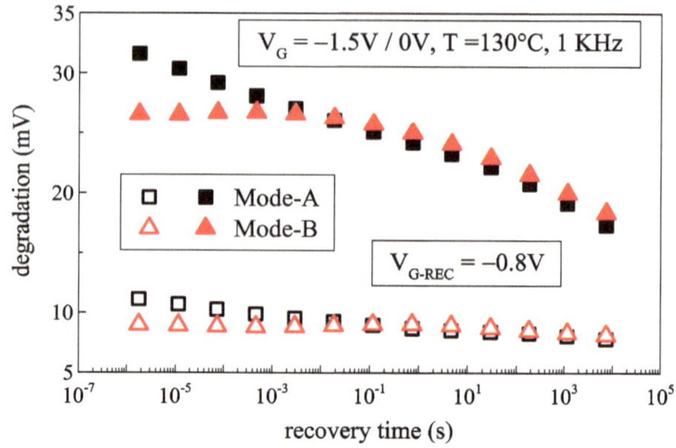

Fig. 6.34 Recovery of ΔV_T after Mode-A and Mode-B AC NBTI stress in HKMG devices at different PDC measured using UF-OTF method

ΔV_{HT} is calculated using the cyclostationary trapping expression shown in Table 6.5. However, the ΔV_{HT} subcomponent reduces at higher f for Mode-A stress, and both Mode-A and Mode-B stress at higher f is dominated by ΔV_{IT} as shown.

As an additional proof, Fig. 6.34 shows recovery of ΔV_T measured in HKMG p-MOSFETs using UF-OTF method, after AC Mode-A and Mode-B NBTI stress at low and high PDC. Note that a relatively higher $|V_{G\text{-}REC}|$ is used to minimize fast electron capture, which ensures recovery of generated traps primarily by the slower H_2 back diffusion induced passivation process. Note that Mode-B stress shows a delayed start of recovery, which is consistent with the trap passivation process and signifies the absence of hole detrapping. In contrast, recovery following Mode-A stress starts early due to detrapping of trapped holes. However, once the available holes get de-trapped, Mode-A ΔV_T recovery merges with that for Mode-B at longer time, and the recovery for both modes are governed by passivation of generated traps as per RD model.

6.5 Compact Model for DC and AC Degradation

The framework consisting of TTOM enabled RD model and cyclostationary hole trapping, discussed earlier in this chapter, is capable of providing the time evolution of ΔV_T for different AC stress conditions. It calculates degradation at the end of each on and off half cycles. It can provide ΔV_T at any t_{STR} within a particular on or off cycle of interest, which is done by noting degradation at the end of previous two half cycles for stress and previous three half cycles for recovery calculations. Although the framework is comprehensive and adheres to the underlying physical process, calculations can become computationally taxing especially for high f AC simulation over long t_{STR}.

In many situations, it is just sufficient to calculate ΔV_T at a fixed t_{STR} during AC stress, which can be done using the compact model shown in Table 6.6. ΔV_{IT} (DC) and ΔV_{HT} (DC) for a particular time can be calculated using the compact

Table 6.6 Compact model equations to calculate ΔV_{IT} and ΔV_{HT} at fixed t_{STR} for Mode-A and Mode-B AC stress

$\Delta V_{IT1} = F_{FAST}{}^{*}k_{DUTY}{}^{*}\Delta V_{HT,DC}{}^{*}(1 - \exp\{-(t_{OFF}/\tau_{EC})^{\beta_{EC}}\})*(\exp\{-(t_{ON}/\tau_{EE})^{\beta_{EE}}\})$	(6.22)
$\Delta V_{IT2} = (1 - F_{FAST})*k_{DUTY}{}^{*}\Delta V_{IT,DC}$	(6.23)
$\Delta V_{HT} = \Delta V^{*}_{HT,DC}(1 - \exp\{-(t_{ON}/\tau_{STR})^{\beta_{STR}}\})*(\exp\{-(t_{OFF}/\tau_{REC})^{\beta_{REC}}\})$	(6.24)
$\Delta V_{T} = \Delta V_{IT1} + \Delta V_{IT2} + \Delta V_{HT}$	(6.25)

model of Chap. 4, Table 4.1. For AC stress, ΔV_{IT} is divided into $\Delta V_{IT\text{-}1}$ and $\Delta V_{IT\text{-}2}$ depending on recovery via fast electron capture and H_2depassivation, respectively. Equation (6.22) calculates $\Delta V_{IT\text{-}1}$ by using stretched exponential relation for electron emission and capture, using parameters τ_{EC}, β_{EC} and τ_{EE}, β_{EE}, respectively. The on (t_{ON}) and off (t_{OFF}) time is calculated using $1/f$ * PDC and $1/f$ * (1 – PDC), respectively, f and PDC being the frequency and duty of applied gate pulse. Equation 6.23 calculates $\Delta V_{IT\text{-}2}$, the parameter k_{DUTY} is the PDC dependent DC to AC scaling factor for generated traps, is given by RD model solution, and can be approximated by using the analytical expression shown in (6.23). As before, F_{FAST} determines the fraction of traps that undergo recovery by electron capture and depends on $V_{G\text{-}LOW}$. Finally, hole trapping and detrapping (6.24) is calculated using stretched exponential equations, with parameters τ_{STR}, β_{STR} and τ_{REC}, β_{REC}, respectively.

Figure 6.35 plots UF-MSM measured ΔV_T at fixed t_{STR} for Mode-A and Mode-B AC stress in HKMG p-MOSFETs for different (a) PDC and (b) frequency. All other stress conditions are kept fixed. The dotted line shows compact model

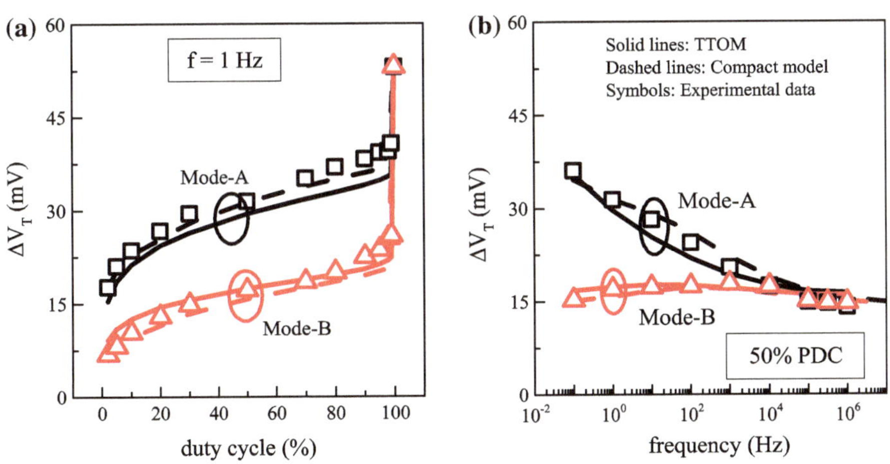

Fig. 6.35 UF-MSM measured ΔV_T at fixed t_{STR} for Mode-A and Mode-B AC NBTI stress in HKMG devices (*symbols*) as a function of **a** duty cycle and **b** frequency. *Lines* represent full transient simulation (*solid*) and compact model calculation (*dashed*)

calculation, while the solid line shows calculation using the TTOM enabled RD model with cyclostationary hole trapping. Note that both models agree quiet well, and the compact model can successfully predict the PDC and *f* dependence of AC Mode-A and Mode-B degradation. It is important to remark that the time constant related parameters for electron capture/emission and hole trapping/detrapping are different between the two models.

Real life circuits often work under random AC activity and hence the capability of the compact model needs to be tested for arbitrary pulse on and off conditions. In particular, the occupancy calculation of generated traps is of interest, as the trap generation subcomponent dominates overall degradation at long time and for AC stress. Figure 6.36 illustrates two different examples, with (a) larger total pulse on than off and (b) smaller total pulse on than off time. The solid line represents time evolution of ΔV_{IT} calculated using the composite framework consisting of TTOM enabled RD model. In this framework, calculation is performed over sequential on and off periods of the gate pulse to exactly track the gate activity. Note that a single pulse of equivalent duty can be used to represent the random gate activity as shown. The time evolution of ΔV_T calculated using the TTOM enabled RD model framework for this single effective pulse is also shown, and good agreement has been achieved at identical total simulation time. Finally, the dots represent compact model calculated DC and AC ΔV_{IT} at end of simulation time. Compact model calculation agrees well with the actual and effective composite model calculation as shown.

Figure 6.37 shows the time evolution of UF-MSM measured ΔV_T for (a) DC and (b) Mode-B AC NBTI stress in HKMG p-MOSFETs. Experiments are done at higher $V_{G\text{-}STR}$ (or $V_{G\text{-}HIGH}$) and T for a short time, and at lower $V_{G\text{-}STR}$ (or $V_{G\text{-}HIGH}$)

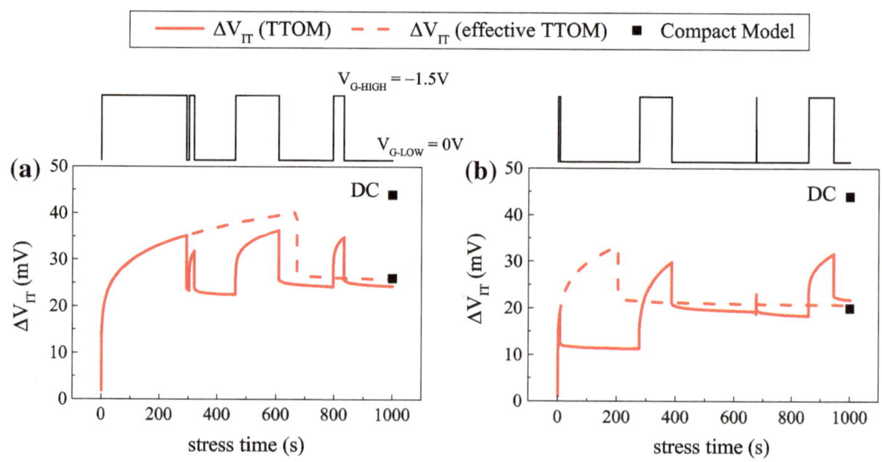

Fig. 6.36 Time evolution of simulated ΔV_{IT} using the transient occupancy enabled RD model for arbitrary stress and recovery sequences. *Data* are shown from exact time tracking (*solid*) and effective duty cycle (*dashed*) calculation. The fixed-time compact model calculation is also shown (*symbols*)

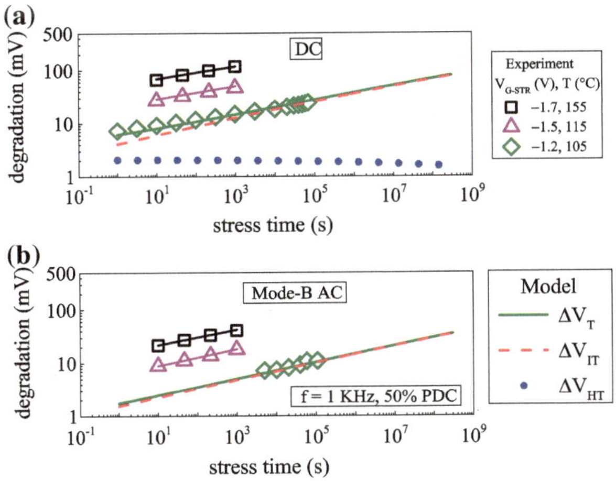

Fig. 6.37 Time evolution of UF-MSM measured ΔV_T for **a** DC and **b** Mode-B AC NBTI stress in HKMG devices (*symbols*). *Solid lines* represent model calculation. Parameters are calibrated by fitting shorter t_{STR} data at high $V_{G\text{-}STR}$ and high stress T, and predictive capability of the calibrated model is tested against longer t_{STR} data at lower $V_{G\text{-}STR}$ and lower stress T. The underlying ΔV_{IT} and ΔV_{HT} subcomponents are shown

and T for moderately long time, for both DC and AC stress. The DC compact model shown in Chap. 4, Table 4.1 is calibrated using measured short-time DC data at higher $V_{G\text{-}STR}$ and T, and the calibrated compact model is used to predict long time measured data at lower $V_{G\text{-}STR}$ and T, and also extrapolated to end-of-life time of 10 years. The calculated ΔV_{IT} and ΔV_{HT} contributions are also shown for the lower $V_{G\text{-}STR}$ and T condition.

Note that time evolution of ΔV_T for Mode-B AC stress is dominated by ΔV_{IT} as discussed earlier in this chapter. Also note that the time evolution of ΔV_{IT} shows power-law dependence with identical $n \sim 1/6$ for both DC and AC stress. The AC to DC scaling factor for ΔV_{IT}, governed by fast electron capture and trap passivation, can be calculated using the compact model shown in Table 6.6 for a particular AC f and PDC. The DC compact model calculated time evolution of ΔV_{IT} is scaled by this factor to calculate the time evolution of ΔV_T for AC stress at different $V_{G\text{-}HIGH}$ and T, since ΔV_{HT} is negligible for Mode-B AC stress. The model agrees well with measured data as shown in Fig. 6.37b. The extrapolated end-of-life degradation is also calculated.

The above analysis has been performed on UF-MSM measured time evolution of ΔV_T for DC and Mode-B AC stress in different HKMG devices D1 through D4 shown in Chap. 3, Fig. 3.2. Figure 6.38 shows the impact of EOT scaling on (a) extrapolated ΔV_T at 10 years, and on (b) time to reach a particular ΔV_T, for DC and AC stress. Note that Mode-B AC stress is dominated by ΔV_{IT}, while both ΔV_{IT} and ΔV_{HT} contributes to DC stress. As a reference, the 10 year extrapolated ΔV_{IT} value is also shown for DC stress in Fig. 6.38a. Note that although ΔV_{HT} contribution is important for shorter time DC stress, end-of-life ΔV_T is completely

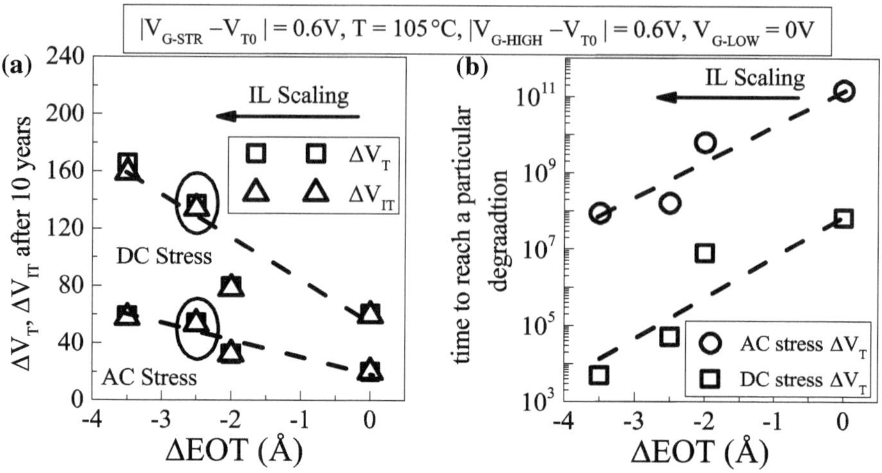

Fig. 6.38 **a** Estimated ΔV_T and ΔV_{IT} at end of life (10 years) using the calibrated compact model for DC and Mode-B AC NBTI stress in HKMG devices having different IL thickness leading to different EOT. **b** Time to reach a particular degradation obtained from calculations shown in (**a**)

dominated by ΔV_{IT} as shown. Therefore, ΔV_{IT} governs NBTI limited lifetime for both DC and AC stress. The degradation increases and corresponding time to reach a particular degradation reduces as EOT is scaled, calculations are done at fixed overdrive and T for different devices. As explained in Chap. 4, enhanced degradation for scaled EOT devices is due to increase in stress E_{OX}, and also due to the presence of higher N content in the gate stack for deeply scaled devices. A relief is obtained for Mode-B AC stress as shown.

Before concluding, it is important to remark that the AC to DC ratio for ΔV_{IT} is governed by trap passivation and electron capture/emission processes. Note that electron capture is a fast process having very short τ_{EC}. However, in certain situations such as very large f and high enough PDC, it is indeed possible that the pulse off phase is not long enough for complete electron capture. Therefore, electron capture will not be efficient if ΔV_T is measured immediately after the ultra-short pulse off phase without the gate spending additional time at lower V_G before measurement. In such situation, ΔV_T is likely to increase at very high f, which is consistent with recently reported results [95]. Therefore, very high f AC measurements need careful attention.

6.6 Summary

A comprehensive modeling framework is proposed to explain DC and AC NBTI degradation. The framework uses generation of interface traps calculated using the Reaction-Diffusion model, together with hole trapping in pre-existing bulk insulator traps and generation of new bulk insulator traps, calculated using empirical expressions, to calculate overall degradation. Model calculation agrees well with experimental time evolution of degradation, measured by using ultra-fast techniques, for DC stress in differently processed SiON and HKMG p-MOSFETs. The model predicts measured degradation from very short to long stress time under different stress conditions, with fixed and process dependent parameters that are consistent with the compact model parameters listed in Chap. 4.

Detrapping of trapped holes in pre-existing and generated bulk insulator traps, as well as reduction of charges associated with generated interface traps, is used to model recovery of NBTI degradation after DC stress. Hole detrapping is computed using analytical expressions. Generated interface trap recovers either by electron capture if the trap goes below the Fermi level as stress bias is reduced, or via trap passivation if the trap stays above the Fermi level. Electron capture is empirically computed, while trap passivation is calculated using the modified RD model that takes into account stochastic effects. The framework is used to explain measured recovery data obtained using ultra-fast methods in differently processed SiON devices for different stress conditions. It is also used to predict ultra-fast measured data in HKMG devices, a wide variety of stress and recovery conditions. Calculated recovery using consistent set of model parameters (as stress) agrees well with measurements for diverse set of experimental conditions.

The DC stress and recovery framework is extended to calculate degradation for successive DC stress and recovery phases and also for Mode-A and Mode-B AC stress. A Transient Trap Occupancy Model is used to calculate the occupancy of generated interface traps obtained using RD model simulation, and a cyclostationary empirical model is used for hole trapping. The framework is used to predict measured degradation under multi-cycle DC and AC stress in HKMG devices for different experimental conditions. Multi-cycle DC data for arbitrary stress and recovery phases and AC data for different pulse frequency and duty is successfully explained, with model parameters that are fully consistent with single DC stress and recovery experiments. The accuracy of trap occupancy calculation is verified using measured data at different recovery or pulse low biases.

Finally, a compact model has been proposed and verified against the composite modeling framework for AC stress under diverse experimental conditions including arbitrary gate activity. The compact model predicts identical DC to AC ratio as the comprehensive model, and has been used to predict end-of-life degradation for HKMG devices having different EOT. It has been shown that EOT scaling results in enhanced NBTI, and moreover, generation of interface traps completely dominates degradation at end-of-life for both DC and AC stress.

Acknowledgment The authors would like to acknowledge Sandip De, Rajan Pandey and K.V.R. M. Murali for DFT calculations, Vrajesh Maheta for NBTI measurements in SiON devices, Subhadeep Mukhopadhyay for DCIV and Bijesh Rajamohanan for flicker noise measurements in HKMG devices, Tejas Naphade for his contribution and useful discussion, Applied Materials for devices used in this study, and Ankush Chaudhary for editorial support.

References

1. B.E. Deal, M. Sklar, A.S. Grove, E.H. Snow, Characteristics of the surface state charge (Qss) of thermally oxidized silicon. J. Electrochem. Soc. **114**, 266 (1967)
2. N. Kimizuka, K. Yamaguchi, K. Imai, T. Iizuka, C.T. Liu, R.C. Keller, T. Horiuchi, NBTI enhancement by nitrogen incorporation into ultrathin gate oxide for 0.10-μm gate CMOS generation, in *Symposium on VLSI Technology: Digest of Technical Papers*, p. 92 (2000)
3. S. Tsujikawa, T. Mine, K. Watanabe, Y. Shimamoto, R. Tsuchiya, K. Ohnishi, T. Onai, J. Yugami, S. Kimura, Negative bias temperature instability of pMOSFETs with ultra-thin SiON gate dielectrics, in *IEEE International Reliability Physics Symposium Proceedings*, p. 183 (2003)
4. V. Huard, M. Denais, F. Perrier, N. Revil, C. Parthasarathy, A. Bravaix, E. Vincent, A thorough investigation of MOSFETs NBTI degradation. Microelectron. Reliab. **45**, 83 (2005)
5. S. Mahapatra, K. Ahmed, D. Varghese, A.E. Islam, G. Gupta, L. Madhav, D. Saha, M.A. Alam, On the physical mechanism of NBTI in silicon oxynitride p-MOSFETs: can differences in insulator processing conditions resolve the interface trap generation versus hole trapping controversy?, in *IEEE International Reliability Physics Symposium Proceedings*, p. 1 (2007)
6. Y. Mitani, H. Satake, A. Toriumi, Influence of nitrogen on negative bias temperature instability in Ultrathin SiON. IEEE Trans. Device Mater. Reliab. **8**, 6 (2008)
7. V.D. Maheta, C. Olsen, K. Ahmed, S. Mahapatra, The impact of nitrogen engineering in silicon oxynitride gate dielectric on negative-bias temperature instability of p-MOSFETs: a study by ultrafast on-the-fly I_{DLIN} technique. IEEE Trans. Electron Devices **55**, 1630 (2008)
8. M. Rafik, X. Garros, G. Ribes, G. Ghibaudo, C. Hobbs, A. Zauner, M. Muller, V. Huard, C. Ouvard, Impact of TiN Metal gate on NBTI assessed by interface states and fast transient effect characterization, in *IEEE International Electron Devices Meeting Technical Digest*, p. 825 (2007)
9. S. Pae, M. Agostinelli, M. Brazier, R. Chau, G. Dewey, T. Ghani, M. Hattendorf, J. Hicks, J. Kavalieros, K. Kuhn, M. Kuhn, J. Maiz, M. Metz, K. Mistry, C. Prasad, S. Ramey, A. Roskowski, J. Sandford, C. Thomas, J. Thomas, C. Wiegand, J. Wiedemer, BTI reliability of 45 nm high-K + metal-gate process technology, in *IEEE International Reliability Physics Symposium Proceedings*, p. 352 (2008)
10. S. Deora, V.D. Maheta, G. Bersuker, C. Olsen, K.Z. Ahmed, R. Jammy, S. Mahapatra, A Comparative NBTI study of HfO_2/HfSiOx, and SiON p-MOSFETs using UF-OTF I_{DLIN} technique. IEEE Electron Device Lett. **30**, 152 (2009)
11. X. Garros, M. Casse, C. Fenouillet-Beranger, G. Reimbold, F. Martin, C. Gaumer, C. Wiemer, M. Perego, F. Boulanger, Detrimental impact of technological processes on BTI reliability of advanced high-K/metal gate stacks, in *IEEE International Reliability Physics Symposium Proceedings*, p. 362 (2009)
12. S. Pae, A. Ashok, T. Ghani, K. Lemay, M. Liu, R. Lu, P. Packan, C. Parker, R. Purser, A. St. Amour, B. Woolery, Reliability characterization of 32 nm high-K and Metal-Gate logic transistor technology, in *IEEE International Reliability Physics Symposium Proceedings*, p. 287 (2010)
13. E. Cartier, A. Kerber, T. Ando, M.M. Frank, K. Choi, S. Krishnan, B. Linder, K. Zhao, F. Monsieur, J. Stathis, V. Narayanan, Fundamental aspects of HfO_2-based high-k metal gate stack reliability and implications on tinv-scaling, in *IEEE International Electron Devices Meeting Technical Digest*, p. 18.4.1 (2011)
14. S. Krishnan, V. Narayanan, E. Cartier, D. Ioannou, K. Zhao, T. Ando, U. Kwon, B. Linder, J. Stathis, M. Chudzik, A. Kerber, K. Choi, Bias temperature instability in High-κ/metal gate transistors—Gate stack scaling trends, in *IEEE International Reliability Physics Symposium Proceedings*, p. 5A.1.1 (2012)
15. K. Joshi, S. Hung, S. Mukhopadhyay, V. Chaudhary, N. Nanaware, B. Rajamohnan, T. Sato, M. Bevan, A. Wei, A. Noori, B. McDougal, C. Ni, G. Saheli, C. Lazik, P. Liu, D. Chu, L. Date, S. Datta, A. Brand, J. Swenberg, S. Mahapatra, HKMG process impact on N, P BTI:

Role of thermal IL scaling, IL/HK integration and post HK nitridation, in *IEEE International Reliability Physics Symposium Proceedings*, p. 4C.2.1 (2013)

16. W. McMahon, C. Tian, S. Uppal, H. Kothari, M. Jin, G. LaRosa, T. Nigam, A. Kerber, B. P. Linder, E. Cartier, W.L. Lai, Y. Liu, R. Ramachandran, U. Kwon, B. Parameshwaran, S. Krishnan, V. Narayanan, Intrinsic dielectric stack reliability of a high performance bulk planar 20 nm replacement gate high-k metal gate technology and comparison to 28 nm gate first high-k metal gate process, in *IEEE International Reliability Physics Symposium Proceedings*, p. 4C.4.1 (2013)
17. S. Ramey, A. Ashutosh, C. Auth, J. Clifford, M. Hattendorf, J. Hicks, R. James, A. Rahman, V. Sharma, A. St. Amour, C. Wiegand, Intrinsic transistor reliability improvements from 22 nm tri-gate technology, in *IEEE International Reliability Physics Symposium Proceedings*, p. 4C.5.1 (2013)
18. A.T. Krishnan, V. Reddy, S. Chakravarthi, J. Rodriguez, S. John, S. Krishnan, NBTI impact on transistor and circuit: models, mechanisms and scaling effects [MOSFETs], in *IEEE International Electron Devices Meeting Technical Digest*, p. 14.5.1 (2003)
19. V. Reddy, J. Carulli, A. Krishnan, W. Bosch, B. Burgess, Impact of negative bias temperature instability on product parametric drift, in *International Test Conference*, p. 148 (2004)
20. A. Haggag, G. Anderson, S. Parohar, D. Burnett, G. Abeln, J. Higman, M. Moosa, Understanding SRAM high-temperature-operating-life NBTI: statistics and permanent vs recoverable damage, in *IEEE International Reliability Physics Symposium Proceedings*, p. 452 (2007)
21. V. Huard, R. Chevallier, C. Parthasarathy, A. Mishra, N. Ruiz-Amador, F. Persin, V. Robert, A. Chimeno, E. Pion, N. Planes, D. Ney, F. Cacho, N. Kapoor, V. Kulshrestha, S. Chopra, N. Vialle, Managing SRAM reliability from bitcell to library level, in *IEEE International Reliability Physics Symposium Proceedings*, p. 655 (2010)
22. A.T. Krishnan, F. Cano, C. Chancellor, V. Reddy, Z. Qi, P. Jain, J. Masin, S. Zuhoski, S. Krishnan, J. Ondrusek, Product drift from NBTI : guardbanding, circuit and statistical effects, in *IEEE International Electron Devices Meeting Technical Digest*, p. 4.3.1 (2010)
23. D.P. Ioannou, S. Mittl, D. Brochu, Burn-in stress induced BTI degradation and post-burn-in high temperature anneal (Bake) effects in advanced HKMG and oxynitride based CMOS ring oscillators, in *IEEE International Reliability Physics Symposium Proceedings*, p. 5C.2.1 (2012)
24. S. Mahapatra, V. Huard, A. Kerber, V. Reddy, S. Kalpat, A. Haggag, Universality of NBTI—From devices to circuits and products, in *IEEE International Reliability Physics Symposium Proceedings*, p. 3B.1.1 (2014)
25. S. Mahapatra, N. Goel, S. Desai, S. Gupta, B. Jose, S. Mukhopadhyay, K. Joshi, A. Jain, A.E. Islam, M.A. Alam, A comparative study of different physics-based NBTI models. IEEE Trans. Electron Devices **60**, 901 (2013)
26. S. Rangan, N. Mielke, E.C. C. Yeh, Universal recovery behavior of negative bias temperature instability [PMOSFETs], in *IEEE International Electron Devices Meeting Technical Digest*, p. 14.3.1 (2003)
27. H. Reisinger, O. Blank, W. Heinrigs, A. Muhlhoff, W. Gustin, C. Schlunder, Analysis of NBTI Degradation- and Recovery-Behavior Based on Ultra Fast V_T-Measurements, in *IEEE International Reliability Physics Symposium Proceedings*, p. 448 (2006)
28. T. Grasser, W. Gos, V. Sverdlov, B. Kaczer, The Universality of NBTI relaxation and its implications for modeling and characterization, in *IEEE International Reliability Physics Symposium Proceedings*, p. 268 (2007)
29. B. Kaczer, T. Grasser, J. Roussel, J. Martin-Martinez, R. O'Connor, B.J. O'Sullivan, G. Groeseneken, Ubiquitous relaxation in BTI stressing—New evaluation and insights, in *IEEE International Reliability Physics Symposium Proceedings*, p. 20 (2008)
30. C. Shen, M.-F. Li, C.E. Foo, T. Yang, D.M. Huang, A. Yap, G.S. Samudra, Y.-C. Yeo, Characterization and physical origin of fast Vth transient in NBTI of pMOSFETs with SiON dielectric, in *IEEE International Electron Devices Meeting Technical Digest*, doi:10.1109/IEDM.2006.346776 (2006)

31. E.N. Kumar, V.D. Maheta, S. Purawat, A.E. Islam, C. Olsen, K. Ahmed, M.A. Alam, S. Mahapatra, Material dependence of NBTI physical mechanism in silicon oxynitride (SiON) p-MOSFETs: a comprehensive study by ultra-fast on-the-fly (UF-OTF) I_{DLIN} technique, in *IEEE International Electron Devices Meeting Technical Digest*, p. 809 (2007)
32. N. Goel, N. Nanaware, S. Mahapatra, Ultrafast AC–DC NBTI characterization of deep IL scaled HKMG p-MOSFETs. IEEE Electron Device Lett. **34**, 1476 (2013)
33. A.E. Islam, H. Kufluoglu, D. Varghese, S. Mahapatra, M.A. Alam, Recent issues in negative-bias temperature instability: initial degradation, field dependence of interface trap generation, hole trapping effects, and relaxation. IEEE Trans. Electron Devices **54**, 2143 (2007)
34. D. Varghese, D. Saha, S. Mahapatra, K. Ahmed, F. Nouri, M. Alam, On the dispersive versus arrhenius temperature activation of NBTI time evolution in plasma nitrided gate oxides: measurements, theory, and implications, in *IEEE International Electron Devices Meeting Technical Digest*, p. 684 (2005)
35. S. Mahapatra, P. Bharath Kumar, M.A. Alam, Investigation and modeling of interface and bulk trap generation during negative bias temperature instability of p-MOSFETs. IEEE Trans. Electron Devices **51**, 1371 (2004)
36. C.L. Chen, Y.M. Lin, C.J. Wang, K. Wu, A new finding on NBTI lifetime model and an investigation on NBTI degradation characteristic for 1.2 nm ultra thin oxide, in *IEEE International Reliability Physics Symposium Proceedings*, p. 704 (2005)
37. A. Islam, G. Gupta, S. Mahapatra, A.T. Krishnan, K. Ahmed, F. Nouri, A. Oates, M.A. Alam, Gate leakage vs. NBTI in plasma nitrided oxides: characterization, physical principles, and optimization, in *IEEE International Electron Devices Meeting Technical Digest*, doi:10.1109/IEDM.2006.346775 (2006)
38. H. Aono, E. Murakami, K. Okuyama, A. Nishida, M. Minami, Y. Ooji, K. Kubota, Modeling of NBTI degradation and its impact on electric field dependence of the lifetime, in *IEEE International Reliability Physics Symposium Proceedings*, p. 25 (2004)
39. A. Haggag, K. Forbes, G. Anderson, D. Burnett, P. Abramowitz, M. Moosa, Product failures: power-law or exponential voltage dependence? in *IEEE International Reliability Physics Symposium Proceedings*, p. 125 (2010)
40. A.T. Krishnan, C. Chancellor, S. Chakravarthi, P.E. Nicollian, V. Reddy, A. Varghese, R.B. Khamankar, S. Krishnan, Material dependence of hydrogen diffusion: implications for NBTI degradation, in *IEEE International Electron Devices Meeting Technical Digest*, p. 688 (2005)
41. B. Kaczer, V. Arkbipov, R. Degraeve, N. Collaert, G. Groeseneken, M. Goodwin, Disorder-controlled-kinetics model for negative bias temperature instability and its experimental verification, in *IEEE International Reliability Physics Symposium Proceedings*, p. 381 (2005)
42. N. Goel, K. Joshi, S. Mukhopadhyay, N. Nanaware, S. Mahapatra, A comprehensive modeling framework for gate stack process dependence of DC and AC NBTI in SiON and HKMG p-MOSFETs. Microelectron. Reliab. **54**, 491 (2014)
43. Y.S. Tsai, N.K. Jha, Y.H. Lee, R. Ranjan, W. Wang, J.R. Shih, M.J. Chen, J.H. Lee, K. Wu, Prediction of NBTI degradation for circuit under AC operation, in *IEEE International Reliability Physics Symposium Proceedings*, p. 665 (2010)
44. R. Fernández, B. Kaczer, A. Nackaerts, S. Demuynck, R. Rodríguez, M. Nafría, G. Groeseneken, AC NBTI studied in the 1 Hz–2 GHz range on dedicated on-chip CMOS circuits, in *IEEE International Electron Devices Meeting Technical Digest*, doi:10.1109/IEDM.2006.346777 (2006)
45. T. Grasser, B. Kaczer, W. Goes, An energy-level perspective of bias temperature instability, in *IEEE International Reliability Physics Symposium Proceedings*, p. 28 (2008)
46. A.E. Islam, S. Mahapatra, S. Deora, V.D. Maheta, M.A. Alam, On the differences between ultra-fast NBTI measurements and Reaction-Diffusion theory, in *IEEE International Electron Devices Meeting Technical Digest*, doi:10.1109/IEDM.2009.5424236 (2009)

47. H. Reisinger, T. Grasser, W. Gustin, C. Schlunder, The statistical analysis of individual defects constituting NBTI and its implications for modeling DC- and AC-stress, in *IEEE International Reliability Physics Symposium Proceedings*, p. 7 (2010)
48. S. Mahapatra, A.E. Islam, S. Deora, V.D. Maheta, K. Joshi, A. Jain, M.A. Alam, A critical re-evaluation of the usefulness of R-D framework in predicting NBTI stress and recovery, in *International Reliability Physics Symposium Proceedings*, p. 6A.3.1 (2011)
49. W.J. Liu, Z.Y. Liu, D. Huang, C.C. Liao, L.F. Zhang, Z.H. Gan, W. Wong, C. Shen, M.-F. Li, On-the-fly interface trap measurement and its impact on the understanding of NBTI mechanism for p-MOSFETs with SiON gate dielectric, in *IEEE International Electron Devices Meeting Technical Digest*, p. 813 (2007)
50. A. Neugroschel, G. Bersuker, R. Choi, Applications of DCIV method to NBTI characterization. Microelectron. Reliab. **47**, 1366 (2007)
51. V. Huard, Two independent components modeling for Negative Bias Temperature Instability, in *IEEE International Reliability Physics Symposium Proceedings*, p. 33 (2010)
52. G. Chen, M.-F. Li, C.H. Ang, J.Z. Zheng, D.L. Kwong, Dynamic NBTI of p-MOS transistors and its impact on MOSFET scaling. IEEE Electron Device Lett. **23**, 734 (2002)
53. J.H. Stathis, G. La Rosa, A. Chou, Broad energy distribution of NBTI-induced interface states in p-MOSFETs with ultra-thin nitrided oxide, in *IEEE International Reliability Physics Symposium Proceedings*, doi:10.1109/RELPHY.2004.1315292 (2004)
54. J.P. Campbell, P.M. Lenahan, A.T. Krishnan, S. Krishnan, NBTI: an atomic-scale defect perspective, in *IEEE International Reliability Physics Symposium Proceedings*, p. 442 (2006)
55. S. Mukhopadhyay, K. Joshi, V. Chaudhary, N. Goel, S. De, R.K. Pandey, K.V.R.M. Murali, S. Mahapatra, NBTI: an atomic-scale defect perspective, in *IEEE International Reliability Physics Symposium Proceedings*, p. GD.3.1 (2014)
56. G. Kapila, N. Goyal, V.D. Maheta, C. Olsen, K. Ahmed, S. Mahapatra, A comprehensive study of flicker noise in plasma nitride SiON p-MOSFETs: process dependence of pre-existing and NBTI stress generated trap distribution profiles, in *IEEE International Electron Devices Meeting Technical Digest*, doi:10.1109/IEDM.2008.4796625 (2008)
57. S. Mahapatra, S. De, K. Joshi, S. Mukhopadhyay, R.K. Pandey, K.V.R.M. Murali, Understanding process impact of hole traps and NBTI in HKMG p-MOSFETs using measurements and atomistic simulations. IEEE Electron Device Lett. **8**, 963 (2013)
58. B. Kaczer, T. Grasser, J. Martin-Martinez, E. Simoen, M. Aoulaiche, P.J. Roussel, G. Groeseneken, NBTI from the perspective of defect states with widely distributed time scales, in *IEEE International Reliability Physics Symposium Proceedings*, p. 26 (2009)
59. D.S. Ang, S. Wang, Recovery of the NBTI-stressed ultrathin gate p-MOSFET: the role of deep-level hole traps. IEEE Electron Device Lett. **27**, 914 (2006)
60. D. Ielmini, M. Manigrasso, F. Gattel, G. Valentini, A new NBTI model based on hole trapping and structural relaxation in MOS dielectrics. IEEE Trans. Electron Device **56**, 1943 (2009)
61. H. Reisinger, T. Grasser, K. Ermisch, H. Nielen, W. Gustin, C. Schlunder, Understanding and modeling AC BTI, in *IEEE International Reliability Physics Symposium Proceedings*, p. 6A.1.1 (2011)
62. T. Grasser, B. Kaczer, H. Reisinger, P.J. Wagner, M. Toledano-Luque, On the frequency dependence of the bias temperature stability, in *IEEE International Reliability Physics Symposium Proceedings*, p. XT.8.1 (2012)
63. T. Grasser, Stochastic charge trapping in oxides: from random telegraph noise to bias temperature instabilities. Microelectron. Reliab. **52**, 39 (2012)
64. T. Grasser, B. Kaczer, W. Goes, T. Aichinger, P. Hehenberger, M. Nelhiebel, A two stage model for negative bias temperature instability, in *IEEE International Reliability Physics Symposium Proceedings*, p. 33 (2009)
65. K. Sakuma, D. Matsushita, K. Muraoka, Y. Mitani, Investigation of nitrogen-originated NBTI mechanism in SiON with high-nitrogen concentration, in *IEEE International Reliability Physics Symposium Proceedings*, p. 454 (2006)

66. J.H. Lee, W.H. Wu, A.E. Islam, M.A. Alam, A.S. Oates, Separation method of hole trapping and interface trap generation and their roles in NBTI Reaction-Diffusion model, in *IEEE International Reliability Physics Symposium Proceedings*, p. 745 (2008)
67. S. Mahapatra, V.D. Maheta, A.E. Islam, M.A. Alam, Isolation of NBTI stress generated interface trap and hole-trapping components in PNO p-MOSFETs. IEEE Trans. Electron Devices **56**, 236 (2009)
68. M. Jo, M. Chang, S. Kim, H.S. Jung, R. Choi, H. Hwang, Contribution of interface states and oxide traps to the negative bias temperature instability of high-k pMOSFETs. IEEE Electron Device Lett. **30**, 291 (2009)
69. T. Grasser, P. Wagner, H. Reisinger, T. Aichinger, G. Pobegen, M. Nelhiebel, B. Kaczer, Analytic modeling of the bias temperature instability using capture/emission time maps, in *IEEE International Electron Devices Meeting Technical Digest*, p. 27.4.1 (2011)
70. D. Varghese, PhD Dissertation, *Multi-probe experimental and 'bottom-up' computational analysis of correlated defect generation in modern nanoscale transistors* (Purdue University, West Lafayette, 2009)
71. J. Yang, M. Masuduzzaman, K. Joshi, S. Mukhopadhyay, J. Kang, S. Mahapatra, M.A. Alam, Intrinsic correlation between PBTI and TDDB degradations in nMOS HK/MG dielectrics, in *IEEE International Reliability Physics Symposium Proceedings*, p. 5D.4.1 (2012)
72. K. Joshi, S. Mukhopadhyay, N. Goel, S. Mahapatra, A consistent physical framework for N and P BTI in HKMG MOSFETs, in *IEEE International Reliability Physics Symposium Proceedings*, p. 5A.3.1 (2012)
73. K. Joshi, S. Mukhopadhyay, N. Goel, N. Nanaware, S. Mahapatra, A detailed study of gate insulator process dependence of NBTI using a compact model. IEEE Trans. Electron Devices **61**, 408 (2014)
74. S. Tsujikawa, J. Yugami, Positive charge generation due to species of hydrogen during NBTI phenomenon in pMOSFETs with ultra-thin SiON gate dielectrics. Microelectron. Reliab. **45**, 65 (2005)
75. T. Grasser, K. Rott, H. Reisinger, M. Waltl, P. Wagner, F. Schanovsky, W. Goes, G. Pobegen, B. Kaczer, Hydrogen-related volatile defects as the possible cause for the recoverable component of NBTI, in *IEEE International Electron Devices Meeting Technical Digest*, p. 9 (2013)
76. T. Grasser, M. Waltl, W. Goes, Y. Wimmer, A.M. El-Sayed, A.L. Shluger, B. Kaczer, On the volatility of oxide defects: activation, deactivation, and transformation', in *IEEE International Reliability Physics Symposium Proceedings*, p. 5A.3 (2015)
77. K.O. Jeppson, C.M. Svensson, Negative bias stress of MOS devices at high electric-fields and degradation of MNOS devices. J. Appl. Phys. **48**, 2004 (1977)
78. M.A. Alam, A critical examination of the mechanics of dynamic NBTI for PMOSFETs, in *IEEE International Electron Devices Meeting Technical Digest*, p. 14.4.1 (2003)
79. S. Chakravarthi, A. Krishnan, V. Reddy, C.F. Machala, S. Krishnan, A comprehensive framework for predictive modeling of negative bias temperature instability, in *IEEE International Reliability Physics Symposium Proceedings*, p. 273 (2004)
80. M.A. Alam, S. Mahapatra, A comprehensive model of PMOS NBTI degradation. Microelectron. Reliab. **45**, 71 (2005)
81. M. Alam, H. Kufluoglu, D. Varghese, S. Mahapatra, A comprehensive model for PMOS NBTI degradation: recent progress. Microelectron. Reliab. **47**, 853 (2007)
82. F. Schanovsky, T. Grasser, On the microscopic limit of the modified reaction-diffusion model for the negative bias temperature instability, in *IEEE International Reliability Physics Symposium Proceedings*, p. XT.10.1 (2012)
83. T. Naphade, N. Goel, P.R. Nair, S. Mahapatra, Investigation of stochastic implementation of reaction diffusion (RD) models for NBTI related interface trap generation, in IEEE international reliability physics symposium proceedings, p. XT.5.1 (2013)
84. T. Grasser, K. Rott, H. Reisinger, M. Waltl, F. Schanovsky, B. Kaczer, NBTI in nanoscale MOSFETs—The ultimate modeling benchmark. IEEE Trans. Electron Devices **61**, 358 (2014)

85. S. Desai, S. Mukhopadhyay, N. Goel, N. Nanaware, B. Jose, K. Joshi, S. Mahapatra, A comprehensive AC/DC NBTI model: stress, recovery, frequency, duty cycle and process dependence, in *IEEE International Reliability Physics Symposium Proceedings*, p. XT.2.1 (2013)
86. N. Goel, S. Mukhopadhyay, N. Nanaware, S. De, R.K. Pandey, K.V.R.M. Murali, S. Mahapatra, A comprehensive DC/AC model for ultra-fast NBTI in deep EOT scaled HKMG p-MOSFETs, in *IEEE International Reliability Physics Symposium Proceedings*, p. 6A.4.1 (2014)
87. N. Goel, T. Naphade, S. Mahapatra, Combined trap generation and transient trap occupancy model for time evolution of NBTI during DC multi-cycle and AC atress, in *IEEE International Reliability Physics Symposium Proceedings*, p. 4A.3 (2015)
88. S. Gupta, B. Jose, K. Joshi, A. Jain, M. A. Alam, S. Mahapatra, A comprehensive and critical re-assessment of 2-stage energy level NBTI model, in *IEEE International Reliability Physics Symposium Proceedings*, p. XT.3.1 (2012)
89. W. Tsai, L.-A. Ragnarsson, L. Pantisano, P.J. Chen, B. Onsia, T. Schram, E. Cartier, A. Kerber, E. Young, M. Caymax, S. De Gendt, M. Heyns, Performance comparison of sub 1 nm sputtered TiN/HfO_2 nMOS and pMOSFETs, in *IEEE International Electron Devices Meeting Technical Digest*, p. 13.2.1 (2003)
90. M.J. Bevan, R. Curtis, T. Guarini, W. Liu, S.C.H. Hung, H. Graoui, Ultrathin SiO_2 interface layer growth, in *International Conference on Advanced Thermal Processing of Semiconductors (RTP)*, p. 154 (2010)
91. P.A. Kraus, K.Z. Ahmed, C.S. Olsen, F. Nouri, Physical models for predicting plasma nitrided Si–O–N gate dielectric properties from physical metrology. IEEE Electron Device Lett. **24**, 559 (2003)
92. C. Olsen, Two-step Post Nitridation Annealing for Lower EOT Plasma Nitrided Gate Dielectrics, in *WO2004081984* p. A2 (2004)
93. M.L. Reed, J.D. Plummer, Chemistry of Si–SiO_2 interface trap annealing. J. Appl. Phys. **63**, 5776 (1988)
94. A.E. Islam, E.N. Kumar, H. Das, S. Purawat, V. Maheta, H. Aono, E. Murakami, S. Mahapatra, M.A. Alam, Theory and practice of on-the-fly and ultra-fast V_T measurements for NBTI degradation: challenges and opportunities, in *IEEE International Electron Devices Meeting Technical Digest*, p. 805 (2007)
95. M.-H. Hsieh, D. Maji, Y.-C. Huang, T.-Y. Yew, W. Wang, Y.-H. Lee, J.R. Shih, K. Wu, Frequency dependence of NBTI in high-k/metal-gate technology, in *IEEE International Reliability Physics Symposium Proceedings*, p. XT.3.1 (2014)

Erratum to: Fundamentals of Bias Temperature Instability in MOS Transistors

Souvik Mahapatra

Erratum to: S. Mahapatra (ed.), *Fundamentals of Bias Temperature Instability in MOS Transistors*, Springer Series in Advanced Microelectronics 139, DOI 10.1007/978-81-322-2508-9

Volume number in the front matter as well as in the chapter opening pages of each chapter is appearing incorrectly as 139.
The correct volume number of this book is 52.

The online version of the original book can be found under
DOI 10.1007/978-81-322-2508-9

S. Mahapatra (✉)
Department of Electrical Engineering, Indian Institute of Technology Bombay,
Mumbai 400076, India
e-mail: souvik@ee.iitb.ac.in; mahapatra.souvik@gmail.com

S. Mahapatra (ed.), *Fundamentals of Bias Temperature Instability in MOS Transistors*, Springer Series in Advanced Microelectronics 52,
DOI 10.1007/978-81-322-2508-9_7

Index

S. Mahapatra (ed.), *Fundamentals of Bias Temperature Instability in MOS Transistors*, Springer Series in Advanced Microelectronics 139,
DOI 10.1007/978-81-322-2508-9

P

R

S

T